Supplements to the 2nd Edition of

RODD'S CHEMISTRY OF CARBON COMPOUNDS

ELSEVIER SCIENCE PUBLISHERS B.V.
Molenwerf 1
P.O. Box 211, 1000 AE Amsterdam, The Netherlands

Distributors for the United States and Canada:

ELSEVIER SCIENCE PUBLISHING COMPANY INC.
52, Vanderbilt Avenue
New York, N.Y. 10017

ISBN 0-444-42236-6

Printed in The Netherlands

Supplements to the 2nd Edition of

RODD'S CHEMISTRY OF CARBON COMPOUNDS

VOLUME I

ALIPHATIC COMPOUNDS
★

VOLUME II

ALICYCLIC COMPOUNDS
★

VOLUME III

AROMATIC COMPOUNDS
★

VOLUME IV

HETEROCYCLIC COMPOUNDS
★

VOLUME V

MISCELLANEOUS
GENERAL INDEX
★

Supplements to the 2nd Edition (Editor S. Coffey) of

RODD'S CHEMISTRY OF CARBON COMPOUNDS

A modern comprehensive treatise

Edited by
MARTIN F. ANSELL
Ph.D., D.Sc. (London) F.R.S.C. C. Chem.
Department of Chemistry, Queen Mary College,
University of London (Great Britain)

Supplement to

VOLUME I ALIPHATIC COMPOUNDS

Part E:

Unsaturated Acyclic Hydrocarbons; Trihydric Alcohols,
Their Oxidation Products and Derivatives

ELSEVIER
Amsterdam – Oxford – New York – Tokyo   1983

CONTRIBUTORS TO THIS VOLUME

BRIAN J. COFFIN, B.Sc., Ph.D., C.Chem., M.R.S.C.

Department of Chemistry, Polytechnic of the South Bank,

London, SEl OAA

ROY H. GIGG, B.Sc., Ph.D., D.Sc., C.Chem., F.R.S.C.

National Institute for Medical Research,

Mill Hill, London, NW7 lAA

DAVID R. TAYLOR, M.A., Ph.D., C.Chem., F.R.S.C.

Department of Chemistry,

University of Manchester,

Institute of Science and Technology,

Manchester M60 lQD

RAYMOND E. FAIRBAIRN, B.Sc., Ph.D., F.R.S.C.

Formerly of Research Department,

Dyestuffs Division,

I.C.I. (INDEX)

PREFACE TO SUPPLEMENT IE

In this supplement will be found not only supplements to the Chapters in volume IE of the second edition but also the supplement to Chapter 2 of volume 1A covering unsaturated hydrocarbons. The original author to the supplement to Chapter 2 was unable to provide a manuscript. I am extremely grateful to Dr. David Taylor for providing the present very useful survey of the chemistry of unsaturated hydrocarbons, covering development since the publication of Volume I of the second edition. There have been great advances in this area and the developments in organometallic chemistry are only briefly surveyed and their chemistry will be dealt with in greater detail in a separate supplement.

I am grateful to Dr. Brian Coffin for supplementing not only the Chapters to which he contributed in the second edition but also Chapter 18; and I am very pleased that Dr. Roy Gigg was willing to supplement his previous contribution, Chapter 21 of volume IE.

At a time when there are many specialist reviews, monographs and reports available, there is still in my view an important place for a book such as "Rodd", which gives a broader coverage of organic chemistry. One aspect of the value of this work is that it allows the expert in one field to quickly find out what is happening in other fields of chemistry. On the other hand a chemist looking for the way into a field of study will find in "Rodd" an outline of the important aspects of that area of chemistry together with leading references to other works to provide more detailed information.

This volume, like the other supplements to Volume III, has been produced by direct reproduction of manuscripts. I am most grateful to the contributors for all the care and effort which they and their secretaries have put into the production of their manuscripts, including in most cases the diagrams. I am confident that readers will find this presentation acceptable. I also wish to thank the staff at Elsevier for all the help they have given me and for seeing the transformation of authors' manuscripts to published work.

May 1983   Martin F. Ansell

CONTENTS

VOLUME I E

Aliphatic Compounds; Unsaturated Acyclic Hydrocarbons; Trihydric Alcohols, Their Oxidation Products and Derivatives

Chapter 2. Unsaturated Acyclic Hydrocarbons
by D.R. TAYLOR

*Chapter 18. Trihydric Alcohols, Their Analogues and Derivatives
and Their Oxidation Products: Trihydric Alcohols to Triketones
by B.J. COFFIN*

*Chapter 19. Trihydric Alcohols and Their Oxidation Products (continued).
Dihydroxycarboxylic and Trihydroxycarboxylic Acids and
Related Compounds
by B.J. COFFIN*

Chapter 20. Trihydric Alcohols and Their Oxidation Products (continued)
by B.J. COFFIN

Chapter 21. Phospholipids and Glycolipids
by R.H. GIGG

OFFICIAL PUBLICATIONS

B.P.	British (United Kingdom) Patent
F.P.	French Patent
G.P.	German Patent
Sw.P.	Swiss Patent
U.S.P.	United States Patent
U.S.S.R.P.	Russian Patent
B.I.O.S.	British Intelligence Objectives Sub-Committee Reports
F.I.A.T.	Field Information Agency, Technical Reports of U.S. Group Control Council for Germany
B.S.	British Standards Specification
A.S.T.M.	American Society for Testing and Materials
A.P.I.	American Petroleum Institute Projects
C.I.	Colour Index Number of Dyestuffs and Pigments

SCIENTIFIC JOURNALS AND PERIODICALS

With few obvious and self-explanatory modifications the abbreviations used in references to journals and periodicals comprising the extensive literature on organic chemistry, are those used in the World List of Scientific Periodicals.

LIST OF COMMON ABBREVIATIONS AND
SYMBOLS USED

A	acid
Å	Ångström units
Ac	acetyl
a	axial; antarafacial
as, asymm.	asymmetrical
at	atmosphere
B	base
Bu	butyl
b.p.	boiling point
C, mC and μC	curie, millicurie and microcurie
c, C	concentration
C.D.	circular dichroism
conc.	concentrated
crit.	critical
D	Debye unit, 1×10^{-18} e.s.u.
D	dissociation energy
D	dextro-rotatory; dextro configuration
DL	optically inactive (externally compensated)
d	density
dec. or decomp.	with decomposition
deriv.	derivative
E	energy; extinction; electromeric effect; Entgegen (opposite) configuration
E1, E2	uni- and bi-molecular elimination mechanisms
ElcB	unimolecular elimination in conjugate base
e.s.r.	electron spin resonance
Et	ethyl
e	nuclear charge; equatorial
f	oscillator strength
f.p.	freezing point
G	free energy
g.l.c.	gas liquid chromatography
g	spectroscopic splitting factor, 2.0023
H	applied magnetic field; heat content
h	Planck's constant
Hz	hertz
I	spin quantum number; intensity; inductive effect
i.r.	infrared
J	coupling constant in n.m.r. spectra; joule
K	dissociation constant
kJ	kilojoule

LIST OF COMMON ABBREVIATIONS

k	Boltzmann constant; velocity constant
kcal	kilocalories
L	laevorotatory; laevo configuration
M	molecular weight; molar; mesomeric effect
Me	methyl
m	mass; mole; molecule; *meta*-
ml	millilitre
m.p.	melting point
Ms	mesyl (methanesulphonyl)
[M]	molecular rotation
N	Avogadro number; normal
nm	nanometre (10^{-9} metre)
n.m.r.	nuclear magnetic resonance
n	normal; refractive index; principal quantum number
o	*ortho*-
o.r.d.	optical rotatory dispersion
P	polarisation, probability; orbital state
Pr	propyl
Ph	phenyl
p	*para*-; orbital
p.m.r.	proton magnetic resonance
R	clockwise configuration
S	counterclockwise config.; entropy; net spin of incompleted electronic shells; orbital state
S_N1, S_N2	uni- and bi-molecular nucleophilic substitution mechanisms
S_Ni	internal nucleophilic substitution mechanisms
s	symmetrical; orbital; suprafacial
sec	secondary
soln.	solution
symm.	symmetrical
T	absolute temperature
Tosyl	*p*-toluenesulphonyl
Trityl	triphenylmethyl
t	time
temp.	temperature (in degrees centigrade)
tert.	tertiary
U	potential energy
u.v.	ultraviolet
v	velocity
Z	zusammen (together) configuration

LIST OF COMMON ABBREVIATIONS

α	optical rotation (in water unless otherwise stated)
$[\alpha]$	specific optical rotation
α_A	atomic susceptibility
α_E	electronic susceptibility
ε	dielectric constant; extinction coefficient
μ	microns (10^{-4} cm); dipole moment; magnetic moment
μB	Bohr magneton
μg	microgram (10^{-6} g)
λ	wavelength
ν	frequency; wave number
χ, χ_d, χ_μ	magnetic, diamagnetic and paramagnetic susceptibilities
$\sim$	about
$(+)$	dextrorotatory
$(-)$	laevorotatory
$(\pm)$	racemic
$\ominus$	negative charge
$\oplus$	positive charge

Chapter 2

UNSATURATED ACYCLIC HYDROCARBONS

D.R. TAYLOR

Nearly twenty years have elapsed since the previous chapter
on this topic was written (2nd Edn. Vol. IA, p.398). In view
of the enormous advances made since 1964 and the quite
prolific publications related to this topic, certain aspects
have had to be savagely curtailed, notably reactions catalysed
by transition metal complexes.

1. Carbenes

The ground state of methylene is now considered to be a bent
triplet (H–C–H angle *ca.* 136°), and the first excited state a
singlet with an even more acute angle (102°) (J.F. Harrison,
Acc.chem.Res., 1974, 7, 378). Linear singlets are predicted
to arise only if the carbon bears substituents less electro-
negative than itself (W.W. Schoeller, Chem.Comm., 1980, 124).
Controversy has raged around the difference in energies
between the ground and first excited states. For methylene,
experimental values have risen from a base of 4–8 kJ/mol (R.W.
Carr, T.W. Eder, and M.G. Toper. J.chem.Phys., 1970, 53, 4716)
through a peak value of *ca.* 82 kJ/mol (W.C. Lineberger *et al.*
J.Amer.chem.Soc., 1976, 98, 3731), before declining to values
around 33 kJ/mol (R.K. Lengel and R.N. Zare, *ibid.*, 1978, 100,
7495) which are more in accord with theoretical calculations
(e.g. J.F. Harrison, R.C. Liedtke, and J.F. Libman, *ibid.*,
1979, 101, 7162).

  Much of the fascination of carbene chemistry stems from the
relative stabilities of the singlet and triplet states and the
speed of intersystem crossing between them, both aspects which
are affected by structural changes in the carbene. Singlets
can be stabilised by either donor or acceptor interaction
between a substituent's orbital and one of the carbene's non-

bonding orbitals, so that although diphenylcarbene has a triplet ground state, halogenocarbenes such as ClCH: and F_2C: (C.W. Bauslicher, Jr., J.F. Schaefer, and P.S. Bagus, *ibid.*, 1977, 99, 7106) and other heteroatom-substituted- and acyl-carbenes, all have singlet ground states. There are several techniques for experimentally probing whether a carbene is formed in the singlet or the triplet state. Triplets generated in the presence of nitric oxide are converted into iminoxyl radicals ($R_2C=N-O^{\bullet}$) which are longlived enough to be detectable by ESR spectroscopy (A.R. Forrester and J.S. Sadd, Chem.Comm., 1976, 631). Another important new technique is CIDNP (review: H.D. Roth, Acc.chem.Res., 1977, 10, 85).

(a) *Generation of carbenes*

An excellent concise summary of the main methods now in use has been published (J.T. Sharp, "Comprehensive Organic Chemistry", Vol. 1, ed. D.H.R. Barton, W.D. Ollis, and J.F. Stoddart, Pergamon, Oxford, 1979, p. 455).

(i) *From diazocompounds*

Carbenes can be generated by thermolysis and by photolysis of diazocompounds (H. Dürr, Topics in Current Chem., 1975, 55, 87 and W.J. Baron *et al.*, "Carbenes", Vol. 1, ed. M. Jones, Jr., and R.A. Moss, Wiley-Interscience, New York, 1973, p. 1). Photolysis usually produces the singlet state unless a triplet sensitizer is added to ensure that photolysis occurs *via* the triplet diazoalkane.

$$R_2CN_2 \xrightarrow{\text{u.v.}} R_2CN_2)_{S_1} \xrightarrow{-N_2} R_2C{\uparrow\downarrow})_{S_1}$$

$$\downarrow {Ph_2CO)_{T_1}}$$

$$Ph_2CO)_{S_o} + R_2CN_2)_{T_1} \xrightarrow{-N_2} R_2C\uparrow)_{T_1}$$

If spontaneous intersystem crossing between the excited states of the diazoalkane occurs in competition with their decomposition, both the singlet and the triplet carbene will be produced; this can be effected by the presence of diluents. Gas phase photolyses are additionally complicated by the excess energy with which methylene is generated.

Thermal decomposition of diazoalkanes in the presence of, for example, an olefin to trap the carbene produced may be accompanied by side-reactions such as direct cycloaddition of the diazoalkane to the alkene. Aprotic conditions are advisable, to avoid the possibility of acid-catalysed carbenium ion (R_2CH^+) formation. Interestingly, tetraphenyl-ethylene has itself been implicated as a catalyst for diazo-alkane decomposition (C.-T. Ho, R.T. Conlin and P.P. Gaspar, J.Amer.chem.Soc., 1974, 96, 8109).

Copper and its salts, e.g. Cu(I) triflate (R.G. Salomon and J.K. Kochi, *ibid.*, 1973, 95, 3300), are well-known promoters of methylene transfers from diazoesters and diazoketones, but changes in the observed product distribution compared to that from routes *via* free carbenes indicate that copper-containing carbenoids are actually involved. Copper-catalysed decompositions of diazoalkanes have also been reported (for a review see W. Kirmse, "Carbene Chemistry", 2nd Edn., Academic Press, New York, 1971, p. 85).

(ii) *Cycloeliminations*

The dibenzonorcaradiene (1) has been proposed as a shelf-stable source of methylene. It is made from phenanthrene by the Simmons-Smith procedure, and on photolysis the yields of methylene-based produces are very high (90%) (D.B. Richardson *et al.*, J.Amer.chem.Soc., 1965, 87, 2763). More common are examples of the generation of a wide range of alkyl-, aryl-, and arylalkyl-carbenes by the photochemical cleavage of heterocyclics such as oxirans (2) and diaziridines (3) (review: G.W. Griffin and N. Bertonière, "Carbenes", Vol. 1, ed. Jones and Moss, *loc.cit.*, p. 305).

(1)   (2)   (3)

In general, thermal cycloeliminations parallel photolytic

ones and similar starting materials can be used, although
fewer examples are to be found in the literature (R.W.
Hoffmann, Angew.Chem.internat.Edn., 1971, 10, 529 and Kirmse,
loc.cit., p. 9).

(iii) *Base-catalysed α-eliminations*

Alkylcarbenes are seldom generated in this way, but it is a
widely used technique for generating aryl- and halogenocarbenes.

$$CH_2X_2 \xrightarrow{\text{n-BuLi or}}_{\text{t-BuOK}} X_2CHM \xrightarrow{\hspace{1cm}} XCH: + MX$$

Such reactions can usually be shown to proceed *via* carbenoids
containing the base's counter-cation; however, in such cases
the addition of a crown ether such as 18-crown-6 leads to
product distributions consistent with the presence of free
carbenes as the main intermediates (R.A. Moss, M.A. Joyce,
and F.G. Pilkiewicz, Tetrahedron Letters, 1975, 2425). Such
α-eliminations can even be induced simply by treating
halogenoalkanes with aqueous bases, provided that a phase
transfer catalyst is present (M. Makosza, Pure appl.Chem.,
1975, 43, 439).

Base-promoted α-eliminations from vinyl triflates may
now be regarded as a standard technique for the generation of
alkylidenecarbenes (P.J. Stang, Chem.Reviews, 1978, 78, 383).

$$R_2C=CHOT_f \xrightarrow{\text{t-BuOK}} R_2C=C:$$

The procedure has been shown to be capable of extension to
allenylcarbenes ($R_2C=C=C:$) and to higher cumulenic carbenes,

$$Me_2C=C(OT_f)C\equiv CH \xrightarrow{\text{t-BuOK}} Me_2C=C=C=C:$$

(P.J. Stang and T.E. Fisk, J.Amer.chem.Soc., 1980, 102, 6813;
P.J. Stang and M. Ladika, *ibid.*, 5406), the energies and
favoured **structures** of which had already been predicted by
ab initio calculations (C.E. Dykstra, C.A. Parsons, and C.L.
Oates, *ibid.*, 1979, 101, 1962).

(iv) *Simmons–Smith and related methylene transfers*

Cyclopropanes are obtained, usually in good yield, when alkenes are treated with diiodomethane in the presence of activated zinc (a zinc–copper couple is most often used), a procedure termed the Simmons–Smith reaction (H.E. Simmons *et al.*, Org.Reactions, 1973, 20, 1).

$$R_2C=CR_2 \quad + \quad CH_2I_2 \quad + \quad Zn(Cu) \longrightarrow \quad \overset{CH_2}{\underset{R_2C \text{------} CR_2}{\triangle}} \quad + \quad ZnI_2$$

Several variations of the technique are known, including the use of diethylzinc in place of the couple (S. Miyano and H. Hashimoto, Bull.chem.Soc.Japan, 1973, 46, 892) and the use of 5-15 μm copper powder (N. Kawabata, I. Kamemura, and M. Naka, J.Amer.chem.Soc., 1979, 101, 2139). Carbenoids, involving the metal, are believed to be the true intermediates rather than free methylene in all such procedures.

(b) *Reactions of carbenes*

(i) *Cycloadditions*

Skell first proposed that the stereochemical outcome of the cycloaddition of a carbene to either the *trans-* or *cis-*isomer of an olefin could be used to distinguish singlets (which react concertedly and hence stereospecifically) from triplets (which react non-concertedly and hence with loss of stereochemical integrity) (P.S. Skell and R.C. Woodworth, *ibid.*, 1956, 78, 4496).

Since then much further work has been reported on this very useful cyclopropane-forming reaction (G.L. Closs, Topics in Stereochem., 1968, 3, 193 and A.P. Marchand, "The Chemistry of

Double-Bonded Functional Groups", Supplement A, ed. S. Patai, Wiley, New York, 1977, Part 1, ch. 7). The concerted process is theoretically interesting because two alternative topologies are possible, only one of which (the "non-least motion" approach) is symmetry allowed; no experimental test of this feature has yet been devised. Theoretical calculations, perhaps not too surprisingly, favour the symmetry allowed approach (N. Bodor, M.J.S. Dewar, and J.S. Wasson, J.Amer.chem.Soc., 1972, 94, 9095 using MINDO/2; R. Hoffmann, D.M. Hayes, and P.S. Skell, J.phys.Chem., 1972, 76, 664 using extended Hückel; and N.G. Rondan, K.N. Houk and R.A. Moss, J.Amer.chem.Soc., 1980, 102, 1770, using *ab initio* STO-3G).

The relative reactivities (selectivities) of various carbenes towards a standard set of alkenes of differing electrophilicity have been measured and interpreted on the basis of frontier orbital theory. It was concluded that electrophilic carbenes will be characterised by smaller LUMO(carbene)-HOMO(alkene) energy difference relative to LUMO(alkene)-HOMO(carbene), whereas the reverse order of differential orbital energies would characterise a nucleophilic carbene (R.A. Moss, Acc.chem.Res., 1980, 13, 58).

Cycloadditions between carbenes and 1,3-dienes rarely proceed in a 1,4-addition manner, with the main exception of the addition of dihalogenocarbenes to norbornadiene. The usual outcome of the reaction of acyclic dienes with alkylcarbenes is 1,2-addition to give vinyl cyclopropanes (R.A. Moss and M. Jones, Jr., "Reactive Intermediates", ed. M. Jones, Jr., and R.A. Moss, Vol. 1, Wiley-Interscience, New York, 1978, p. 69).

(ii) *Insertions*

Carbenes insert intra- and inter-molecularly into a variety of single bonds, notably C-H, Si-H, N-H, O-H, S-H, C-halogen, and C-Si (Kirmse, *loc.cit.*, p. 209). Singlet methylene inserts with retention of configuration at a chiral C-H or Si-H centre, although the insertion shows a rather low selectivity between primary, secondary, and tertiary C-H bonds. For isopentane the relative rates per CH bond are 1:1.2:1.4 when methylene is generated by gas phase photolysis (B.M. Herzog and R.W. Carr, Jr., J.phys.Chem., 1967, 71, 2688). The geometry of this evidently concerted process has been examined by *ab initio* methods (R.C. Dobson, D.M. Hayes, and R. Hoffmann, J.Amer.chem.Soc., 1971, $\underline{93}$, 6188). Triplet methylene inserts non-concertedly by a mechanism firmly established as an abstraction-recombination sequence by CIDNP (H.D. Roth, Acc.chem.Res., 1977, $\underline{10}$, 85).

$$\underset{R^2}{\overset{R^1}{>}}\underset{R^3}{\overset{|}{C}}\text{-H} \quad \xrightarrow{X_2C\uparrow\downarrow} \quad \underset{R^2}{\overset{R^1}{>}}\underset{R^3}{\overset{|}{C}}\text{-CHX}_2$$

$$\xrightarrow{X_2C\uparrow\uparrow} \quad [R^1R^2R^3C^\cdot \quad H\overset{\cdot}{C}X_2] \longrightarrow R^1R^2R^3CCHX_2$$

(racemate)

This explains why triplet carbenes cause racemisation during insertion at a chiral centre; even singlet carbenes may react by this route with certain bonds, e.g. with C-halogen.

(iii) *Rearrangements*

The intramolecular α- and β-CH insertions which commonly occur in alkylcarbenes may conveniently be regarded as rearrangements.

$$R^1{}_2CH\text{-}CHR^2\text{-}\overset{\cdot\cdot}{C}R^3 \quad \xrightarrow{\quad\quad} \quad R^1{}_2CHCR^2\text{=}CHR^3 \quad + \quad \underset{R^1{}_2 \quad\quad R^2}{\triangle}{}^{R^3}$$

Although acyclic carbenes are less prone than their cyclic analogues to other types of rearrangements, known examples include 1,2-alkyl shifts and double-bond migrations which effectively produce isomeric carbenes.

$$R^1{}_2C\text{=}CR^2\text{-}\overset{\cdot\cdot}{C}R^3 \quad \rightleftharpoons \quad \underset{R^2 \quad\quad R^3}{\overset{R^1{}_2}{\triangle}} \quad \rightleftharpoons \quad R^2\overset{\cdot\cdot}{C}R^3\text{=}CR^1{}_2$$

(For reviews see W.M. Jones, Acc.chem.Res., 1977, $\underline{10}$, 353 and C. Wentrup, Topics in Current Chem., 1976, $\underline{62}$, 173).

(iv) *Transition metal complexes*

Complexes in which carbenes are attached as ligands to transition metals are believed to play an important role as reaction intermediates in processes such as alkene metathesis (D.J. Cardin, B. Cetinkaya, M.J. Doyle, and M.F. Lappert, Quart.Rev., 1973, 2, 99; A. Nakamura, Pure appl.Chem., 1978, 50, 37; and H.C. Clark, *ibid.*, p.43).

2. Hydrocarbons containing one double bond (alkenes)

(a) *Introduction*

The chemistry of alkenes is covered in detail in Houben-Weyl, "Methoden der Organischen Chemie", ed. E. Müller, Vol. 5/1b, 4th Edn., Thieme-Verlag, Stuttgart, 1972; "The Chemistry of Alkenes", Wiley-Interscience, London, Vol. 1, ed. S. Patai, 1964, and Vol. 2, ed. J. Zabicky, 1970; and G.H. Whitham in "Comprehensive Organic Chemistry", Vol. 1, ed. D.H.R. Barton, W.D. Ollis and J.F. Stoddart, Pergamon, Oxford, 1979, p. 121. There are also two useful monographs dealing mainly with industrial aspects (F. Asinger, "Mono-olefins: Chemistry and Technology", Pergamon, London, 1968; E.G. Hancock "Propylene and its Industrial Derivatives", Wiley, New York, 1973).

The stereochemistry of olefins should now be described by the prefix *E* (entgegen = opposite) or *Z* (zusammen = together) according to the proximity of the two substituents of highest priority attached to the C=C bond. Priority is defined by the Cahn-Ingold-Prelog rule (R.S. Cahn, C.K. Ingold, and V. Prelog, Experientia, 1956, 12, 81). Thus (4) should be referred to as (*E*)-, and (5) as (*Z*)-, 2-bromopent-2-ene, respectively, while for simple non-terminal alkenes (*E*) replaces *trans*, and (*Z*) replaces *cis* (J.E. Blackwood *et al.*, J.Amer.chem.Soc., 1968, 90, 509; for the IUPAC Rules on alkene nomenclature, see J.org.Chem., 1970, 35, 2849).

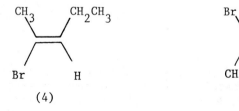

(4)   (5)

(b) *Synthetic methods*

The synthesis of olefins has attracted a consistently high
level of attention since the previous edition appeared.
Several entirely new preparative procedures have emerged, but
significant improvements have also been effected in the older
techniques, and this progress will where appropriate be
included. Particularly powerful methods have been developed
for stereoselective synthesis of alkenes (for reviews see J.
Reucroft and P.G. Sammes, Quart.Rev., 1971, 25, 135; A.S.
Arora and I.K. Ugi, Houben–Weyl, *loc.cit.*, p. 728; D.J.
Faulkner, Synthesis, 1971, 175; C.A. Henrick, Tetrahedron,
1977, 33, 1845). The mechanisms and stereochemical features
of olefin-forming elimination reactions have been reviewed
several times (J. Sicher, Angew.Chem.internat.Edn., 1972, 11,
200; W.H. Saunders and A.F. Cockerill, "Mechanisms of
Elimination Reactions", Wiley-Interscience, New York, 1973;
W.H. Saunders, Jr., in "The Chemistry of Alkenes", Vol. 1,
loc.cit., ch. 2; J.L. Coke, Selective Organic Transformations,
1972, 2, 269; E. Baciocchi, Acc.chem.Res., 1979, 12, 430;
C.J.M. Stirling, *ibid.*, p. 198; R.A. Bartsch, *ibid.*, 1975, 8,
239; W.T. Ford, *ibid.*, 1973, 6, 410; A. Fry, Quart.Rev., 1972,
1, 163).

   (i) *Eliminations of HX*

This group of procedures includes acid-catalysed dehydration
of alcohols and base-catalysed dehydrohalogenation of alkyl
halides. New developments in these older methods have led to
the use of milder dehydrating agents than strong acids, which
frequently cause rearrangements or other complications;
typical examples of these new reagents are the methyl-
(carboxysulphamoyl)triethylammonium hydroxide inner salt (6)
and the sulphurane (7), both of which are suitable for use
with secondary or tertiary alcohols (E.M. Burgess, H.R. Penton
and E.A. Taylor, J.Amer.chem.Soc., 1970, 92, 5224; R.J. Arhart
and J.C. Martin, *ibid.*, 1972, 94, 5003).

$$MeO_2\overset{-}{C}NSO_2\overset{+}{N}Et_3 \qquad\qquad Ph_2S[OC(CF_3)_2Ph]_2$$

$$(6) \qquad\qquad\qquad\qquad (7)$$

Dehydrations effected by thionyl chloride-pyridine are also
less prone than acid-catalysed ones to Wagner-Meerwein
rearrangements. Thus, 1,1-di-t-butylethylene is obtained in

high yield from (8) (J.S. Lomas, D.D. Sagatys, and J.E. Dubois, Tetrahedron Letters, 1971, 599), although rearrangement may still occur with severely crowded alcohols: tris(t-butyl)ethylene is not obtained by similar treatment of (9).

$$(t-Bu)_2C(OH)CH_3 \quad \xrightarrow{\text{SOCl}_2, \text{ pyr.}} \quad (t-Bu)_2C=CH_2$$

(8)

$$(t-Bu)_2C(OH)CH_2CMe_3 \quad \longrightarrow \quad t-BuCMe(CH_2CMe_3)CMe=CH_2$$

(9)

Alcohols can alternatively be dehydrated by treatment with a Grignard reagent followed by thermal decomposition of the magnesium alkoxide (E.C. Ashby, G.F. Willard, and A.B. Goel, J.org.Chem., 1979, 44, 1221.

The well-established base catalysed bimolecular *anti*-elimination of hydrogen halide from an alkyl halide proceeds stereoselectively to give the alkene with the most bulky groups in an *E*-orientation, and with a regioselectivity embodied in Saytzeff's Rule.

A particularly useful modification is to employ a bulky base to ensure formation of the least substituted (*i.e.* contrary to Saytzeff) alkene; such a procedure is termed a Hofmann elimination.

$$EtCH=CHCH_3 \quad \xleftarrow{\text{Saytzeff}} \quad n-PrCHBrCH_3 \quad \xrightarrow{\text{Hofmann}} \quad n-PrCH=CH_2$$

Primary halides require especially non-nucleophilic bases such as Hünig's base [(i-Pr)$_2$NEt] (S. Hünig and M. Kiessel, Ber., 1958, 91, 380) or potassium t-butoxide in DMSO (D.E. Pearson and C.A. Buehler, Chem.Rev., 1974, 74, 45). Competing

nucleophilic substitution can also be minimised by choice of
bases such as the bicylic amidines DBN (10, n=3) or, better
DBU (10, n=5) (H. Oediger, F. Möller and K. Eiter, Synthesis,
1972, 591), although these do not give 1-enes preferentially.
Crown ethers can be used to promote dehydrohalogenations by
potassium t-butoxide in non-polar solvents (E.V. Dehmlow and
M. Lissel, Synthesis, 1979, 372).

$$n\text{-BuCH}_2\text{CHBrCH}_3 \quad \xrightarrow[80-90^\circ C]{\text{DBU}}$$

(E)-n-BuCH=CHMe   (80%)

\+

n-BuCH$_2$CH=CH$_2$ (20%)

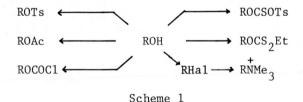

(10)

Considerable variation in this general procedure is possible,
since the leaving group (e.g., OH$^-$ or Br$^-$), and if required its
stereochemistry at the site of attachment, can be modified by
inter-conversions such as those shown in Scheme 1. Major
objectives achievable by such interconversions are (i) less
stringent conditions for the elimination after conversion into
a more labile leaving group (e.g. OTs), (ii) preference for
Hofmann regioselectivity after conversion into a bulkier
leaving group (e.g. -NMe$_3^+$), and (iii) formation of a substrate
suitable for pyrolytic *syn*-elimination (e.g. ROAc, ROCS$_2$Et,
ROCOCl, or RNMe$_2$-O). Gas-phase pyrolytic eliminations have
been reviewed (A. Maccoll in "The Chemistry of Alkenes",
Wiley-Interscience, London, Vol. 1, ed. S. Patai, 1964, ch. 3).

ROTs ←   ↗ → ROCSOTs

ROAc ← —— ROH —→ ROCS$_2$Et

ROCOCl ← ——/   ↘ +
                  RHal —→ RNMe$_3$

Scheme 1

Particularly useful substrates for *syn*-eliminations are
selenoxides generated by peroxide-oxidation of selenides,

themselves obtainable either from halides or tosylates by substitution (K.B. Sharpless, M.W. Young, and R.F. Lauer, Tetrahedron Letters, 1973, 1979) or by selenide induced opening of epoxides (K.B. Sharpless and R.F. Lauer, J.Amer. chem.Soc., 1973, 95, 2697).

$$MeCH_2CHMe \xrightarrow[\text{(ii)}]{\text{(i)}} MeCH\text{---}CHMe \longrightarrow MeCH=CHMe$$

(with OTs group below first structure; H, Se$^+$, Ph, and O groups in the middle structure)

$$EtCH_2 \overset{O}{\triangle} CH_2Et \xrightarrow{\text{(i)}} EtCH_2\overset{OH}{CH}\text{-}CHCH_2Et \xrightarrow{\text{(ii)}}$$

PhSe

EtCH=CHCH(OH)CH$_2$Et

Reagents: (i) PhSeNa; (ii) H$_2$O$_2$

(ii) *Eliminations of XY*

Since such methods in principle afford precise control of the site at which the double bond is formed, eliminations from substrates of the type >CX-CY< are a valuable group of reactions which have continued to receive a great deal of attention.

(1) *Dehalogenation*

Zinc-promoted elimination from a vicinal dihalide is one of the oldest known methods of alkene synthesis. It has been adapted to modern requirements by using a hot suspension of zinc powder in ethylene glycol (C.A. Brown, J.org.Chem., 1975, 40, 3154) and by the use of new initiators such as lithium aluminium hydride, potassium iodide, or a catalytic amount of sodium iodide in the presence of a phase transfer agent (D. Landini, S. Quici, and F. Rolla, Synthesis, 1975, 397).

(2) *Decarboxylation*

Decarboxylation-dehydration of hydroxy acids and decarboxylation-dehalogenation of halogeno acids are two well-established techniques.

$$R^1CH-\!\!\!-CHR^2$$

with structure showing CO_2^- and X groups, arrow labeled $X = OH, Hal$ giving $R^1CH=CHR^2$

An improvement of this technique is to heat β-hydroxyacids (easily obtained by the Reformatsky reaction) with the dimethyl acetal of dimethylformamide (S. Hara et al., Tetrahedron Letters, 1975, 1545).

RCH—CHMe with OH and CO_2H groups, reacting with $Me_2NCH(OMe)_2$ $(R = n-C_9H_{19})$, giving a cyclic intermediate (with Me, R, O, NMe_2), leading to RCH=CHMe

Another, highly stereoselective, synthesis of both (E) and (Z) olefins uses threo-3-hydroxy acids as the key intermediates (J. Mulzer et al., Chem.Comm., 1979, 52).

Reagents: (i) $EtO_2CN=NCO_2Et$, Ph_3P;   (ii) $-CO_2$, $-Ph_3PO$;
           (iii) $PhSO_2Cl$;   (iv) $-CO_2$, $-PhSO_3H$.

vic-Dicarboxylic acids can be oxidatively decarboxylated using copper(I) oxide in quinoline (R.A. Snow, C.R. Degenhardt, and L.A. Paquette, Tetrahedron Letters, 1976, 4447).

(3) *Wittig reaction and related procedures*

Although the Wittig procedure, culminating in *syn*-elimination from a phosphonium betaine, places the double bond precisely between the carbon atom of the ylide and that of the carbonyl group, the stereochemistry of the alkene is seldom so clearly controlled.

$$Ph_3\overset{+}{P}-\overset{-}{C}HR^1 \xrightarrow{R^2CHO} \quad (11) \xrightarrow{80\ ^{\circ}C} \quad Ph_3P{=}0 \quad + \quad (12)$$

When R^1 and R^2 are both uncrowded alkyl groups, the (*Z*)-isomer predominates under salt-free conditions which can best be ensured by avoiding the use of alkyllithium bases. Thus, for the synthesis of the pheromone precursor (*Z*)-2-methyloctadec -7-ene [12; $R^1 = Me_2CH(CH_2)_4$, $R^2 = n{-}C_{10}H_{21}$], successive reductions in the proportion of the unwanted (*E*)-isomer are obtained by changing from n-BuLi in DMSO (88:12; B.A. Bierl, M. Beroza, and C.W. Collier, Science, 1970, 170, 87) to potassium in hexamethylphosphoric triamide (HMPT) (94:6; H.J. Bestmann and O. Vostrowsky, Tetrahedron Letters, 1974, 207), and finally to sodium bis(trimethylsilyl)amide in THF at -78 °C (99:1; H.J. Bestmann, O. Vostrowsky, and W. Stransky, Ber., 1976, 109, 3375).

Equilibration between the diastereoisomeric betaines (11) and (13) is accelerated by added lithium salts, giving mixtures of olefins containing substantial amounts of the (*E*)-isomer (H.O. House, V.K. Jones, and G.A. Frank, J.org.Chem., 1964, 29, 3327), but the *threo*-betaine (13) completely dominates the equilibrium if α-lithiation is induced by the addition of an alkyl-lithium, providing a useful procedure which is specific for the synthesis of pure (*E*)-isomers (M. Schlosser, Topics Stereochem., 1970, 5, 1).

$$(11) \xrightarrow{\text{RLi}} \quad \begin{array}{c} Ph_3\overset{+}{P}-\overset{R^1}{\underset{|}{\underset{O^-}{C}}}\overset{\text{....}}{\underset{|}{\underset{}{}}} \\ \overset{-}{O}-\overset{\text{....}}{\underset{R^2}{C}}\overset{H}{} \end{array} \quad \xrightarrow[\text{t-BuOH}]{\overset{+}{H}} \quad \begin{array}{c} Ph_3\overset{+}{P}-\overset{\text{....}R^1}{\underset{|}{\underset{}{C}}}\overset{}{\underset{}{}}H \\ \overset{-}{O}-\overset{\text{....}}{\underset{R^2}{C}}\overset{H}{} \end{array}$$

$$(13)$$

$$\begin{array}{cc} R^1 & H \\ \diagdown & \diagup \\ & C=C \\ \diagup & \diagdown \\ H & R^2 \end{array} \quad \xleftarrow{-Ph_3PO}$$

This Wittig-Schlosser modification can be extended by
alkylation of the intermediate anion to provide more highly
substituted alkenes, though this does not overcome the other
main limitation of the Wittig reaction, its failure in the
synthesis of tetra-substituted olefins. Polymer-bound
phosphoranes (S.V. McKinley and J.W. Rakshys, Jr., Chem.Comm.,
1972, 124) and the generation and use of Wittig reagents in a
two-phase system (e.g., CH_2Cl_2-H_2O) assisted by quaternary
ammonium salts (G. Märkl and A. Merz, Synthesis, 1973, 295)
or crown ethers (R.M. Boden, *ibid.*, 1975, 784), are two other
promising modifications of the conventional technique.

The Wittig-Horner modification, which uses ylides derived
from phosphonates so that a neighbouring P=O bond stabilises
the carbanion, has also been adapted to a two-phase system (M.
Mikolajczyk *et al.*, *ibid.*, 1975, 278). The Horner procedure
gives mainly (E)-isomers, though it is seldom applied to the
synthesis of non-aromatic alkenes, since one of its main uses
is with stabilised ylides. Aliphatic esters can be converted
directly into branched alkenes by treatment with an excess of
an alkylidenephosphorane in DMSO (A.J. Uijttewaal, F.L. Jonkers
and A. van der Gen, J.org.Chem., 1979, 44, 3157), presumably
via intermediate ketones which react promptly with additional
phosphorane.

$$R^1CO_2R^2 \xrightarrow{Ph_3\overset{+}{P}-\overset{-}{C}HR^3} R^1COCH_2R^3 \xrightarrow{Ph_3\overset{+}{P}-\overset{-}{C}HR^3} R^1C(CH_2R^3)=CHR^3$$

$$(E + Z)$$

Industrial applications of the Wittig reaction have been
reviewed (H. Pommer, Angew.Chem.internat.Edn., 1977, 16, 423).

(4) Eliminations of HO-X.

One of the earliest stereoselective syntheses of di- and tri-alkyl-olefins, the Cornforth synthesis (2nd Edn., Vol IA, p. 407) is based upon stereoselective formation and ring-opening of an epoxide.

Reagents: (i) R^3MgX, $-78°C$; (ii) NaOH; (iii) NaI; (iv) $SnCl_2$, $POCl_3$, pyridine.

This sequence has been elegantly shortened to an essentially one-pot procedure by the simple expedient of adding lithium to the initially-formed magnesium alkoxide. Metallation occurs at $-60°C$; elimination to the alkene follows on warming to $20°C$ (J. Barluenga, M. Yus, and P. Bernad, Chem.Comm., 1978, 847).

The philosophy of the Cornforth synthesis has recently been extended to include eliminations from β-hydroxy-silanes, -selenides, -sulphides, and -phosphine oxides.

$(M = SiMe_3, SePh, SR, or P(O)Ph_2)$

The Peterson reaction (D.J. Peterson, J.org.Chem., 1968, <u>33</u>, 780; T.-H. Chan, Acc.Chem.Res., 1977, <u>10</u>, 442) is based on this type of elimination from β-hydroxysilanes and is currently the most important of the group, owing to the variety of possible ways for generating the required silanes. The usual starting point is the treatment of an aldehyde or a ketone with an α-silyl-carbanion, a route which closely resembles the

Wittig reaction.

$$R^1COR^2 \xrightarrow{\quad R^3_3SiCR^4R^5 \quad}$$

The reaction is irreversible and non-stereospecific up to the last step, however; the last stage is a selective *syn*-elimination under base-catalysed conditions, but can be turned into an *anti*-elimination under acidic conditions. Various techniques have emerged for the stereoselective synthesis of β-hydroxysilanes (P.F. Hudrlik and D.J. Peterson, J.Amer.chem. Soc., 1975, 97, 1464; M. Obayashi, K. Utimoto, and H. Nozaki, Bull.chem.Soc.Japan, 1979, 52, 1760; Chan, *loc.cit.*) and for the formation of α-silylcarbanions (T.H. Chan, E. Chang, and E. Vinokur, Tetrahedron Letters, 1970, 1137; T.H. Chan and E. Chang, J.org.Chem., 1974, 39, 3264). Routes to alkenylsilanes, which can be converted into silylcarbanions by addition of an alkyllithium or Grignard reagent, have been reviewed (T.H. Chan and I. Fleming, Synthesis, 1979, 761), together with a discussion of their conversion to alkenes. Examples of these reactions are shown in Scheme 2.

Acid-catalysed eliminations from β-hydroyselenides are very facile and proceed with *anti*-stereospecificity, so that although such selenides are toxic and generally less easy to handle than organosilanes, the procedure may come to be regarded as preferable to the Peterson reaction (J. Rémion, W. Dumont, and A. Krief, Tetrahedron Letters, 1976, 1385):

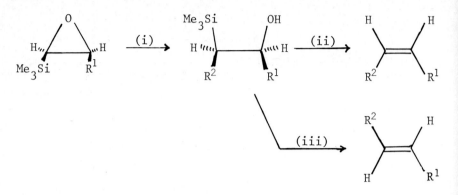

$Me_3SiCHR^1CHR^2OH$ $\xrightarrow{\text{(iv)}}$ $Me_3SiCHR^1COR^2$ $\xrightarrow{\text{(v)}}$

RCHO $\xrightarrow{\text{(vi)}}$ $RCH(OH)C(SiMe_3)=CH_2$ $\xrightarrow{\text{(vii)}}$

Reagents: (i) R_2^2CuLi; (ii) KH, THF, 25 °C;

(iii) $BF_3 \cdot OEt_2$, 0 °C; (iv) CrO_3; (v) $i\text{-}Bu_2AlH$,

−120 °C; (vi) $Me_3SiC(Li)=CH_2$; (vii) $SOCl_2$;

(viii) acid.

Scheme 2

$$R^1CHO \xrightarrow{\text{PhSeH}} R^1CH(SePh)_2 \xrightarrow[\text{(ii) } R^2CHO]{\text{(i) } Bu^nLi} R^1\underset{\underset{PhSe}{|}}{CH}\!\!-\!\!\underset{\underset{OH}{|}}{CHR^2}$$

$$\xrightarrow{\text{acid}} \quad \underset{H}{\overset{R^1}{\Big\diagdown}}C\!=\!C\underset{H}{\overset{R^2}{\Big\diagup}}$$

The route can be extended with complete stereospecificity to
tri- and tetra-substituted olefins by commencing with a
selenide-induced opening of an epoxide (J. Rémion and A.
Krief, *ibid.*, p. 3743). The same workers have shown that
P_2I_4, PI_3, or $SOCl_2$ effect the conversion of β-hydroxysulphides
into alkenes, including tetra-substituted alkenes, in
excellent yields (J.N. Denis, W. Dumont, and A. Krief, *ibid.*,
1979, 4111):

$$R^1SR^2R^3CLi \xrightarrow{R^4R^5CO} R^1SCR^2R^3CR^4R^5OH \xrightarrow{P_2I_4} R^2R^3C=CR^4R^5$$

Closely resembling both the Wittig and the Peterson reactions
is the base-catalysed *syn*-elimination of phosphinate from a
β-hydroxyphosphine oxide (A.J. Bridges and G.H. Whitham, Chem.
Comm., 1974, 142).

(iii) Ring fragmentations

This last group of eliminations is analogous to the final stage
stages of the Wittig and the base-catalysed Peterson reaction,
in that it involves extrusion of a segment of, usually, a
heterocyclic ring and therefore *syn*-stereochemistry. The
ring which breaks may vary considerably in size: commonly used
methods are, however, largely restricted to those involving 3-
and 5-membered rings.

Extrusion of sulphur dioxide from episulphones is a key step
in the Ramberg-Backlund reaction (N.P. Neureiter, J.Amer.chem.
Soc., 1966, 88, 558). Episulphones can be made in other ways,
which enlarges the scope of the method.

$$RCH_2SO_2CHClR \xrightarrow[-HCl]{base} \quad \underset{R \quad\quad R}{\triangle} \overset{O=S=O}{} \xrightarrow[-SO_2]{heat} RCH=CHR$$

Deamination of aziridines is brought about by nitrosation (R.M. Carlson and S.Y. Lee, Tetrahedron Letters, 1969, 4001). The required aziridines are most often made from oximes or Δ^2-1,2-oxazolines by reduction, or by addition of a Grignard reagent to an azirine (A.S. Arora and I.K. Ugi, Houben-Weyl, "Methoden der Organischen Chemie", ed. E. Müller, Vol. 5/1b, 4th Edn., Thieme-Verlag, Stuttgart, 1972, p. 754).

$$\underset{\text{n-Am} \quad\quad \text{Me}}{\triangle} \xrightarrow{\text{n-BuONO}} \underset{\text{n-Am} \quad\quad \text{Me}}{\overset{H \quad\quad H}{C=C}}$$

An important route to hindered olefins was developed through the discovery that stereospecifically prepared episulphides lose sulphur without loss of stereochemical integrity when treated with a phosphine or an alkyllithium. A key step is the formation of Δ^3-1,3,4-thiadiazolines (14), which lose nitrogen in a conrotatory extrusion when heated: the overall sequence is now termed Barton's double extrusion (D.H.R. Barton, F.S. Guziec, and I. Shahak, J.chem.Soc.Perkin I, 1974, 1794). Quite extraordinarily hindered olefins such as adamantylidene-adamantane have now been made by this route. The intermediate Δ^3-1,3,4-thiadiazolines (14) can also be made by the dipolar cyclo-addition of diazoalkanes to thiones, and selenium may replace sulphur throughout (D.H.R. Barton *et al.*, *ibid.*, 1976, 2079).

$$t\text{-BuCHO} \xrightarrow{N_2H_4, H_2S} \qquad \xrightarrow[(E=CO_2Et)]{EN=NE} \qquad (14)$$

$$(14) \xrightarrow[-N_2]{\triangle} \qquad \xrightarrow{Ph_3P} \qquad$$

Stereospecific deoxygenation of epoxides can be achieved in a variety of ways, including treatment with
(i) triphenylphosphineselenide-trifluoroacetic acid (D.L.J. Clive and C.V. Denyer, Chem.Comm., 1973, 253),
(ii) the anion [CpFe(CO)$_2$]$^-$ followed by acid for retention of configurations or thermal decomposition for inversion of configuration (M. Rosenblum, M.R. Saidi, and M. Madhavarao, Tetrahedron Letters, 1975, 4009),
(iii) potassium selenocyanate (J.M. Behan, R.A.W. Johnstone, and M.J. Wright, J.chem.Soc.Perkin I, 1975, 1216),
(iv) ferric chloride-n-butyllithium (T. Fujisawa, K. Sugimoto, and M. Ohta, Chem.Letters, 1974, 883),
(v) sodium iodide-trifluoroacetic acid in acetonitrile-THF (P.E. Sonnet, J.org.Chem., 1978, 43, 1841), or
(vi) phosphorus iodides (A. Krief et al., Nouveau J.Chem., 1979, 3, 705).
These routes complement the Cornforth and Peterson reactions, which also involve stereospecific epoxide openings (for a useful summary see P.E. Sonnet, Tetrahedron, 1980, 36, 557).

Cyclo-eliminations from five-membered heterocycles, many of which are derived from 1,2-diols, are also frequently used for *syn*-stereospecific eliminations. Some typical examples are shown in Scheme 3 (for details see A.S. Arora and I.K. Ugi, Houben-Weyl, "Methoden der Organischen Chemie", ed. E. Müller, Vol. 5/1b, 4th Edn., Thieme-Verlag, Stuttgart, 1972, p. 747).

Comparable eliminations occur *via* 2-phenyl-1,3-dioxalans and 2-phenyl-1,3-oxathiolans on treatment with phenyllithium or lithium diethylamide (G.H. Whitham et al., J.chem.Soc.Perkin I, 1973, 2332; 1974, 433); these are essentially retro-1,3-dipolar cycloadditions. A frequently selected target molecule for such procedures is *trans*-cyclo-octene (G.H. Whitham and M. Wright, J.chem.Soc.(C), 1971, 886).

(iv) Additions to acetylenes

This group of reactions is particularly attractive for stereospecific syntheses, and is all the more powerful as a result of developments in alkyne synthesis and in organoborane chemistry.

(1) Partial reduction

Catalysts for the selective *cis* hydrogenation of alkynes

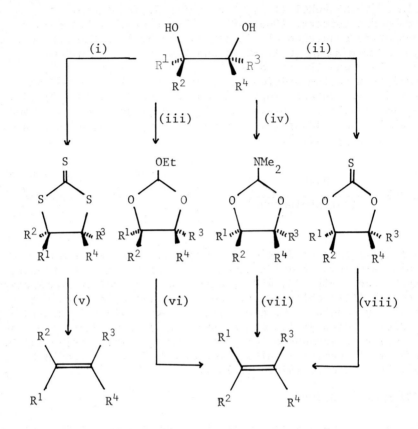

Reagents: (i) several steps; (ii) n-BuLi, CS_2, MeI;
(iii) $HC(OEt)_3$; (iv) $Me_2NCH(OMe)_2$;
(v) $(RO)_3P$, 135 °C; (vi) H^+(catalytic), 160 °C;
(vii) Ac_2O, 90 °C; (viii) $(MeO)_3P$, 110 °C.

Scheme 3

include $(Ph_3P)_3RuHCl$ (I. Jardine and F.J. McQuillin, Tetrahedron Letters, 1966, 4871), "P-2 nickel" (obtained by borohydride reduction of nickel(II) acetate) in the presence of ethylenediamine (C.A. Brown and V.K. Ahuja, Chem.Comm., 1973, 553), and the so-called "complex reducing agents", formed from sodium hydride and iron or nickel salts and an alkoxide (P. Caubère, Topics in Current Chem., 1978, $\underline{73}$, 105). The same result is achieved by the acid cleavage of an alkyne-catecholborane adduct (H.C. Brown and S.K. Gupta, J.Amer.chem. Soc., 1975, $\underline{97}$, 5249).

$$R^1C{\equiv}CR^2 \xrightarrow[\text{(catecholborane)}]{R_2BH} \underset{\underset{R_2B\qquad\quad H}{}}{\overset{\overset{R^1\qquad\quad R^2}{}}{C{=}C}} \xrightarrow{H^+} \underset{\underset{H\qquad\quad H}{}}{\overset{\overset{R^1\qquad\quad R^2}{}}{C{=}C}}$$

Selective *trans*-reduction is frequently achieved using lithium aluminium hydride alone (E.F. Magoon and L.H. Slaugh, Tetrahedron, 1967, $\underline{23}$, 4509), although when used in combination with nickel(II) chloride it, too, gives essentially pure *cis*-addition products (E.C. Ashby and J.J. Lin, J.org.Chem., 1978, $\underline{43}$, 2567). High yields of *cis*-hydrogenation products are also obtained by treatment of alkynes with magnesium hydride in the presence of copper(I) salts (E.C. Ashby, J.J. Lin, and A.B. Goel, J.org.Chem., 1978, $\underline{43}$, 757).

(2) Hydroboration

There are numerous procedures for converting alkynes into di- and tri-substituted alkenes stereospecifically, and such applications of organoboranes have been reviewed (P.F. Hudrlik and A.M. Hudrlik in "The Chemistry of the Carbon-carbon Triple Bond", ed. S. Patai, Vol. 1, Wiley-Interscience, New York, 1978, ch. 7; G.M.L. Cragg "Organoboranes in Organic Synthesis", Dekker, New York, 1973; H.C. Brown "Organic Synthesis *via* Boranes", Wiley-Interscience, New York, 1975). Sterically hindered dialkylboranes such as catecholborane, di-isoamylborane (Sia_2BH), and dicyclohexylborane, are particularly useful reagents in this context, because they add with high regioselectivity to unsymmetrical alkynes, e.g. (G. Zwiefel, G.M. Clark, and N.L. Polston, J.Amer.chem.Soc., 1971, $\underline{93}$, 3395). Such additions are invariably found to occur with *syn*-stereochemistry.

$$R^1C{\equiv}CR^2 \xrightarrow{\text{Sia}_2\text{BH}}$$

R^1 = Ph		81%	19%
R^1 = n-Pr	R^2 = Me	67	33
R^1 = i-Pr	R^2 = Me	93	7
R^1 = t-Bu	R^2 = Me	97	3

Apart from the simple replacement of boron by proton, a reaction effectively catalysed by palladium(II) acetate (H. Yatagai, Y. Yamamoto, and K. Maruyama, Chem.Comm., 1978, 702), migration of an alkyl group from boron to vinylic carbon is readily effected by iodine. This boron to carbon transfer is accompanied by an inversion of the alkene configuration so that the alkyne substituents are *trans* in the product (G. Zweifel, H. Arzoumanian, and C.C. Whitney, J.Amer.chem.Soc., 1967, 89, 3652):

$$EtC{\equiv}CEt \xrightarrow[R = c\text{-}C_6H_{11}]{R_2BH} \quad \xrightarrow{I_2,\text{NaOH}}$$

The opposite result is achieved if a 1-bromoalkyne is used, or if an adduct from a terminal alkyne and a dialkylborane is treated with palladium(II) acetate and triethylamine; these two modifications only appear to operate for the preparation of disubstituted alkenes (G. Zweifel *et al.*, *ibid.*, 1971, 93, 6309; Yatagai *et al.*, *loc.cit.*).

$$R^1C{\equiv}CBr \xrightarrow{R^2{}_2BH}$$

$$R^1C{\equiv}CH \xrightarrow{R^2{}_2BH}$$

Alkynylborates, formed from a trialkylborane and a lithium acetylide, and dilithium ethynylbis(trialkylborates), from dilithium acetylide, are powerful alternative synthons for the preparation of 1,1-dialkyl-, tri-alkyl-, and tetra-alkyl-olefins (Scheme 4) (H.C. Brown, A.B. Levy, and M.M. Midland, J.Amer.chem.Soc., 1975, $\underline{97}$, 5017; A. Pelter et $al.$, Tetrahedron Letters, 1975, 1633 and 3327; G. Zweifel and R.P. Fisher, Synthesis, 1975, 376; and N. Miyaura et $al.$, $ibid.$, 1975, 669).

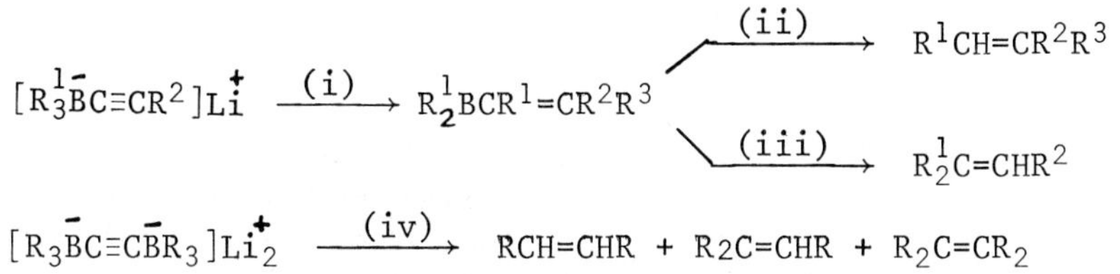

$$[R_3\bar{B}C{\equiv}CR^2]L\overset{+}{i} \xrightarrow{\ (i)\ } R_2^1BCR^1{=}CR^2R^3$$

$$\xrightarrow{\ (ii)\ } R^1CH{=}CR^2R^3$$

$$\xrightarrow{\ (iii)\ } R_2^1C{=}CHR^2$$

$$[R_3\bar{B}C{\equiv}\bar{C}R_3]L\overset{+}{i}_2 \xrightarrow{\ (iv)\ } RCH{=}CHR + R_2C{=}CHR + R_2C{=}CR_2$$

Reagents: (i) R^3Hal or $R_2^3SO_4$; (ii) acid; (iii) $I_2(R^3{=}H)$; (iv) BrCN, NaOMe

Scheme 4

(3) Addition of organometallic reagents.

Since Wurtz-type coupling reactions between vinylcoppers and alkyl halides occur in good yield and without loss of stereochemistry, the addition to alkynes of alkylcoppers, usually in combination with a magnesium salt or an alkyllithium, has proved a valuable route to alkenes (G.H. Posner, Org. Reactions, 1975, $\underline{22}$, 253; J.F. Normant, "New Applications of Organometallic Reagents in Organic Synthesis", ed. D. Seyferth, Elsevier, Amsterdam, 1976, p. 219; for use of lithium dialkylcuprates, see A. Alexakis, G. Cahiez, and J.F. Normant, Synthesis, 1979, 826). A useful extension of this procedure follows from the smooth Michael additions of the intermediate alkenylcoppers to α,β-unsaturated carbonyl compounds (A. Marfat, P.R. McGuirk, and P. Helquist, J.org.Chem., 1979, $\underline{44}$, 3888).

$$n\text{-}BuC{\equiv}CH \xrightarrow{\text{EtCuMgBr}_2} \underset{\substack{\text{Et} \quad\quad \text{CuMgBr}_2}}{\overset{\substack{\text{n-Bu} \quad\quad \text{H}}}{C=C}} \xrightarrow{\text{RBr}} \underset{\text{Et} \quad\quad \text{R}}{\overset{\text{n-Bu} \quad\quad \text{H}}{\diagup\diagdown}}$$

$$\Big\downarrow \begin{array}{l}\text{(i)} \ R^1CH{=}CR^2COR^3 \\[6pt] \text{(ii)} \ NH_4Cl\end{array}$$

Et   R¹

n-Bu —COR³

R²

A similar technique which is more versatile in that either geometrical isomer of a trialkylolefin can be obtained by adjusting the reaction conditions, is the treatment of an alkyne di-isobutyl-aluminium hydride adduct with an alkyl-lithium, followed by alkylation (S. Baba, D.E. van Horn, and E.-i. Negishi, Tetrahedron Letters, 1976, 1927; J.J. Eisch and G.A. Damasevitz, J.org.Chem., 1976, 41, 2214; K. Uchida, K. Utimoto, and H. Nozaki, *ibid.*, p. 2215).

$$R^1C{\equiv}CR^2 \xrightarrow{\text{(i)}} \begin{cases} \xrightarrow{n\text{-}C_7H_{16}\text{-}Et_2O} \underset{\substack{H \quad\quad AlBu^i_2}}{\overset{\substack{R^1 \quad\quad R^2}}{C=C}} \xrightarrow[\text{(iii)}]{\text{(ii)}} \underset{\substack{H \quad\quad R^3}}{\overset{\substack{R^1 \quad\quad R^2}}{C=C}} \\[24pt] \xrightarrow{n\text{-}C_7H_{16}} \underset{\substack{H \quad\quad R^2}}{\overset{\substack{R^1 \quad\quad AlBu^i_2}}{C=C}} \xrightarrow[\text{(iii)}]{\text{(ii)}} \underset{\substack{H \quad\quad R^2}}{\overset{\substack{R^1 \quad\quad R^3}}{C=C}} \end{cases}$$

Reagents: (i) i-Bu₂AlH; (ii) MeLi; (iii) R³Hal

Rather similar results have been reported using vinylmercurials produced by mercuration of alkynes, a procedure which is claimed to have a greater degree of tolerance for other functional groups than either hydroalanation or hydroboration (R.C. Larock, J.org.Chem., 1975, 40, 3237).

Mixed-metal catalysis appears to be particularly fruitful in promoting alkene formation by additions to alkynes. An important reagent in this connection is bis(cyclopentadienyl)-zirconium dichloride: this, when used in conjunction with trialkylalanes, promotes alkyl addition to the triple bond (E.-i. Negishi and D.E. Van Horn, J.Amer.chem.Soc., 1978, $\underline{100}$, 2252).

$$R^1C\equiv CR^2 \xrightarrow[\text{0-50 }^\circ\text{C}]{R^3_3Al/Cl_2ZrCp_2} \quad \begin{array}{c} R^1 \\ \diagdown \\ R^3 \end{array} \Big/ \hspace{-0.5em} = \hspace{-0.5em} \Big\backslash \begin{array}{c} R^2 \\ \\ ML_n \end{array} \xrightarrow{H_2O} \quad \begin{array}{c} R^1 \\ \diagdown \\ R^3 \end{array} \Big/ \hspace{-0.5em} = \hspace{-0.5em} \Big\backslash \begin{array}{c} R^2 \\ \\ H \end{array}$$

$$(15)$$

Further alkylation of the alkenylmetal intermediate (15) is achieved by treatment with an alkyl bromide in the presence of a Pd(0) or Ni(0) phosphine complex in combination with zinc or cadmium chlorides (E.-i. Negishi *et al.*, *ibid.*, p. 2254).

$$(15) \xrightarrow[\text{Pd(PPh}_3)_4,\text{ZnCl}_2]{R^4\text{Hal}} \quad \begin{array}{c} R^1 \\ \diagdown \\ R^3 \end{array} \Big/ \hspace{-0.5em} = \hspace{-0.5em} \Big\backslash \begin{array}{c} R^2 \\ \\ R^4 \end{array}$$

(v)   Other methods.

This section briefly discusses some of the many other methods available for the synthesis of olefins, several of which commence with ketonic starting materials. Thus, the Bamford-Stevens procedure utilises the elimination of sulphinite anion from sulphonyl-hydrazones of ketones containing an α-CH group, and is initiated by strong bases (R.H. Shapiro, Org.Reactions, 1976, $\underline{23}$, 405). The reaction can be terminated using electrophiles other than water (A.R. Chamberlin and F.T. Bond, Synthesis, 1979, 44).

$$R^1CH_2CR^2=NNHTs \xrightarrow{MeLi} R^1CH_2CR^2=N\bar{N}Ts \xrightarrow[-Ts^-]{MeLi}$$

$$R^1CH=CR^2N=NLi \xrightarrow[-N_2]{H_2O} R^1CH=CHR^2$$

In the Bamford-Stevens reaction, where two alternative enolisable sites exist for double-bond incorporation, selectivity is often observed under conditions of kinetic control, and in some cases a preference for the formation of the *cis* isomer has also been reported.

The second technique, developed by McMurry and co-workers (J.E. McMurry, Acc.chem.Res., 1974, $\underline{7}$, 281), involves the reductive coupling of ketones or aldehydes with low valency titanium species generated by reduction of titanium(III) or titanium(IV) halides.

$$R_2CO \xrightarrow[\text{or } TiCl_3-LiAlH_4]{TiCl_4-Zn} R_2C=CR_2$$

This method seems likely to find applications mainly in the synthesis of tetra-substituted alkenes, for which the Wittig carbonyl olefination is generally inadequate. Since the reaction involves a pinacol-type reductive dimerization, it is not obvious that unsymmetrically substituted alkenes can be obtained, but by using an excess of a small ketone in the presence of a much larger one, or a 1:1 proportion of acetone with a diaryl ketone, excellent yields of the mixed product have been obtained (J.E. McMurry and L.R. Krepski, J.org. Chem., 1976, $\underline{41}$, 3929). This procedure has already proved exceptionally useful for preparing highly hindered olefins like tetrakis(neopentyl)ethylene (G.A. Olah and G.K.S. Prakash, *ibid.*, 1977, $\underline{42}$, 580).

Terminal olefins may be obtained by methylenation of ketones with methylene iodide/magnesium (F. Bertini *et al.*, Tetrahedron, 1970, $\underline{26}$, 1281) or by the use of methylene iodide or bromide with zinc activated by trimethylaluminium or titanium tetrachloride (K. Oshima *et al.*, Tetrahedron Letters, 1978, 2417), or by treating Eschenmoser's salt (16) with a Grignard reagent and then proceeding *via* an amine oxide elimination (J.L. Roberts, P.S. Borromeo, and C.D. Poulter, Tetrahedron Letters, 1977, 1299).

$$R^1COR^2 \xrightarrow{CH_2I_2/Mg} R^1R^2C=CH_2$$

$$\uparrow \text{(i) } H_2O_2 \quad \text{(ii) heat}$$

$$CH_2=NMe_2I \xrightarrow{R^1R^2CHMgX} R^1R^2CHCH_2NMe_2$$

(16)

The interconversions of mono-olefins by base-catalysed

isomerization have been reviewed (H. Pines and W.M. Stalick, "Base Catalysed Reactions of Hydrocarbons and Related Compounds", Academic Press, New York, 1977, ch. 2).

(c) Physical properties of alkenes

Important advances have been made in several areas, notably in the use of spectroscopic data to demonstrate the location of double bonds and the relative positions of the surrounding substituents. Correlations of the ^{1}H nuclear magnetic resonance (nmr) spectroscopic shifts and coupling constants in alkenes have been reviewed (V.S. Watts and J.H. Goldstein in "The Chemistry of Alkenes" Vol. 2, ed. J. Zabicky, Wiley-Interscience, London, 1970, ch. 1). Proton to carbon-13 coupling constants can also be used to distinguish *cis* from *trans* isomers (J.E. Anderson, Tetrahedron Letters, 1975, 4079; J.W. de Haan and L.J.M. van den Ven *ibid.*, 1971, 3965). The use of nmr spectra for the recognition of the stereochemistry of alkenes has been reviewed (G.J. Martin and M.L. Martin, Progress in NMR Spectroscopy, 1972, 8, 163), and the regularities in the nmr shifts of 200 olefins have been analysed (O. Yamamoto *et al.*, **Analyt.Chem.**, 1972, 44, 2180). Nmr studies of congested tetra-substituted alkenes suggest that the conformational arrangement of their substituents may lead to non-equivalence (D.S. Bomse and T.H. Morton, Tetrahedron Letters, 1975, 781; R.F. Langler and T.T. Tidwell, *ibid.*, p. 777). Amongst very crowded alkenes, tetrakis(t-butyl)ethylene still eludes synthesis, although close analogues are now known (A. Krebs and W. Rüger, *ibid.*, 1979, 1305).

Vertical ionization potentials (IP's) of alkenes can be measured precisely and rapidly by photoelectron spectroscopy (PES); the values found have been related to the calculated MO energies (K.B. Wiberg *et al.*, J.Amer.chem.Soc., 1976, 98, 7179; G. Mouvier *et al.*, Bull.soc.Chim.France, 1974, 567 and J. electron Spec. Related Phenomena, 1973, 2, 225). *cis* and *trans*-Isomers have the same π-IP's, but display differences in their σ-IP's. Increased alkylation lowers both π- and σ-IP's in a regular fashion (D.A. Krause, J.W. Taylor, and R.F. Fenske, J.Amer.chem.Soc., 1978, 100, 718). PES has also been used to study the conformations of normal and congested olefins (T.H. Morton *et al.*, *ibid.*, 1976, 98, 4732), an area being explored by *ab initio* calculations (O. Ermer and S. Lifson, Tetrahedron, 1974, 30, 2425). The wave mechanical

properties of the π-bond have been discussed by several
authors (C.A. Coulson and E.T. Stewart, in "The Chemistry of
Alkenes", Vol. 1, ed. S. Patai, Wiley-Interscience, London,
1964, ch. 1).

The ultraviolet spectra and excited state structures of
ethylene and its simple alkyl-derivatives have been reviewed
and the experimental data related to MO calculations (A.J.
Merer and R.S. Mulliken, Chem.Rev., 1969, 69, 639). The
Cotton effects of many olefins have been correlated with Mills'
and Brewster's rules (A.I. Scott and A.D. Wrixon, Tetrahedron,
1971, 27, 4787) and the sign of the Cotton effect has been
related to the chirality of a series of π-complexes of olefins
with tetracyanoethylene (A.I. Scott and A.D. Wrixon, *ibid.*,
1972, 28, 933). Rotational isomerism about the sp^2-sp^3 bonds
in alkenes has been reviewed (G.J. Karabatsos and D.J.
Fenoglio, Topics **Stereochem.** 1970, 5, 167).

(d)   Chemical reactions of alkenes

The enormous literature on the chemical reactions of alkenes,
which reflects their importance as industrial and laboratory
intermediates and products, can only be dealt with in a very
selective fashion. The material dealt with below has been
chosen to highlight the most significant developments; the
reader is referred to the major sources cited earlier for more
thorough treatments.

(i) Electrophilic additions

Several important reviews of this topic have appeared (R.C.
Fahey, Topics in Stereochem., 1968, 3, 237; G.H. Schmid and
D.G. Garratt in "The Chemistry of Double-bonded Functional
Groups", Wiley-Interscience, ed. S. Patai, London, 1977, ch.
9; R. Bolton in "Comprehensive Chemical Kinetics", ed. C.H.
Bamford and C.F.H. Tipper, Vol. 9, Elsevier, London, 1973,
p. 1). Ingold's notation (Ad$_E$n, where n indicates the
molecularity of the rate-determining step) is now generally
adopted, and several variations of the main Ad$_E$2 and Ad$_E$3
mechanisms are well-established for acyclic alkenes. The
most important are considered below.
(i) Concerted bimolecular additions, such as are thought to
occur with the boron hydrides; ostensibly forbidden on
orbital symmetry grounds, the evidence for a four-centre
mechanism is not yet soundly based on kinetic data (Bolton,

loc.cit., p. 49), though stereochemical and relative rate data are in accord with a transition state such as (17) (G.M.L. Cragg, "Organoboranes in Organic Synthesis", Dekker, New York, 1973, ch. 3).

(17)

(ii) Rate-determining bimolecular formation of bridged 'onium ions, thought to occur during bromination in solutions containing bromide ions; there is evidence in certain cases for the prior formation of π-complexes.

(iii) Rate-determining bimolecular formation of open-chain cations, thought to be the **pathway** followed in acid-catalysed hydration, for which rate-determining proton-transfer has been convincingly demonstrated.

(iv) An Ad_E3 (i.e., third order) pathway, as for additions of hydrogen halides proceeding *via* transition states such as (18) (*anti*-addition route) or (19) (*syn*-addition route).

In many reactions there may be competing mechanisms and reliable broadly-based evidence is often lacking, so that the exact reaction pathway(s) may be in doubt. Ion pairs are formed during many electrophilic additions to alkenes, particularly in non-polar solvents, and their formation may complicate the stereochemistry and kinetic effects of added nucleophiles.

(1) Addition of H-X

This group of reactions includes addition of hydrogen halides, of water by acid-catalysed hydration and *via* hydroboration-oxidation, of boranes, and of carboxylic acids. Addition of hydrogen chloride to simple alkenes in organic solvents proceeds *via* carbocation-containing ion pairs; rearrangement and solvent incorporation often occur (R.C. Fahey and C.A.A. McPherson, J.Amer.chem.Soc., 1969, 91, 3865).

$$Me_3CCH=CH_2 \xrightarrow{HCl, \; AcOH} Me_3CCHClCH_3 \; (37\%) + Me_3CCH(OAc)CH_3$$

$$(19\%) + Me_2CClCHMeCH_3 \; (44\%)$$

Kinetic data for such reactions vary from solvent to solvent, the reaction being second order in acetic acid but third order (second order in HCl) in nitromethane or ether, rising to almost fourth order (nearly third order in HCl) in n-heptane (Y. Pocker, K.D. Stevens, and J.J. Champoux, *ibid.*, p. 4199; Y. Pocker and R.F. Buckholz, *ibid.*, 1970, 92, 4033). The most commonly ascribed pathway is Ad_E3 with rate-determining ion-pair formation occurring through proton transfer from HCl molecule, accompanied in the transition state by a second HCl molecule or a molecule of solvent, or if one is present by an ionic additive such as lithium perchlorate. Under conditions in which no solvent is available, as in the gas phase or in surface reactions, evidence has been obtained for rate-determining collisions of $(HCl)_2$ with an alkene/HCl complex (M.J. Haugh and D.R. Dalton, *ibid.*, 1975, 97, 5674).

$$R_2C=CR_2 \xrightarrow[\text{slow}]{\text{HCl (XY)}} [\ R_2\overset{+}{\underset{H}{C}}-CR_2\ ][\ ClXY]^- \quad (XY = HCl, AcOH, LiClO_4, \text{etc.})$$

In reactions where the steric course of the addition can be determined, it has been found that *anti* HCl addition occurs in nitromethane, but *syn* addition competes in acetic acid suggesting that both Ad_E2 and Ad_E3 pathways may be occurring in that solvent (Y. Pocker and K.D. Stevens, *ibid.*, 1969, 91, 4205; R.C. Fahey and M.W. Monahan, *ibid.*, 1970, 92, 2816).

Hydrogen bromide additions are complicated by the facility of the alternative free-radical chain addition pathway, which can only be reliably arrested by carrying out the addition in triply-degassed, dark, solutions under an inert gas (D.J. Pasto, G.R. Meyer and B. Lepeska, *ibid.*, 1974, 96, 1858). When these precautions are observed, the evidence obtained suggests an Ad_E3 mechanism in acetic acid, the rate equation taking the form:

$$-d[C=C]/dt = k_{RBr}[C=C][HBr]^2 + k_{ROAc}[C=C][HBr].$$

This indicates that either a solvent molecule or a second HBr molecule is present in the transition state of the cation-forming step.

Structural effects in acid-catalysed hydration have been studied in depth (V.J. Nowlan and T.T. Tidwell, Acc.chem.Res., 1977, 10, 252). Reaction rates can be correlated with the substituents' σ^+-constants in 4-substituted styrenes, and similar behaviour in 1,1-dialkyl- and 1,2-dialkyl-ethylenes indicates rate-determining C-protonation (T.T. Tidwell *et al.*, J.Amer.chem.Soc., 1977, 99, 3395, 3401, and 3408) leading to open-chain carbocations.

$$R_2C=CR_2 \xrightarrow[H^+]{slow} R_2\overset{+}{C}-CHR_2 \xrightarrow[H_2O]{fast} R_2C(OH)CHR_2$$

The transition state for the first step appears to have much double bond character, since the relative rates of hydration of the (E) and the (Z) isomers of crowded alkenes are not very different (W.K. Chwang and T.T. Tidwell, J.org. Chem., 1978, 43, 1904). The stereochemistry of the hydration of simple acyclic alkenes is not well-established.

Perhaps the most synthetically useful type of electrophilic addition to alkenes is hydroboration, i.e., the addition of borane or a substituted borane. Numerous reviews have appeared (H.C. Brown, "Organic Synthesis *via* Boranes", Wiley-Interscience, New York 1975; G.M.L. Cragg, "Organoboranes in Organic Synthesis", Dekker, New York, 1973). Anti-Markownikov hydration is best achieved by hydroboration followed by oxidation. A number of reagents for highly regioselective hydroboration have been developed, particularly notable being 9-borabicyclo[3,3,1]nonane (9BBN) (20) (H.C. Brown, E.F. Knights, and C.G. Scouten, J.Amer.chem.Soc., 1974, 96, 7765), t-hexyl- (termed "thexyl") borane (21) (H.C. Brown, E.-i. Negishi, and J.J. Katz, *ibid.*, 1975, 97, 2791, and catecholborane (22) (H.C. Brown and S.K. Gupta, *ibid.*, p. 5249). Diborane itself is conveniently handled as its complex

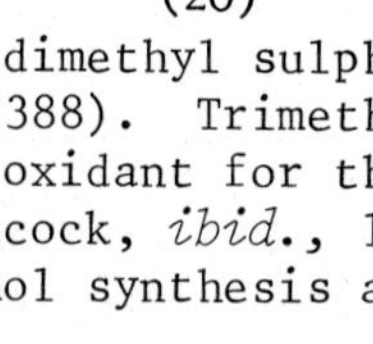

$$Me_2CHCMe_2BH_2$$

(20) (21) (22)

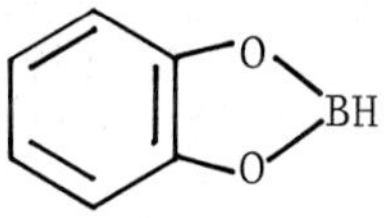

with dimethyl sulphide (R.M. Adams *et al.*, J.org.Chem., 1971, 36, 2388). Trimethylamine N-oxide dihydrate is useful as a mild oxidant for the B-C cleavage step (G.W. Kabalka and H.C. Hedgecock, *ibid.*, 1975, 40, 1776). Typical examples of alcohol synthesis are shown below.

$$n\text{-BuCH=CH}_2 \xrightarrow{R_2BH} n\text{-BuCH}_2CH_2BR_2 \xrightarrow[NaOH]{H_2O_2} n\text{-BuCH}_2CH_2OH$$

$$\text{i-PrCH=CHMe} \xrightarrow[\text{(ii) } H_2O_2, \text{ NaOH}]{\text{(i) } 9BBN, \text{ THF}} \text{i-PrCH}_2\text{CH(OH)Me} \quad (>99\%)$$

Anti-Markownikov addition of hydrogen bromide and hydrogen iodide is achieved by treatment of hydroboration products with bromine/sodium methoxide and iodine/sodium hydroxide, respectively (H.C. Brown *et al.*, J.Amer.chem.Soc., 1968, 90, 5038 and 1970, 92, 6660), e.g.

$$CH_2=CHPh \xrightarrow{\text{i-Am}_2BH} \text{i-Am}_2BCH_2CH_2Ph \xrightarrow{I_2, \text{NaOH}} ICH_2CH_2Ph$$

Amination of boranes is best achieved using mesitylene-sulphonylhydroxylamine (Y. Tamura *et al.*, Synthesis, 1974, 196). Anti-Markownikov addition of carboxylic acids results from a high yield sequence involving mercuration of organo-boranes (R.C. Larock, J.org.Chem., 1974, 39, 834).

$$RCH=CH_2 \xrightarrow{Me_2SBH_3} (RCH_2CH_2)_3B \xrightarrow[\text{(ii) } I_2]{\text{(i) } Hg(OAc)_2} RCH_2CH_2OAc$$

Overall addition of an aldehyde to a carbon-carbon double bond is achieved by hydroboration followed by carbonylation (see for example, H.C. Brown, J.L. Hubbard, and K. Smith, Synthesis, 1979, 701).

(2) *Addition of halogens and interhalogens*

Non-stereospecific addition of fluorine can be achieved controllably using xenon difluoride in the presence of hydrogen fluoride (M. Zupan and A. Pollak, Tetrahedron Letters, 1974, 1015), presumably *via* an open-chain carbocation.

Very complex mechanisms have been proposed to account for the experimental data accumulated over the years pertaining to ionic chlorination (great care must be taken to exclude free-radical chain pathways: see for example M.L. Poutsma, J.Amer. chem.Soc., 1965, 87, 4293). Established facts include a clear second order rate law (G.H. Schmid *et al.*, *ibid.*, 1973, 95, 160) and the exclusive formation of *anti*-adducts (except in highly polar solvents), suggesting the involvement of bridged chloronium ion intermediates (M.L. Poutsma and J.L. Kartch, *ibid.*, 1967, 89, 6595) in an Ad$_E$2 mechanism. Both alkene-chlorine charge-transfer complexes (G.A. Olah *et al.*,

ibid., 1974, **96**, 3581) and chloronium ions (G.A. Olah and
J.M. Bollinger, *ibid.*, 1967, **89**, 4744) have been observed
by nmr spectroscopy in suitable solvents, but the possible
intervention of open-chain carbenium ions (trivalent
carbocations, R_3C^+) and ion pairs are still matters of
dispute (see for example, P.B.D. de la Mare, M.A. Wilson, and
M.J. Rosser, J.chem.Soc. Perkin II, 1973, 1480). As the
examples below show, nucleophilic solvents lead to solvent-
incorporated products and carbenium ion rearrangements may
occur competitively in suitable cases. Added chloride ion
does not appear to change the product distribution, but
catalysis by HCl is well-established (M.C. Cabaleiro *et al.*,
J.chem.Soc.(B), 1968, 1026).

$$\text{t-BuCH=CHR} \xrightarrow[\text{CCl}_4]{\text{Cl}_2} \text{t-BuCHClCHClR}$$

$$+$$

$$\text{CH}_2\text{=CMeCHMeCHClR}$$

(R = t-Bu)

syn-Chlorination of acyclic alkenes can be achieved using
molybdenum pentachloride (S. Uemura, A. Onoe, and M. Okano,
Bull.chem.Soc.Japan, 1974, **47**, 3121).

Bromination follows a more complex rate law than
chlorination

$$-d[Br_2]/dt = k_2[Br_2][C=C] + k_3[Br_2]^2[C=C] + k'_3[Br_3^-][C=C]$$

although the above expression simplifies to overall second
order kinetics at low bromine, high bromide, concentrations
(J.-E. Dubois and G. Mouvier, Bull.Soc.chim.France, 1968,
1426). A possible explanation of this and other data is that
the rate-determining step involves solvent-, bromide-, or
bromine-assisted conversion of a charge-transfer complex into
a cyclic bromonium ion.

$$\underset{\text{fast}}{\overset{Br_2}{\rightleftharpoons}} \qquad \longrightarrow Br-Br \qquad \underset{\text{slow}}{\overset{X}{\longrightarrow}} \qquad Br^+ \quad + \quad Br^-$$

$$(X = \text{solvent, } Br_2, \text{ or } Br^-)$$

Good evidence for bromonium ions has been obtained by nmr spectroscopy (G.A. Olah *et al.*, J.Amer.chem.Soc., 1974, 96, 3565). 1,1-Disubstituted alkenes react faster than *cis*-1,2-disubstituted ones, which themselves react faster than the *trans*-isomers, suggesting some asymmetry in the structure of the bromonium ion. The generally accepted *anti*-addition stereochemistry of electrophilic bromination (J.H. Rolston and K. Yates, *ibid.*, 1969, 91, 1477) has been challenged (G.H. Schmid and D.G. Garratt, "The Chemistry of Double-bonded Functional Groups", ed. S. Patai, Wiley-Interscience, London, 1977, ch. 9). There has also been a considerable controversy over the extent of bromonium ion participation, but the current view seems to be in their favour (G.H. Schmid and T.T. Tidwell, J.org.Chem., 1978, 43, 460; E. Bienvenue-Goetz and J.E. Dubois, Tetrahedron, 1978, 34, 2021).

Electrophilic attack by iodine is important mainly in synthetically useful additions of interhalogens and pseudo-halogens (e.g. INCO). These and a number of related procedures are shown in Scheme 5. *anti*-Addition stereo-specificity is observed, suggesting that intermediate bridged iodonium ions are formed (A. Hassner, Acc.chem.Res., 1971, 4, 9).

(3) Other electrophilic additions.

Oxymercuration is the general term given to addition of mercury salts to alkenes. Such reactions lead *via* bridged mercurinium ions (W. Kitching, Organometal.Reactions, 1972, 3, 319; G.A. Olah and P.R. Clifford, J.Amer.chem.Soc., 1971, 93, 2320) to solvent- or added nucleophile-incorporated products which have a stereochemistry consistent with an *anti*-addition pathway.

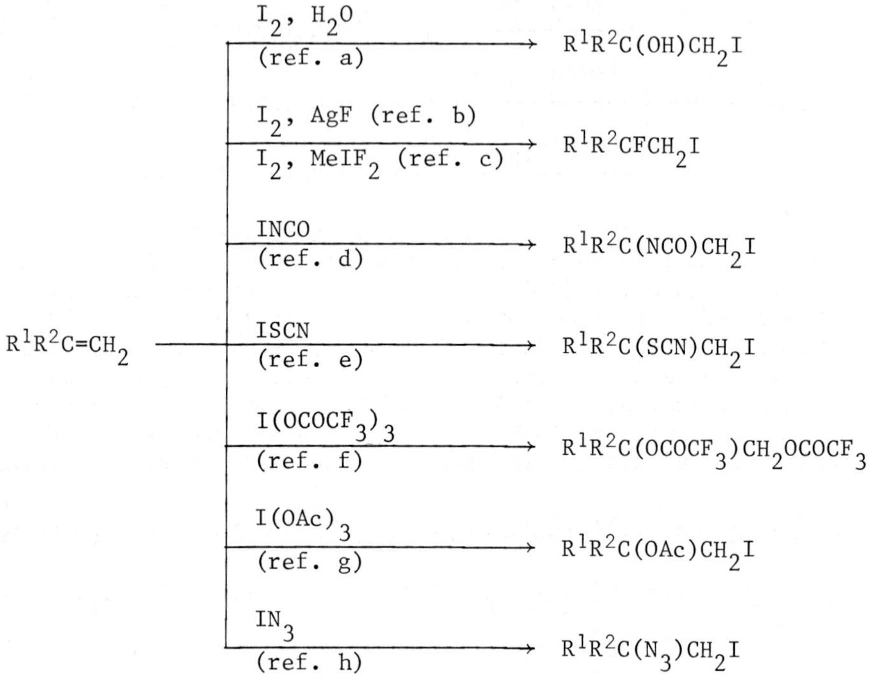

$$R^1R^2C=CH_2 \xrightarrow{}$$

$$\xrightarrow[\text{(ref. a)}]{I_2,\ H_2O} R^1R^2C(OH)CH_2I$$

$$\xrightarrow[\text{I_2, MeIF$_2$ (ref. c)}]{I_2,\ AgF\ \text{(ref. b)}} R^1R^2CFCH_2I$$

$$\xrightarrow[\text{(ref. d)}]{INCO} R^1R^2C(NCO)CH_2I$$

$$\xrightarrow[\text{(ref. e)}]{ISCN} R^1R^2C(SCN)CH_2I$$

$$\xrightarrow[\text{(ref. f)}]{I(OCOCF_3)_3} R^1R^2C(OCOCF_3)CH_2OCOCF_3$$

$$\xrightarrow[\text{(ref. g)}]{I(OAc)_3} R^1R^2C(OAc)CH_2I$$

$$\xrightarrow[\text{(ref. h)}]{IN_3} R^1R^2C(N_3)CH_2I$$

Scheme 5

(a) R.C. Cambie *et al.*, J.chem.Soc. Perkin I, 1977, 226;
(b) L.D. Hall and K.D. Jones, Canad.J.Chem., 1973, 51, 2902;
(c) M. Zupan and A. Pollak, J.chem.Soc. Perkin I, 1976, 1745 and J.org.Chem., 1976, 41, 2179;
(d) A. Hassner *et al.*, J.Amer.chem.Soc., 1970, 92, 1326;
(e) J.C. Hinshaw, Tetrahedron Letters, 1972, 3567;
(f) J. Buddrus, Angew.Chem.internat.Edn., 1973, 12, 163;
(g) R.C. Cambie *et al.*, J.chem.Soc. Perkin I, 1977, 2231;
(h) A. Hassner, Acc.chem.Res., 1971, 4, 9.

$$CH_3CH=CH_2 \quad \begin{array}{l} \xrightarrow{\text{HgCl}_2, \text{ ROH}} CH_3CH(OR)CH_2HgCl \\[2mm] \xrightarrow{\text{Hg(OAc)}_2, \text{ AcOH}} CH_3CH(OAc)CH_2HgOAc \\[2mm] \xrightarrow{\text{Hg(OAc)}_2, \text{ R}_2NH} CH_3CH(NR_2)CH_2HgOAc \end{array}$$

The importance of these reactions lies in the facile demercuration which occurs on treatment of the adducts with sodium borohydride (H.C. Brown and P.J. Geoghegan, J.org. Chem., 1970, 35, 1844) or with lithium in THF followed by hydrolysis (V.G. Aranda *et al.*, Synthesis, 1974, 135). The overall sequence of mercuration-demercuration thereby leads to Markownikov orientated hydration, carboxylation, or ether- or amine-formation: conditions are mild and rearrangements, a frequent complication in acid-catalysed additions, are avoided.

$$Me_3CCH(OH)CH_3 \xleftarrow{\text{(i),(ii)}} Me_3CCH=CH_2 \xrightarrow{\text{(iii),(iv)}} Me_2C(OH)CHMe_2$$

Reagents: (i) $Hg(OAc)_2$, aq. THF; (ii) $NaBH_4$; (iii) H_2SO_4;
           (iv) H_2O

Although lead(IV) and thallium(III) are isoelectronic with mercury(II), their alkene additions proceed differently and overall oxidation with solvent- or metal ligand-incorporation results (A. Lethbridge, R.O.C. Norman, and C.B. Thomas, J. chem.Soc. Perkin I, 1974, 1929; R.M. Moriarty, "Selective Organic Transformations", 1972, 2, 183).

$$CH_2=CH_2 \xrightarrow{\text{TlCl}_3, \text{ H}_3O^+} HOCH_2CH_2OH + CH_3CHO$$

$$Me_2C=CMe_2 \xrightarrow[\text{MeOH}]{\text{Tl(NO}_3)_3} t\text{-BuCOMe} + Me_2C(OMe)C(OMe)Me_2$$
$$+ Me_2C(ONO_2)C(ONO_2)Me_2$$
$$+ Me_2C(OMe)C(ONO_2)Me_2$$

Electrophilic addition of oxygen will be considered under oxidation. Attack by positive sulphur and selenium are both generally regarded as proceeding by an Ad_E2 mechanism, the *anti*-stereochemistry arising as the result of bridged episulphenium ions (23) and episeluranes (24), respectively.

(23)   (24)

The selenium addition reaction has been developed into a
synthetic approach to allyl alcohols and their esters using
the oxidative fragmentation of the adduct selenides (D.L.
Clive, Tetrahedron, 1978, $\underline{34}$, 1049; H.J. Reich, Acc.chem.Res.,
1979, $\underline{12}$, 22). β-Hydroxyselenides are made by the combined
use of PhSeBr and silver trifluoroacetate (Clive, *loc.cit.*),
and β-alkoxy- and acetoxy-selenides are available by obvious
modification of the reaction.

Reagents: (R^1 = n-Pr) (i) PhSeBr, CF_3CO_2Ag, then aq. MeOH,
                              $KHCO_3$;

                   (ii) PhSeBr, AcOH, AcOK.

   Although carbocations certainly attack alkenes readily, the
onset of cationic polymerization limits the utility of this
electrophilic addition. Two notable exceptions are the Prins
reaction (reviewed by D.R. Adams and S.P. Bhatnagar, Synthesis,
1977, 661) (see Section 2d-viii).

$$Me_2C=CHMe \xrightarrow[H_2O]{CH_2O, \ BF_3} Me_2C(OH)CHMeCH_2OH$$

and the acylation of ethylene (J.K. Groves, Quart.Rev., 1972,
$\underline{1}$, 73).

$$CH_2=CH_2 \xrightarrow[AlCl_3, \ 0\ °C]{EtCOCl} EtCOCH_2CH_2Cl$$

(ii) Nucleophilic additions

In the absence of electron-withdrawing functional groups there are few important types of nucleophilic additions to alkenes. Grignard reagents and lithium alkyls react with olefinic hydrocarbons to give adducts and proton-abstraction products (H. Felkin and G. Swierczewski, Tetrahedron, 1975, 31, 2735; J. Klein and A. Medlik-Belan, Chem.Comm., 1975, 877). The regioselectivity of the addition of Grignard reagents to terminal alkenes varies from 99% C_1-attack by t-BuMgCl to 100% C_2-attack by PhMgBr; the orientation pattern can be correlated with the sum of the Taft σ^*-values for the R-substituents in R_3CMgX (H. Lehmkuhl *et al.*, Ann., 1975, 145).

Alkylarenes can be ethylated by ethylene using sodium as the promoter in 2-chlorotoluene; temperatures of 175-200 $^\circ$C and an over-pressure of ethylene are required (B. Stepanović and H. Pines, J.org.Chem., 1969, 34, 2106):

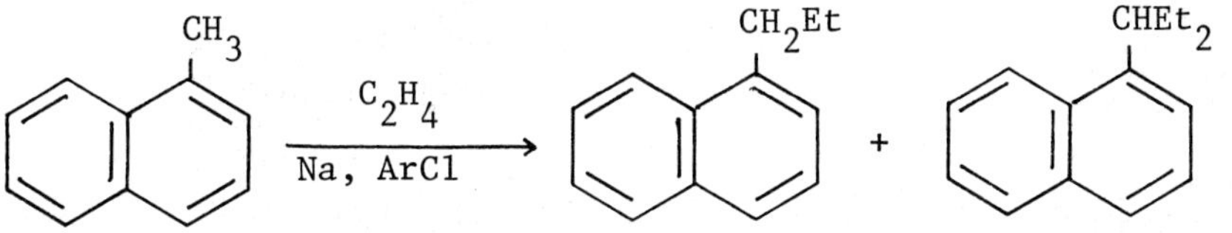

Lower reaction temperatures are possible if potassium replaces sodium, but other reactions besides side-chain alkylation then occur. Higher alkenes, being less susceptible to nucleophilic addition, react less readily (reviews: H. Pines, Acc.chem. Res., 1974, 7, 155 and Synthesis, 1974, 309).

(iii) Free-radical additions

The kinetics and orientation of free-radical additions to alkenes are now much more thoroughly understood, although some interesting problems remain unsolved (J.M. Tedder and J.C. Walton, Acc.chem.Res., 1976, 9, 183; P.I. Abell in "Free Radicals", Vol. 2, ed. J.K. Kochi, Wiley-Interscience, New York, 1973, p. 63; J.I.G. Cadogan and M.J. Perkins in "The Chemistry of Alkenes", ed. S. Patai, Wiley-Interscience, London, 1964, ch. 9). The general reaction scheme for free-radical addition is shown below.

Initiation   A–B $\xrightarrow{\text{heat or light}}$ A$^{\cdot}$ + B$^{\cdot}$

     or   A–B + Init$^{\cdot}$ $\longrightarrow$ A$^{\cdot}$ + B–Init
                                         (Init = initiator radical
                                         Ph$^{\cdot}$, etc.)

Propagation   A$^{\cdot}$ + RCH=CH$_2$ $\longrightarrow$ ACH$_2\overset{\cdot}{\text{C}}$HR

                  ACH$_2\overset{\cdot}{\text{C}}$HR + AB $\longrightarrow$ ACH$_2$CHRB + A$^{\cdot}$

Suitable substrates (A–B) for the formation of 1:1-adducts
are (attacking radical A$^{\cdot}$ first): Br–H, Br–Br, Cl–Cl, CCl$_3$–Br,
CCl$_3$–Cl, (EtO$_2$C)$_2$CH–H, RS–H (review: K. Griesbaum, Angew.Chem.
internat.Edn., 1970, 9, 273), R$_3$Si–H, Me$_2$C(OH)–H, OHCCH$_2$–H,
(RO)$_2$P(O)–H, R$_2$P–H, C$\overline{\text{F}}_3$–I, and Cl–SF$_5$.
In several of these additions, free-radical polymerization
(review: J.C. Bevington and J.R. Ebdon, "Developments in
Polymerization" Vol. 2, ed. R.N. Haward, Applied Science,
London, 1979), or the intermediate stage of such a process
yielding 2:1, 3:1, 4:1 adducts, etc., (or oligomers), may be
a competing process but can be minimised by taking a large
excess of the addendum A–B. The extent of bridging in
suitable intermediate radicals (e.g. RCHCHBrR) is a matter of
debate. *anti*-Addition, its usual consequence, is observed at
low temperatures but not exclusively and *syn*-adducts increase
with elevation of the reaction temperatures.

     The typical *anti*-Markownikov regioselectivity of free-
radical additions to unsymmetrical alkenes has usually been
explained on the basis of faster formation of the more stable
intermediate radical (*i.e.*, RCHCH$_2$A and not RCHACH$_2^{\cdot}$), though
polar, steric and bond-strength factors are also important.
Thus the rate and orientation of addition to a given alkene
may be significantly altered by a change in the attacking
radical (Tedder and Walton, *loc.cit.*).

CF$_3$CHF–$\overset{\cdot}{\text{C}}F_2$ $\xleftarrow{0.031}$ $\overset{\cdot}{\text{C}}F_3$ CHF=CF$_2$ Me$^{\cdot}$ $\xrightarrow{1.0}$ MeCHF–$\overset{\cdot}{\text{C}}F_2$

$\overset{\cdot}{\text{C}}$HF–CF$_2CF_3$ $\xleftarrow{0.016}$ $\xrightarrow[1.90]{}$ $\overset{\cdot}{\text{C}}$HF–CF$_2$Me

(numbers are values of k$_{\text{rel}}$)

(iv) Oxidation

Although an enormous variety of metallic and non-metallic oxidants attack the olefinic π-bond (P.M. Henry and G.M. Lange, "The Chemistry of Double Bonded Functional Groups", ed. S. Patai, Wiley-Interscience, London, 1977, ch. 11), the most important laboratory oxidation modes are hydroxylation, ozonolysis, epoxidation, and allylic oxidation. Important advances have also occurred in industrial oxidation reactions such as the Wacker process and ammoxidation.

(1) Hydroxylation

Alternatives to the use of alkaline permanganate for effecting *syn*-hydroxylation have emerged, the most promising of which involves the use of catalytic amounts of osmium tetroxide in combination with either t-butyl hydroperoxide (K.B. Sharpless and K. Akashi, J.Amer.chem.Soc., 1976, 98, 1986) or a tertiary amine N-oxide (V. van Rheenan, R.C. Kelly, and D.Y. Cha, Tetrahedron Letters, 1976, 1973). *syn*-Oxyamination is also possible using either osmium tetroxide (catalytic) and chloramine-T (K.B. Sharpless, A.O. Chong, and K. Oshima, J.org.Chem., 1976, 41, 177), or alkylamido-osmium trioxides, formed from OsO_4 and an amine, followed by treatment with LAH (K.B. Sharpless *et al.*, J.Amer.chem.Soc., 1975, 97, 2305), e.g.,

(2) Ozonolysis

There has been an informative controversy over the mechanism of ozonolysis, centering upon the nature of the intermediates which precede formation of the isolable secondary ozonide (25); the main contenders are the Criegee mechanism *via* the primary ozonide (26) and the Story-Murray pathway *via* the Staudinger molozonide (27), itself possibly forming *via* a peroxy-epoxide (28):

Challenges to the Criegee mechanism (R. Criegee, Angew.Chem. internat.Edn., 1975, 14, 745) stemmed mainly from the following observations.
(i) Incorporation of ^{18}O mainly into the peroxide bridge of the secondary ozonide when radio-labelled aldehydes or ketones were added to alkene ozonolysis mixtures (R.W. Murray and R. Hagen, J.org.Chem., 1971, 36, 1103).
(ii) Suppression of ozonide formation in so-called "Baeyer-Villiger" solvents such as pinacolone (P.R. Story, E.A. Whited, and J.A. Alford, J.Amer.chem.Soc., 1972, 94, 2143).
(iii) Variations in the yield and cis/trans ratios of the secondary ozonides from certain olefins resulting from changes in solvent and with dilution (V. Ramachandran and R.W. Murray, ibid., 1978, 100, 2197 and R.W. Murray et al., Tetrahedron, 1968, 24, 4347).
These challenges have either been refuted, as was the case for the ^{18}O studies by, for example, G. Klopman and C.M. Joiner (J.Amer.chem.Soc., 1975, 97, 5287) who showed that ^{18}O-exchange between added aldehyde and ozone occurs faster than ozonolysis, and D.P. Higley and R.W. Murray (ibid., 1976, 98, 4526) who reported improved analytical procedures which located the labelled oxygen in the ether, not the peroxide, bridge, or have been incorporated into the Criegee three-step mechanism without requiring the intervention of the molozonide (27) (see G.D. Fong and R.L. Kuczkowski, ibid., 1980, 102, 4763 and references cited therein). Thus, although there is spectroscopic evidence for the formation of π-complexes between ozone and non-aromatic terminal alkenes at low temperature (W.

G. Alcock and B. Mile, Chem.Comm., 1976, 5), the Criegee
mechanism stands - for the present! A book on ozonolysis
gives details of this and other aspects of the reaction (P.S.
Bailey, "Ozonation in Organic Chemistry", Vol. 1, Academic
Press, New York, 1978).

(3) Epoxidation

An important discovery made in the early 1970's was that
the epoxidation of alkenes by t-butyl hydroperoxide is
catalysed by metal complexes, especially those of molybdenum
and vanadium (M.N. Sheng and J.G. Zajacek, J.org.Chem., 1970,
35, 1839; H. Mimoun, I.S. de Roch, and L. Sajus, Tetrahedron,
1970, 26, 37). Sometimes these reactions are very regioselective (K.B. Sharpless and R.C. Michaelson, J.Amer.chem.
Soc., 1973, 95, 6136):

(93%)

Stereoselectivity is also observed in some cases, and this
has led to the development of techniques for asymmetric
epoxidation with high enantiomeric selectivity, e.g. using
titanium(IV) isopropoxide/t-butyl hydroperoxide in combination
with optically-active diethyl tartrate (T. Katsuki and K.B.
Sharpless, ibid., 1980, 102, 5974). These methods represent
a significant improvement over earlier attempts with, for
example, peroxycamphoric acid. Even terminal alkenes have
been converted into optically-active epoxides (C. Döbler and
E. Höft, Z.Chem., 1978, 18, 218).

$$RCH=CH_2 \xrightarrow[\text{(+)-3-CF}_3\text{CO-camphor}]{\text{t-BuOOH, VOSO}_4} \overset{*}{R}CH\!-\!CH_2 \; (O)$$

(ee = enantiomeric excess)   (60-100% ee)

Epoxidations and hydroxylations can also be carried out
using micro-organisms, a field of study of increasingly
promising results (R.A. Johnson in "Oxidation in Organic
Chemistry" Part C, ed. W.S. Trahanovsky, Academic Press,
London, 1978, ch. 2). Some very stereoselective conversions
have been reported, (H. Ziffer and D.T. Gibson, Tetrahedron

Letters, 1975, 2137).

$$H_2C=CH(CH_2)_4CH=CH_2 \xrightarrow[\textit{oleoverans}]{\textit{Pseudomonas}} H_2C\underset{O}{\overset{}{\diagdown}}C\overset{(CH_2)_4CH=CH_2}{\underset{H}{\diagup}}$$

Other new alternatives to the use of peroxycarboxylic acids
are listed below.
(a) Catalysis of hydrogen peroxide epoxidations by
arylselenic acids, presumed to involve perselenic acids (T.
Hori and K.B. Sharpless, J.org.Chem., 1978, $\underline{43}$, 1689).

$$Me_2C=CHBu\text{-}n \xrightarrow[\text{H}_2\text{O}_2,\ \text{CH}_2\text{Cl}_2]{\text{ArSeO}_2\text{H}\ (\frac{1}{2}\%)} Me_2\overset{O}{\triangle}Bu\text{-}n \quad (90\%)$$

(b) Triphenylsilylhydroperoxide (J. Rebek and R. McReady,
Tetrahedron Letters, 1979, 4337).
(c) The hexafluoroacetone-hydrogen peroxide adduct (R.P.
Heggs and B. Ganem, J.Amer.chem.Soc., 1979, $\underline{101}$, 2484).
(d) Photochemical epoxidation using oxygen-benzil mixtures
(N. Shimizu and P.D. Bartlett, *ibid.*, 1976, $\underline{98}$, 4193).
(e) Polymer-bound peroxybenzoic acid (C.R. Harrison and P.
Hodge, J.chem.Soc. Perkin I, 1976, 605).
  Alkene oxidations with peroxides and peroxyacids have been
reviewed (B. Plesničar in "Oxidation in Organic Chemistry",
loc.cit., ch. 3 and V.G. Dryuk, Tetrahedron, 1976, $\underline{32}$, 2855).
Selenium-based reactions, including epoxidation and allylic
functionalisation have also been reviewed (H.J. Reich,
"Oxidation in Organic Chemistry", *loc.cit.*, ch. 1).

4. Allylic oxidation

  Much work has been done on the dye-photosensitized allylic
oxidation of alkenes, regarded as an ene reaction of singlet
oxygen ($^1O_2$) (A.A. Frimer, Chem.Rev., 1979, $\underline{79}$, 359; various
authors in "Singlet Oxygen", ed. H.H. Wasserman and R.W.
Murray, Academic Press, New York, 1978; and "Singlet Oxygen:
Reactions with Organic Compounds and Polymers", ed. B. Rånby
and J.F. Rabek, Wiley-Interscience, Chichester, 1978). The
initially formed allylic hydroperoxides are readily reduced
by borohydride to allylic alcohols. A hotly contested debate
over the exact mechanism has, like ozonolysis research,
centered on the possible involvement of intermediate
perepoxides, zwitterions and 1,2-dioxetanes (Frimer, *loc.cit*)

and continues unabated. The concerted ene insertion process (mechanism A) and the formation of intermediate perepoxides (mechanism B) are the two main contenders on the basis of experimental data such as low activation enthalpies (<42 kJ/mol), absence of rearrangements, negligible substituent directing- and solvent-effects, relatively small isotope effects, and strong preference for *syn*-addition.

Mechanism A

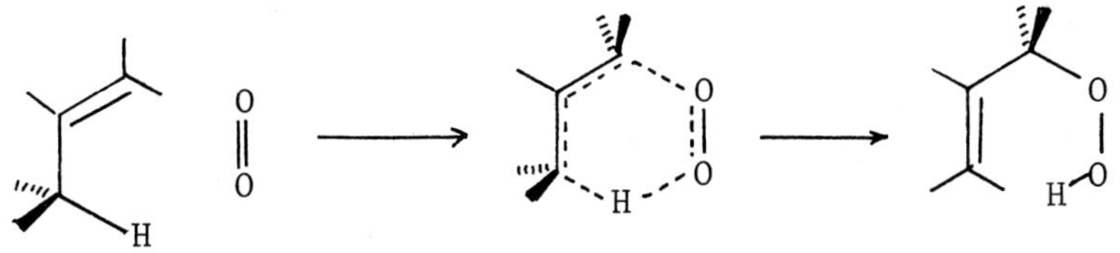
O
O
O
O
H
O
O
H

Mechanism B

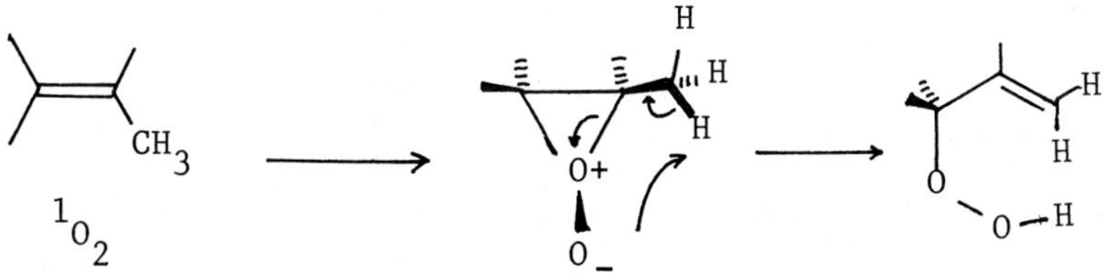
CH3
1O2
H
H
H
O+
O
H
H
O
O—H

The preference for H abstraction from the more congested side of a tri-substituted alkene is difficult to rationalise by the fully concerted ene mechanism (M. Orfanopoulos, Sr. M.B. Grdina, and L.M. Stephenson, J.Amer.chem.Soc., 1979, 101, 275, 3111).

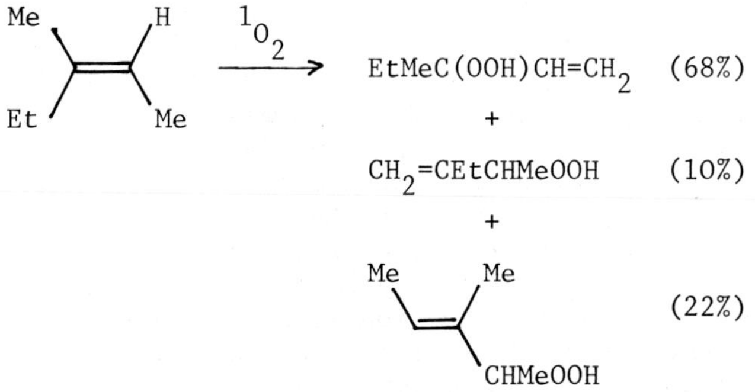
Me
H
Et
Me
1O2
EtMeC(OOH)CH=CH2
(68%)
+
CH2=CEtCHMeOOH
(10%)
+
Me
Me
CHMeOOH
(22%)

A compromise between the two most popular pathways has been proposed (L.M. Stephenson, Tetrahedron Letters, 1980, 1005). It is suggested that a π-complex (29) is formed irreversibly, and that the stereochemical bias is imparted by interactions between oxygen LUMO and alkene HOMO orbitals. However, the distinction between this and a stereoselectively formed perepoxide (30) is a fine one.

(29)   (30)

Of considerable practical importance are the various techniques for the generation of singlet oxygen, either chemically, *via* HOCl–H_2O_2 and by thermal decomposition of anthracene peroxides (31) or the ozone adducts of phosphites (32), or photochemically, using a sensitizer such as Rose

(31)   (32)

Bengal or eosin, which may if required be polymer-supported. Alkene reactivities are in the order: $Me_2C=CMe_2$ > $Me_2C=CHMe$ > MeCH=CHMe, and *cis*-isomers react faster than *trans*-isomers. These and other aspects of photosensitized oxidations have been reviewed (C.S. Foote, Pure appl.Chem., 1971, <u>27</u>, 635).

Singlet oxygen oxidation should not be confused with allylic oxidation (autoxidation) which is a free-radical process and involves triplet molecular oxygen (J.A. Howard, "Free Radicals", Vol. 2, ed. J.K. Kochi, Wiley-Interscience, New

York, 1973, p. 3).

The industrially important catalytic oxidations of ethylene to ethyleneoxide and acetaldehyde and of propene to propylene oxide and acrolein, have been reviewed (D.J. Hucknall, "Selective Oxidation of Hydrocarbons", Academic Press, London, 1974).

(v) Reduction

The low electron affinity of most non-terminal alkenes prevents reduction by dissolving metals, but tetra-alkyl-ethylenes are reduced by sodium in HMPT containing t-butanol (G.M. Whitesides and W.J. Ehman, J.org.Chem., 1970, 35, 3565). A significant development in catalytic hydrogenation is the emergence of polymer-supported and silica-bonded catalysts (e.g. I.V. Howell *et al.*, J.organometal.Chem., 1976, 107, 393; E.S. Chandrasekaren, R.H., Grubbs, and C.H. Brubaker, *ibid.*, 120, 49; and K. Kochloefl, W. Liebelt, and H. Knösinger, Chem.Comm., 1977, 510). Such catalysts are usually based on chemical attachment of a phosphine which then binds the key metal (e.g. rhodium or ruthenium) to the surface. Other improvements in catalysis include the very active form of nickel obtained by borohydride reduction of nickel(II) acetate (C.A. Brown and V.K. Ahuja, J.org.Chem., 1973, 38, 2226).

Another important development concerns the evolution of asymmetric reducing agents for chiral induction using prochiral alkenes. Although attempts have been made to develop an asymmetric form of Raney nickel by doping it with optically-active acids (Y. Izumi, Angew.Chem.internat.Edn., 1971, 10, 871), the optical yields are low and more successes have been achieved with asymmetric Rh(I) complexes (B. Bogdanović, *ibid.*, 1973, 12, 954). Some of the highest optical yields are cited for amino-acid synthesis by hydrogenation of enamino-acids' double bonds (Y. Sugi and W.R. Cullen, Chem.Letters, 1979, 39):

$$R^1CH_2\text{-----}C\begin{array}{c}NHAc\\|\\H\\CO_2R^2\end{array}$$

with H_2, (33) giving product (90% ee)

(33) = $\left\{\left[\begin{array}{c}Ph\\O\\O\\Ph_2PO\\OPPh_2\end{array}\quad OMe\right]Rh(norbornadiene)\right\}^+ PF_6^-$

(ee = enantiomeric excess)

Asymmetric reduction of alkenes has been reviewed (H.B. Kagan and J.C. Fiaud, Topics in Stereochemistry, 1978, 10, 175).

The use and mechanisms of action of homogeneous metal hydrogenation catalysts have been extensively reviewed (R.E. Harmon, S.K. Gupta, and D.J. Brown, Chem.Rev., 1973, 73, 21; R.S. Coffey in "Aspects of Homogeneous Catalysis", Vol. 1, ed. R. Ugo, Carlo Manfredi, Milan, 1970, ch. 1; and A.J. Birch and D.H. Williamson, Org.Reactions, 1976, 24, 1), and these and other aspects of alkene reduction have been reviewed in more general treatments (S. Mitsui and A. Kasahara "The Chemistry of Alkenes" Vol. 2, ed. J. Zabicky, Wiley-Interacience, London, 1970, ch. 4; G. Brieger and T.J. Nestrick, Chem. Rev., 1974, 74, 567; and M. Freifelder, "Catalytic Hydrogenation in Organic Synthesis", Wiley, New York, 1978).

(vi) Photochemical reactions

In this section and the discussion of thermal cycloadditions which follows it, familiarity with the various treatments of orbital symmetries and interactions has been assumed (see for example T.L. Gilchrist and R.C. Storr "Organic Reactions and Orbital Symmetry", Cambridge UP, 2nd Edn.1979). The main point for olefin photochemistry from these rules is that suprafacial [2 + 2] cycloadditions, although thermally 'forbidden', are photochemically 'allowed' (Table 1). This theoretical prediction has stimulated a greater interest in such processes.

Table 1.   Orbital symmetry rules for allowed alkene
            cycloadditions (s = suprafacial;
            a = antarafacial)

Number of π-electrons involved (n = integer)	Thermal processes	Photochemical processes
2+2	s-a or a-s	s-s or a-a
2+4	s-s or a-a	s-a or a-s
2+6	s-a or a-s	s-s or a-a
2+(4n-2)	s-a or a-s	s-s or a-a
2+4n	s-s or a-a	s-a or a-s

The theoretical prediction of photochemically allowed
dimerization was elegantly confirmed for acyclic alkenes by a
study using (E)- and (Z)-isomers of but-2-ene, which dimerize
with the perfect stereospecificity of a concerted process
(H. Yamazaki and R.J. Cvetanović, J.Amer.chem.Soc., 1969, 91,
520).

It should be noted that such cyclisations proceed *via* photoexcitation of one molecule to the first singlet excited state
$(S_0 \rightarrow S_1)$ followed by bimolecular concerted ring closure
between a S_0 ground state molecule and a S_1 excited state
molecule. They must be distinguished from photosensitized
cycloadditions, such as that which occurs when cyclopentadiene
is irradiated in the presence of acetophenone, its triplet
(T_1) state then reacting in a stepwise fashion with an alkene
such as (E) or (Z)-but-2-ene (P.D. Bartlett, Quart.Rev.,
1970, 24, 473):

The apparent occurrence of the forbidden photochemical [4+2] process and the loss of stereochemical integrity in the butene are ample evidence for a triplet diradical intermediate in the photo-sensitized process. Even more complex cyclo-additions arise when the co-reagent is an arene (K.E. Wilzbach and L. Kaplan, J.Amer.chem.Soc., 1971, 93, 2073).

Non-concerted [2+2] cycloadditions can also occur when simple alkenes react with photoexcited (triplet, $n \to \pi*$) aliphatic or aromatic ketones (D.R. Arnold, Adv.Photochem., 1968, 6, 301). Evidence for long-lived 1,4-diradicals as intermediates in this and related processes includes loss of stereochemical integrity by the alkene and regioselective additions, (H.A.J. Carless, Tetrahedron Letters, 1973, 3173).

The excited states of simple alkenes are not as readily accessible as those of dienes or trienes, since their S_0-S_1 ($\pi \to \pi*$) transitions occur at 170-180 nm, i.e. in the vacuum ultraviolet. For an alkene the singlet state is also well above the triplet in energy (by about 350-450 kJ/mol) so that intersystem crossing is relatively slow. In both the singlet and triplet (T_1) states rotation can occur freely; indeed, the lowest energy conformation is thought to be one in which the ends are mutually rotated by 90°. Thus cis –

trans isomerization occurs freely after photoexcitation.
These and related isomerizations of alkenes and dienes have
been reviewed (J. Saltiel *et al.*, Org.Photochem., 1973, 3, 1;
K.J. Crowley and P.H. Mazzocchi in "The Chemistry of Alkenes",
Vol. 2, ed. J. Zabicky, Wiley-Interscience, London, 1970,
ch. 6; for a general review of alkene photochemistry, see
J.D. Coyle, Quart.Rev., 1974, 3, 329).

(vii) Thermal cycloadditions

The cycloadditions of carbenes to mono-olefins, of
dienophiles to conjugated dienes, and the photochemical
dimerization of alkenes have all been dealt with elsewhere.
Such reactions are examples of concerted cycloadditions, now
to be interpreted by the concepts of orbital symmetry (R.B.
Woodward and R. Hoffmann, "The Conservation of Orbital
Symmetry", Verlag Chemie, Weinheim, 1970), frontier orbital
interactions (I. Fleming, "Frontier Orbitals and Organic
Chemical Reactions", Wiley, London, 1976), or by evaluation
of the aromaticity of their transition states (M.J.S. Dewar,
Tetrahedron, 1966, Suppl. 8 part 1, 75; H.E. Zimmerman, Acc.
chem.Res., 1971, 4, 272). Reviews of these three inter-
related treatments abound, as do attempts at more general
treatments (S.-Y. Chu, Tetrahedron, 1978, 34, 645; N.T. Ang
et al., *ibid.*, 1977, 33, 523, 553; N.D. Epiotis, Angew.Chem.
internat.Edn., 1974, 13, 751; and W.C. Herndon, Chem.Rev.,
1972, 72, 157).

Simplified representations of the possible cycloaddition
processes for alkenes are shown in equations 1-5.

$$\| \quad + \quad R_2\ddot{C} \quad \longrightarrow \quad \triangleright R_2 \qquad (1)$$

$$\| \quad + \quad \| \quad \longrightarrow \quad \square \qquad (2)$$

54

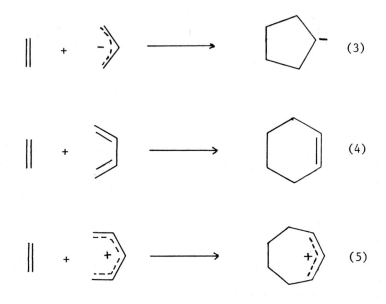

(3)

(4)

(5)

Thus, carbene cycloadditions (Section 1b-i) (equation 1) are predicted to be allowed thermally for an approach geometry in which the empty orbital of a singlet carbene overlaps with the alkene's HOMO (π) orbital. The alternative geometries are forbidden, but the point is difficult to test experimentally.

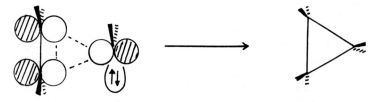

Alkene-alkene cycloadditions are predicted to be thermally forbidden (*i.e.* to require too high an activation energy for the pathway to be probable) unless they proceed suprafacially -antarafacially ($\pi2s + \pi2a$):

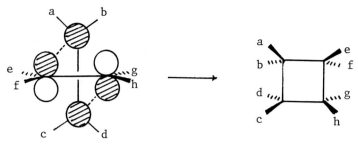

Such an event is unlikely for simple acyclic alkenes, and has
not even been demonstrated for the intermolecular dimerization
of the highly strained (E,Z)-cycloocta-1,4-diene (J. Leitich,
Angew.Chem.internat.Edn., 1969, 8, 909). The only well-
authenticated class of such [π2s + π2a] reactions is that of
ketene-olefin thermal cycloaddition. Evidence in support of
their concertedness includes the stereospecificity observed in
diphenylketene's reactions with (E)- and with (Z)-but-2-ene
(R. Huisgen, L.A. Feiler, and G. Binsch, Ber., 1969, 102,
3460), the kinetic data for substituent effects, and related
activation parameters (reviewed by P. Beltrame in
"Comprehensive Chemical Kinetics", Vol. 9, ed. C.H. Bamford and
C.F.H. Tipper, Elsevier, London, 1973, ch. 2).

Reactions of the type represented by equation 2 occur with
unexpected facility with certain classes of alkenes,
especially allenes (Section 3a-ii/7) and fluoro-olefins, which
dimerize and co-dimerize thermally at $ca.$ 120 ^{o}C, and when an
electron-rich and an electron-deficient alkene co-dimerize
(the latter type of cyclization is termed polar 2+2 cycloaddition). It is claimed that treatments which embody
configuration interactions define the allowed and forbidden
classes of cycloadditions more precisely and that such
treatments then deal successfully with polar cycloadditions
(N.D. Epiotis, J.Amer.chem.Soc., 1976, 98, 453). Polar [2+2]
cycloadditions are considered to be outside the scope of this
chapter; reviews are available (R. Huisgen, Acc.chem.Res.,
1977, 10, 117; D.N. Reinhoudt, Adv.heterocyclic Chem., 1977,
21, 253; T.L. Gilchrist and R.C. Storr, "Organic Reactions and
Orbital Symmetry", Cambridge UP, 2nd Edn., 1979, p. 193).
Fluoro-olefin 1,2-cycloadditions to alkenes and to 1,3-
dienes are generally considered to be non-concerted and to
proceed via intermediate diradicals, since such reactions
lack stereospecificity (P.D. Bartlett et $al.$, J.Amer.chem.
Soc., 1972, 94, 2899).

50%   50%

However, when fluoro-olefins react with 1,3-dienes, concerted 1,4-cycloadditions may compete with stepwise 1,2-addition (P.D. Bartlett, Quart.Rev., 1970, 24, 473).

Cycloadditions of the type represented by equation 3 are unlikely for simple alkenes unless the 4-electron component is a 1,3-dipole such as an azide, a diazoalkane, a nitrone, or a nitrilimine; the 4-electron component may also be an aza-allyl anion (T. Kauffmann and E. Köppelmann, Angew.Chem. internat.Edn., 1972, 11, 290). Interest in dipolar cyclo-additions was greatly stimulated by Huisgen's classification of dipolar reagents (R. Huisgen, *ibid.*, 1963, 2, 565). Nevertheless, simple unstrained alkenes make poor dipolarophiles, the best being styrene.

Solvent effects and kinetic evidence for the concertedness of such reactions, which has not gone unchallenged (R.A. Firestone, Tetrahedron, 1977, 33, 3009), have been reviewed (Beltrame, *loc.cit.*). The rates of such reactions are insensitive to solvent effects, which argues against zwitterionic intermediates, and the concertedness is further supported both by theoretical considerations and by experimental data such as the large negative $\Delta S^{\ddagger}$ and moderate $\Delta H^{\ddagger}$ values, as well as by retention of the alkene (the dipolarophile) stereochemistry (W.J. Linn and R.E. Benson, J.Amer.chem.Soc., 1965, 87, 3657).

Much work has been concentrated upon the successful prediction of the regioselectivities of such additions, with considerable success using frontier orbital treatments (K.N. Houk, Acc.chem.Res., 1975, 8, 361 and I. Fleming, "Frontier Orbitals and Organic Chemical Reactions", Wiley, London, 1976). In a concise form this approach suggests that the strongest interactions for electron-rich dipolarophiles will involve overlap of the dipole-LUMO with the dipolarophile-HOMO, which will occur most rapidly at termini where these orbitals have large atomic orbital coefficients. A knowledge of such coefficients (K.N. Houk *et al.*, J.Amer.chem.Soc., 1973, 95, 7287) is therefore essential to rationalise or predict regioselectivities in such additions. These and other

features of 1,3-dipolar cycloadditions are fully discussed in
a review (G. Bianchi, C. De Micheli, and R. Gandolfi, "The
Chemistry of Double Bonded Functional Groups", ed. S. Patai,
Wiley-Interscience, London, 1977, ch. 6).

Consideration of the Diels-Alder reaction is deferred to
Section 3b-iii.

(viii) Ene reactions

Interest in the ene reaction has focussed on the mechanism
of the concerted thermal process, interpreted on the basis of
the orbital symmetry rules as a $[\pi 2s + \pi 2s + \sigma 2s]$ process:

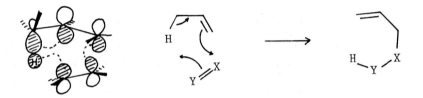

$\pi 2s + \pi 2s + \sigma 2s$

A wide selection of enophiles X=Y can be tolerated (H.M.R.
Hoffmann, Angew.Chem.internat.Edn., 1969, 8, 556(including
benzyne (P. Crews and J. Beard, J.org.Chem., 1973, 38, 522).
One of the most powerful yet discovered is N-phenyltriazol-
indione (34), and other widely used enophiles are maleic
anhydride, electron-deficient acetylenes such as hexafluoro-
but-2-yne and dimethyl acetylenedicarboxylate, azodicarboxylic
esters, and thiones.

(34)

Nitrosoarenes were thought to behave as simple enophiles, but
now appear to react with alkenes by a one-electron transfer
pathway (D. Mulvey and W.A. Waters, J.chem.Soc. Perkin II,
1978, 1059). It has been suggested that (34) may, like singlet

oxygen, react to form a cyclic intermediate initially (C.A.
Seymour and F.D. Greene, J.Amer.chem.Soc., 1980, 102, 6384).
Alkenes do not themselves make good enophiles, and even intramolecular ene insertions in suitable dienes such as (35)
(review: W. Oppolzer and V. Snieckus, Angew.Chem.internat.Edn.,
1978, 17, 476) require much higher temperatures than electron-
deficient dienophiles would normally need.

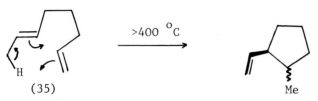

(35) >400 °C Me

The concertedness of uncatalysed ene reactions of simple
alkenes is now accepted to have been established by asymmetric
induction, the prevalence of *endo* over *exo* addition-insertion
pathways, the *cis*-configuration of the newly-formed C-X and
Y-H bonds, and the high entropy of activation and negligible
solvent effects observed (see for example: R.K. Hill *et al.*,
J.Amer.chem.Soc., 1974, 96, 4201; V. Garsky, D.F. Koster, and
R.T. Arnold, *ibid.*, p. 4207; and O. Achmatowicz and J.
Szymoniak, J.org.Chem., 1980, 45, 1228).

Acid-catalysed reactions between simple aldehydes,
especially formaldehyde, and olefinic hydrocarbons have been
known for many years as the Prins reaction (review: D.R. Adams
and S.P. Bhatnagar, Synthesis, 1977, 661) and although they
could be viewed as ene insertions are more sensibly regarded
as electrophilic attack by protonated aldehyde upon the C=C
bond (Section 2d-i/3). However, as a means of lowering the
temperatures necessary for normal ene reactions, Lewis acid
catalysis of the ene insertions of certain classes of carbonyl
compounds has been developed successfully (B.B. Snider *et al.*,
J.org.Chem., 1974, 39, 255; Tetrahedron Letters, 1977, 2831;
1980, 1815; and J.Amer.chem.Soc., 1980, 102, 5872, 5926).

$$(E = CO_2R)$$

Catalysed ene reactions also occur with certain aldehydes such as chloral and t-BuCHO (G.B. Gill and B. Wallace, Chem. Comm., 1977, 380, 382; G.B. Gill et al., J.chem.Soc. Perkin I, 1978, 93 and Tetrahedron Letters, 1979, 4867), and also appear to be stereospecific.

A complication in these acid-catalysed ene reactions is Lewis-acid induced [2+2] cycloaddition, which appears to be particularly facile when the carbonyl group is adjacent to a triple bond (K. Griesbaum and W. Seiter, J.org.Chem., 1976, 41, 937; B.B. Snider et al., ibid., p. 3061; J.H. Lukas, F. Baardman, and A.P. Kouwenhoven, Angew.Chem.internat.Ed., 1976, 15, 369). Such reactions have been found to be highly stereospecific (H. Fienemann and H.M.R. Hoffmann, J.org.Chem., 1979, 44, 2802).

(ix) Rearrangements

Interconversions between (E) and (Z) isomers are termed inversions (review: P.E. Sonnet, Tetrahedron, 1980, 36, 557). They are most simply achieved by photolysis (Section 2d-vi), which gives mixtures containing mainly the isomer with the higher triplet energy, often the *cis* isomer. Inversions can also be induced by heating the alkene with a catalytic amount of a Lewis acid such as iodine or BF_3, a method which is generally used to convert *cis* isomers into mixtures containing mainly the thermodynamically favoured *trans* isomers. A recent innovation is the use of sulphur dioxide and a zeolite (K. Otsuka and A. Morikawa, J.Catalysis, 1977, 46, 71). More elaborate sequences featuring *syn*-addition followed by *anti*-

elimination, or *vice versa*, are clearly possible. Highest
yields of this type of inversion are attained by epoxidation
sequences since several methods exist for stereospecific
epoxide reduction with or without inversion (Section 2b-iii,
and Sonnet, *loc.cit.*).

Isomerization of a 2-ene to a 3-ene, or of a 1-ene to a
2-ene, may be treated by orbital symmetry as potentially
concerted [1,3]sigmatropic shifts. For a hydrogen atom to
migrate suprafacially across the allyl unit (36), access to
the excited state is necessary because a 4-electron Hückel
set indicates an anti-aromatic transition state in a thermal
process. Theoretically an antarafacial hydrogen migration
(37) should be thermally allowed but it is self-evidently an
unlikely event for the small hydrogen atom.

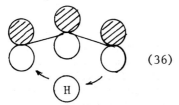   (36)

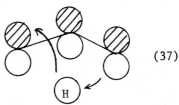   (37)

thermally 'forbidden'   thermally 'allowed'
anti-aromatic   Möbius aromatic
transition state   transition state

In spite of the forbidden nature of the [1,3] prototropic
shift, it is a well-known process provided that acidic or
basic catalysts are present (reviews: A.J. Hubert and H.
Reimlinger, Synthesis, 1969, 97; 1970, 405). Methods based on
transition metal catalysts such as rhodium chlorides and iron
carbonyls have now emerged to challenge more classical
procedures.

$$\text{Fe}_3(\text{CO})_{12} \atop 80^\circ\text{C}$$

PhCHMeCH=CH$_2$ $\xrightarrow[\text{t-BuOH}]{\text{t-BuOK}}$ PhCMe=CHMe

(C.P. Casey and C.R. Cyr, J.Amer.chem.Soc., 1973, <u>95</u>, 2248)
(D.J. Cram and R.T. Uyada, *ibid.*, 1964, <u>86</u>, 5466).

Comprehensive reviews of these reactions are available (K.
Mackenzie, "The Chemistry of Alkenes", Vol. 2, ed. J. Zabicky,
Wiley-Interscience, London, 1970, ch. 3; H. Pines and W.M.
Stalick, "Base-Catalysed Reactions of Hydrocarbons and Related
Compounds", Academic, New York, 1977, ch. 2; R.H. De Wolfe,
"Comprehensive Chemical Kinetics", ed. C.H. Bamford and C.F.H.
Tipper, Elsevier, London, 1973, ch. 5; and for transition metal
catalysts see R. Cramer, Ann.New York Acad.Sci., 1970, 172,
507; and P.N. Rylander, "Organic Synthesis with Noble Metal
Catalysts", Academic, New York, 1973, ch. 5).

(x) Metal and metal complex catalysed reactions

This is now a topic of such importance that it will be the
subject of a separate supplement. To provide a temporary
coverage of the topic, a list of some of the available books
and reviews is given below.

(1) General

D.V. Banthorpe, Chem.Rev., 1970, 70, 295; C.W. Bird, Topics
Lipid Chem., 1971, 2, 247; J.-F. Biellmann, H. Hemmer, and J.
Levisalles, "The Chemistry of Alkenes", ed. J. Zabicky, Wiley-
Interscience, London, 1970, ch. 5; D.St.C. Black, W.R. Jackson,
and J.M. Swan, "Comprehensive Organic Chemistry", ed. D.H.R.
Barton and W.D. Ollis, Pergamon, Oxford, 1979, Vol. 3, p.
1127; J.M. Davidson, M.T.P. Internat.Rev.Sci., Ser. 1, Vol. 6,
ch. 9, ser. 2, Vol. 6, ch. 10, ed. M.J. Mays, Butterworths,
London, 1972-1975; R.D.W. Kemmitt, *ibid.*, Ser. 1, Vol. 6, ch.
6; Ser. 2, Vol. 6, ch. 9; M.M. Taqui Khan and A.E. Martell,
"Homogeneous Catalysis by Metal Complexes", Academic, New
York, 1974; Rylander, *loc.cit.*; J.K. Kochi, "Organometallic
Mechanisms and Catalysis", Academic, New York, 1978; G.W.
Parshall, "Homogeneous Catalysis", Wiley-Interscience, New
York, 1980; various authors in "Transition Metals in
Homogeneous Catalysis: ed. G.N. Schrauzer, Dekker, New York,
1971.

(2) Thermodynamic and structural data on alkene complexes

F.R. Hartley, Chem.Rev., 1973, 73, 163; L.D. Pettit and D.S.
Barnes, Topics Current Chem., 1972, 28, 85.

(3) Organopalladium complexes in synthesis

R.F. Heck, *ibid.*, 1971, 16, 221 and Acc.chem.Res., 1979, 12, 146; J. Tsuji, "Organic Synthesis with Palladium Compounds", Springer-Verlag, Berlin, 1980; and Pure appl.Chem., 1979, 51, 1235; B.M. Trost, Tetrahedron, 1977, 33, 2615.

(4) π-Allyl metal derivatives

R. Baker, Chem.Rev., 1973, 73, 487; M. Cooke, "Organometallic Chemistry", ed. E.W. Abel, Chem.Soc.Spec.Per.Rep., 1975, 3, 327; M.F. Semmelhack, Org.Reactions, 1972, 19, 115; L.S. Hegedus, "New Applications of Organometallic Reagents in Organic Synthesis", ed. D. Seyferth, Elsevier, Amsterdam, 1976, p. 329; P.W. Jolly and G. Wilke, "The Organic Chemistry of Nickel", Vols. 1-2, Academic, New York, 1975.

(5) Hydroformylation and carbonylation

J.K. Stille and D.E. James, "The Chemistry of Double Bonded Functional Groups", ed. S. Patai, Wiley-Interscience, London, 1977, ch. 12; G.P. Chiusoli, Acc.chem.Res., 1973, 6, 422; L. Marko, Aspects homog.Catal., 1974, 2, 3; R.L. Pruett, Adv. organometal.Chem., 1979, 17, 1; various authors in "Organic Synthesis *via* Metal Carbonyls", Vols. 1-2, ed. I Wender and P. Pino, Wiley-Interscience, New York, 1968-1977.

(6) Olefin metathesis

N. Calderon, "The Chemistry of Double Bonded Functional Groups", *loc.cit.*, ch. 10; R.L. Banks, Topics Current Chem., 1972, 25, 39; W.B. Hughes, Organometal.chem.Syn., 1972, 1, 341; R.J. Haines and G.J. Leigh, Quart.Rev., 1975, 4, 155; Proc.Internat.Symp.on Metathesis, Noordwijkerhout, 1977, Rec. trav.chim., 1977, 96, ml; R.H. Grubbs, Prog.Inorg.Chem., 1978, 24, 1; N. Calderon, J.P. Lawrence and E.A. Ofstead, Adv. organometal.Chem., 1979, 17, 449.

(7) Homogeneous hydrogenation catalysts

B.R. James, "Homogeneous Hydrogenation", Wiley-Interscience, New York, 1973, and Adv.organometal.Chem., 1979, 17, 319; G. Brieger and T.J. Nestrick, Chem.Rev., 1974, 74, 567; R.E. Harmon, S.K. Gupta, and D.J. Brown, *ibid.*, 1973, 73, 21; F.J. McQuillin, "Homogeneous Hydrogenation in Organic

Chemistry", Riedel, Boston, 1976 (see also Section 2d-v).

(8) Insertion reactions

G. Henrici-Olivé and S. Olivé, Topics curr.Chem., 1976, 67, 107.

(9) Industrial applications

J.P. Candlin, K.A. Taylor, and D.T. Thompson, Ind.Chim.Belg., 1970, 35, 1085.

(10) Dimerization, oligomerization, and ring formation

B. Bogdanović, B. Henc, and H.-G. Karmann, Ind.eng.Chem., 1970, 62, 34; J. Hetflejs and J. Langova, Chem.Listy, 1973, 67, 590; P. Heimbach, Angew.Chem.internat.Edn., 1973, 12, 975; G. Lefebvre and Y. Chauvin, Aspects homog.Catal., 1970, 1, ch. 3; B. Bogdanović, Adv.organometal.Chem., 1979, 12, 105.

(11) Polymerization

"Coordination Polymerization", ed. J.C.W. Chien, Academic, New York, 1975; T. Keii "Kinetics of Ziegler-Natta Polymerization", Chapman-Hall, London, 1972; L. Reich and A. Schindler, "Polymerization by Organometallic Compounds", Wiley-Interscience, New York, 1966; H. Sinn and W. Kaminski, Adv.organometal.Chem., 1980, 18, 99; P.J. Tait "Developments in Polymerization", ed. R.N. Haward, Applied Science, London, 1979, ch. 3; various authors in "Ziegler Natta Catalysts and Polymerization", ed. J. Boor, Academic Press, New York, 1979; A.D. Caunt, Catalysis, 1977, 1, 234.

(12) Asymmetric syntheses

Y. Izumi, Angew.Chem.internat.Edn., 1971, 10, 871; B. Bogdanović, *ibid.*, 1973, 12, 954; A. Nakamura, Pure appl.Chem., 1978, 50, 37; H.B. Kagan and J.C. Faiud, Top.Stereochem., 1978, 10, 175; D. Valentine and J.W. Scott, Synthesis, 1978, 329; J.W. ApSimon and R.P. Seguin, Tetrahedron, 1979, 35, 2797.

3. Hydrocarbons containing two double bonds

(a) Allenes

The greatly increased interest in the chemistry of compounds containing the C=C=C grouping is reflected in the spate of reviews (H. Fischer in "The Chemistry of Alkenes", Vol. 1, ed. S. Patai, Wiley-Interscience, London, 1964, p. 1025; D.R. Taylor, Chem.Reviews, 1967, 67, 317; T.F. Rutledge, "Acetylenes and Allenes", Reinhold, New York, 1969; and various authors in "The Chemistry of Ketenes, Allenes, and Related Compounds", ed. S. Patai, Wiley-Interscience, New York, 1980).

(i) Preparation of allenes

Apart from the texts cited above there are several which deal mainly with methods of synthesis (S.R. Sandler and W. Karo, "Organic Functional Group Preparations", Vol. 2, Academic Press, New York, 1971, p. 1; M. Murray in Houben-Weyl "Methoden der Organischen Chemie", Vol. 5/2a, 4th Edn., ed. E. Müller, G. Thieme, Stuttgart, 1977, p. 693; and M.V. Mavrov and V.F. Kucherov in "Reaktsui Metody Issled.Organ. Soedin.", ed. B.A. Kazanskii *et al.*, available in English from IFI/Plenum Press, 1973, 21, 93) including one restricted to chiral allenes (R. Rossi and P. Diversi, Synthesis, 1973, 25). The methods discussed below are limited to those most applicable to acyclic allenic hydrocarbons, and the reviews listed above should be consulted for methods appropriate to allenes containing functional groups (especially H. Hopf in "The Chemistry of Ketenes, Allenes, and Related Compounds", *loc.cit.*, p. 779).

(1) From 1,1-dihalogenocyclopropanes

Important contributors to the development of this procedure have been Doering (W.von E. Doering and P.M. LaFlamme, Tetrahedron, 1958, 2, 75), Moore (W.R. Moore and H.R. Ward, J.org.Chem., 1960, 25, 2073) and Skattebøl (L. Skattebøl, Tetrahedron Letters, 1961, 167). The precursors are made by addition of dibromo- or dichloro-carbene to alkenes, most efficiently using phase-transfer techniques (R.F. Heldeweg and H. Hogeveen, J.org.Chem., 1978, 43, 1916), whilst the dehalogenation is best performed using methyllithium at low temperatures to preclude isomerization of the desired product.

$$R^1R^2C=CHR^3 \xrightarrow{X_2C:} \quad \triangle \quad \xrightarrow[<0\ °C]{MeLi} \quad R^1R^2C=C=CHR^3$$

A one-step modification using tetrabromomethane and two molar proportions of methyllithium has been proposed (K.G. Untch, D.J. Martin and N.T. Castellucci, *ibid.*, 1965, **30**, 3572). Bicyclobutanes may be significant by-products, especially when tetra-substituted allenes are desired (L. Skattebøl, Tetrahedron Letters, 1970, 2362) or when the allene would have two geminal bulky groups e.g., t-BuCMe=C=CH$_2$ (D.W. Brown, M.E. Hendrick, and M. Jones, Jr., *ibid.*, 1973, 3951). Dehalogenation of appropriate dihalogeno-cyclopropanes by an alkyllithium/(−)-sparteine complex gives optically-active allenes in low optical purity (H. Nozaki *et al.*, Tetrahedron, 1971, **27**, 905).

(2) By propargylic displacements

Although this general procedure may be depicted as an S$_N$2' displacement:

$$L-C-C\equiv C- + \ \bar{N} \longrightarrow L^- + \quad \mathord{>}C=C=C\mathord{<}$$

the mechanism in specific cases is frequently quite different. The leaving group (L) is commonly acetate, tosylate, sulphinate, or halide; the entering group (N) may be hydride ion or alkyl anion, the latter provided by a lithium dialkyl-cuprate or by a Grignard reagent in the presence of a metal salt such as ferric chloride or copper(I) bromide. Practically speaking the most successful of these combinations is the treatment of acetylenic tosylates (R^{1}C≡CCHR^{2}OTs in which R^1,R^2 may be H or alkyl) with the organocopper(I) species derived from a Grignard reagent and copper(I) bromide (P. Vermeer, J. Meijer, and L. Brandsma, Rec.trav.Chim., 1975, **94**, 112),

$$t\text{-}BuC\equiv CCH(OTs)Bu\text{-}t \xrightarrow[Cu(I)Br]{t\text{-}BuMgCl} t\text{-}Bu_2C=C=CHBu\text{-}t$$

Because tertiary propargylic tosylates are difficult to prepare, the use of methylsulphinates is recommended in their place (P. Vermeer *et al.*, *ibid.*, 1978, **97**, 56), a method which has also been applied successfully to the synthesis of highly substituted silylallenes (H. Westmijze and P. Vermeer, Synthesis, 1979, 390).

The factors affecting the success of the reaction of propargyl halides with Grignard reagents, once the cause of controversy, have been established. For good yields of allenes to be obtained catalytic amounts of transition metal salts, e.g. ferric chloride, must be present (D.J. Pasto *et al.*, J.org.Chem., 1978, 43, 1382, 1385).

$$Me_2CClC\equiv CR^1 \xrightarrow{\text{R}^2\text{MgBr, FeCl}_3} Me_2C=C=CR^1R^2 \ (60-90\%)$$

$$(R^1 = H \text{ or } Me, \quad R^2 = Et, \text{ n-Bu, s-Bu})$$

This reaction does not provide a route to t-butylallenes and is also less successful in avoiding concurrent formation of the isomeric alkyne when the propargyl halide is non-terminal ($R^1 \neq H$). In many cases these difficulties may be overcome by replacing the Grignard/FeCl$_3$ combination by the appropriate lithium dialkylcuprate (D.J. Pasto *et al.*, *loc.cit.*, p. 1389) although, as noted by earlier workers (P. Crabbé *et al.*, Chem. Comm., 1976, 183), reduction can become a serious side-reaction.

$$R^1_2CClC\equiv CR^2 \quad\begin{cases} \xrightarrow{R^3_2CuLi} & R^1_2C=C=CMeR^3 \ (60-65\%) \\ (R^2 = Me) & + \\ & R^1_2C=C=CHMe \\ \xrightarrow{\text{t-BuMeCuLi}} & R^1_2C=C=CHBu\text{-}t \ (80\%) \\ (R^2 = H) \end{cases}$$

(3) By 1,4-additions to enynes

Nucleophilic attack upon the vinylic terminus of a conjugated enyne by lithium aluminium hydride or an alkyllithium leads to an allenic carbanion which may be trapped by an electrophile.

$$\rangle C=\overset{|}{C}-C\equiv C- \xrightarrow{\text{N}^-} \rangle C-\overset{|}{C}=C=\overset{-}{C}- \xrightarrow{\text{E}^+} \rangle \overset{|}{C}-\overset{|}{C}=C=CE-$$

The technique, which has been reviewed (M. Murray, in Houben-Weyl "Methoden der Organischen Chemie", Vol. 5/2, ed. E. Müller, G. Thieme, Stuttgart, 1977, p. 693), is most often applied to the synthesis of allenic alcohols, but if the capturing electrophile is a proton the reaction yields allenic hydrocarbons. The method is exemplified by the key step in the synthesis of a seed-oil constituent (S.R. Landor *et al.*, J.chem.Soc. Perkin I, 1972, 2197).

$$R^1CH=CHC\equiv CR^2 \quad \xrightarrow[\text{(ii) } H^+]{\text{(i) } LiAlH_4} \quad R^1CH_2CH=C=CHR^2$$

$$[R^1 = HOCH_2, \quad R^2 = (CH_2)_8CH \stackrel{t}{=} CHMe]$$

(4) From other allenic derivatives

Propadiene is mono-lithiated by n-butyllithium at -50 °C; the resulting allenyllithium is converted by alkyl halides mainly into the allenic isomer, although the propadienyl anion is capable of various formulations including the propynyl structure $HC\equiv CCH_2-$. This useful result is in accord with *ab initio* calculations (R.J. Bushby *et al.*, J.chem.Soc. Perkin II, 1978, 807) which favour an allene-like geometry for the anion and indicate charge concentration at the less-substituted terminus. The procedure may then be repeated with equal success (G. Linstrumelle and D. Michelot, Chem.Comm., 1975, 561).

$$H_2C=C=CH_2 \quad \xrightarrow{\text{BuLi}} \quad H_2C=C=CH^-Li^+ \quad \longleftrightarrow \quad H_2\bar{C}C\equiv CH \ Li^+$$

$$\downarrow R^1Br$$

$$H_2C=C=CHR \quad (80\%) \quad + \quad R^1CH_2C\equiv CH \quad (10\%)$$

$$\downarrow \begin{array}{l} \text{(i) BuLi} \\ \text{(ii) } R^2X \end{array}$$

$$R^2CH=C=CHR^1 \quad (94\%)$$

1-Bromoallenes may be alkylated by treatment with lithium dialkylcuprates at low temperature, best yields being obtained if di- or tri-alkylallenes are produced (M. Kalli, P.D. Landor, and S.R. Landor, J.chem.Soc. Perkin I, 1973, 1347). This and the previous technique have their marriage in the conversion of allenyllithiums to allenylcuprates at -70 °C, affording reagents suitable for stereospecific addition to alkynes (D. Michelot and G. Linstrumelle, Tetrahedron Letters, 1976, 275).

(5) Other methods

Tetramethylallene is best prepared by the pyrolysis of the lactone (38) obtained by isomerization of the dimer of dimethylketene (J.C. Martin, U.S. Patent 3 131 234, 1964).

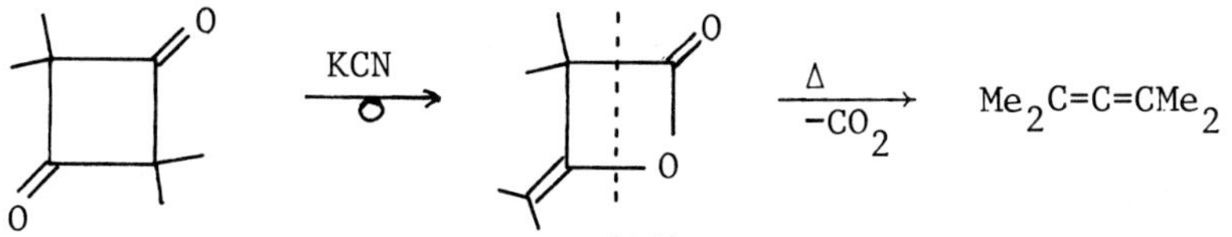

(38)

1,3-Bis(cyclopropyl)allene (W.F. Berkowictz and A.A. Ozorio, J.org.Chem., 1975, 40, 527) and bis(adamantylidene)allene (J. Strating, A.H. Alberts, and H. Wynberg, Chem.Comm., 1970, 818) have been obtained by similar procedures. The latter allene does not apparently qualify as the most sterically hindered allene, since tetrakis(t-butyl)allene is claimed to have exceptionally low reactivity; it was made by a classical dehydration route. however (A. Berndt and R. Bolze, Dissertation of R. Bolze, Marburg, 1980).

A number of new methods have been reported for the synthesis of allenes.

(i) From olefins (F. Sato, K. Oguro, and M. Sato, Chem. Letters, 1978, 805).

$$RCH=CH_2 \xrightarrow{\text{LiAlH}_4, \text{ TiCl}_4} (RCH_2CH_2)_4AlLi \xrightarrow[\text{Cu(I)Cl}]{\text{HC}\equiv\text{CCH}_2\text{Br}}$$

$$RCH_2CH_2CH=C=CH_2$$

(ii) From ketones *via* vinylsilanes (T.H. Chan *et al.*, J.org. Chem., 1978, 43, 1526).

$$R^1R^2CO \xrightarrow[\text{ether, } -78\ ^\circ C]{Me_3SiC(Li)=CH_2} R^1R^2C(OH)C=CH_2 \xrightarrow[\substack{(R^2 = H\ or \\ R^1 = R^2 = Ar)}]{SOCl_2}$$

with $SiMe_3$ below the $C=CH_2$

$$R^1R^2CClC=CH_2 \xrightarrow[\text{acetonitrile}]{Et_4\overset{+}{N}F^-} R^1R^2C=C=CH_2$$

with $SiMe_3$ below

(iii) From ketones *via* vinylsulphoxides (G.H. Posner, P.-W. Tang and J.P. Mallamo, Tetrahedron Letters, 1978, 3995; M. Mikołajzcyk *et al.*,J.org.Chem., 1978, <u>43</u>, 473).

$$R^1R^2CO \xrightarrow{(MeO)_2P(O)CH(Li)S(O)Ar} R^1R^2C=CHS(O)Ar \xrightarrow[\text{CH}_3I]{\text{i-Pr}_2NLi}$$

$$R^1R^2C=C\text{---}SAr \xrightarrow[\text{THF, } -100\ ^\circ C]{R_2NLi} R^1R^2C=C=CH_2$$

with CH_3 and O below

(iv) From ketones *via* enol triflates (P.J. Stang and R.J. Hargrove, *ibid.*, 1975, <u>40</u>, 657).

$$R^1R^2CHCOCH_3 \xrightarrow[\text{pyridine}]{T_fOT_f,\ CCl_4} R^1R^2C=C(OT_f)CH_3 \xrightarrow[100\ ^\circ C]{\text{quinoline}}$$

$$R^1R^2C=C=CH_2$$

Another very promising method, the mechanism for which is not so far established, affords allenes by simple homologation of terminal alkynes by treatment with formaldehyde and di-isopropylamine in the presence of copper(I) bromide (P. Crabbé *et al.*, Chem.Comm., 1979, 860).

$$RC\equiv CH \xrightarrow[\text{CuBr}]{CH_2O,\ \text{i-Pr}_2NH} RCH=C=CH_2 \quad (26\text{-}97\%)$$

(ii) Structure and physical properties of allenes

The resolution of optically-active 1,3-disubstituted allenes (R. Rossi and P. Diversi, Synthesis, 1973, 25) has established beyond doubt their orthogonal structure in the ground state:

$$\underset{H}{\overset{R}{>}}C=C=C\underset{R}{\overset{\cdots H}{<}} \quad \neq \quad \underset{R}{\overset{H_{\cdots}}{<}}C=C=C\underset{H}{\overset{R}{>}}$$

The application of *ab initio* SCF calculations to allenes, their stable isomers, and their various excited states has been reviewed (C.E. Dykstra and H.F. Schaefer III, "The Chemistry of Ketenes, Allenes and Related Compounds", ed. S. Patai, Wiley-Interscience, New York, 1980, p. 1) though without inclusion of a recent study of the cyclopropylidene-allene interconversion (D.J. Pasto, M. Haley, and D.M. Chipman, J.Amer.chem.Soc., 1978, 100, 5272). The proposal that the excited planar state of propadiene is non-linear (C-C-C bond angle 136°) is now accepted by most theoreticians (J.A. Pople *et al.*, *ibid.*, 1977, 99, 7103).

The photoelectron spectra of propadiene, its 1,1-d_2- and d_4-derivatives (R.K. Thomas and H.R. Thompson, Proc.roy.Soc., 1974, A339, 29) and various methylallenes (F. Brogli *et al.*, J.electron.Spec.Related Phenomena, 1973, 2, 455) display unusual fine structure in the band for the lowest ionization (*ca.* 10 eV), ascribed to a Jahn-Teller effect. A considerable body of data such as bond lengths and angles and rotational barriers have been compared with theoretical predictions (W. Runge, "The Chemistry of Ketenes, Allenes, and Related Compounds", *loc.cit.*, p. 45) with which they are usually in good agreement. Rotational barriers are around 180 kJ/mol, sufficiently higher than activation barriers for racemisation not to occur; however, racemisation can be induced by some reagents, including organocuprates (A. Claesson and L.-I. Olsson, Chem.Comm., 1979, 524).

The spectroscopic properties of allenes have been thoroughly tabulated (J.W. Munson, "The Chemistry of Ketenes, Allenes and Related Compounds", *loc.cit.*, p. 165). The use of the characteristic asymmetric stretching mode at *ca.* 1950 cm^{-1} in qualitative analysis is well-known, and use can also be made of the 850 cm^{-1} absorption of terminal allenes. The ^{13}C nmr spectra of allenes are splendidly characterised by the extremely low chemical shift of the central sp-carbon (δ 200-220 ppm) (R. Steur *et al.*, Tetrahedron Letters, 1971, 3307). The mass spectra of allenes do not reveal whether on ionization they interconvert with alkynes, but the McLafferty rearrangement occurs if the alkyl residue is sufficiently long (J.R. Wiersig, A.N.H. Yeo and C. Djerassi, J.Amer.chem.Soc.,

1977, 99, 532).

(iii) Reactions of allenes

(1) Isomerization

This topic has also been exhaustively reviewed (W.D. Huntsman, "The Chemistry of Ketenes, Allenes and Related Compounds", *loc.cit.*, p. 522). Most of the detailed work has involved strongly base-catalysed reactions, using sodamide in ethylenediamine (J.H. Wotiz *et al.*, J.org.Chem., 1966, 31, 2069; 1973, 38, 489) or potassium t-butoxide in dimethyl-sulphoxide (W. Smadja, C. Prevost and C. Georgoulis, Bull.Soc. chim.Fr., 1974, 925), although instances of prototropy are known when suitably substituted compounds are merely eluted through basic alumina (T.L. Jacobs, D. Dankner, and S. Singer, Tetrahedron, 1964, 20, 2177). The usual outcome of the equilibration of alkylallenes with their isomeric alkynes and conjugated dienes is the formation of the energetically favoured conjugated isomers, unless kinetic acidities or precipitation of insoluble salts (e.g. RC≡C- M+) intervene. Careful mechanistic studies using hexa-1,2-diene and its isomers (M.D. Carr, L.H. Gan, and I. Reid, J.chem.Soc.Perkin II, 1973, 668, 672) have shown that sodamide or t-butoxide induce the isomerization of hexa-1,2-diene much faster than that of other isomers, the first-formed product being hex-2-yne.

$$RCH_2CH=C=CH_2 \xrightarrow{\text{base}} RCH_2C{\equiv}CCH_3$$

Acid-catalysed prototropic isomerizations have also been studied (M.D. Carr *et al.*, Chem.Comm., 1973, 177), and in sulpholane containing strong acids such as HBF_4 the allenic isomers are again the fastest to isomerize.

(2) Electrophilic additions

This subject has been well reviewed (M.C. Caserio, Selective Organic Transformations, 1970, 1, 299; P.J. Stang, Prog.phys. org.Chem., 1973, 10, 205; and P. Blake, "The Chemistry of Ketenes, Allenes, and Related Compounds", *loc.cit.*, p. 348). Most of the typical electrophilic additions to olefins have also been studied with allenes. Halogenation (T. Okuyama *et al.*, J.org.Chem., 1974, 39, 2255) and addition of HOBr (W.R.

Dolbier and B.H. Al-Sader, Tetrahedron Letters, 1975, 2159)
both appear to proceed *via* bridged halogenonium ions (39).

$$\overset{+}{\underset{R}{\bigtriangleup}}{\overset{X}{}}{=}CH_2 \qquad (39) \quad R = H \text{ or } Ph, \quad X = Br, Cl$$

Mercuration, studied using oxymercuration of optically-
active acyclic allenes (W.H. Pirkle and C.W. Boeder, J.org.
Chem., 1977, 42, 3697) and by ion cyclotron resonance
examination of the gas phase transfer of MeHg$^+$ to propadiene
(R.D. Bach, J. Patane, and L. Kevan, *ibid.*, 1975, 40, 257),
involves similar bridged complex ions. Bridged episulphonium
ions formed by addition to the least substituted π-bond have
been proposed as intermediates in the addition of aryl-
sulphenylhalides to alkylallenes (T.L. Jacobs. and R.C.
Kammerer, J.Amer.chem.Soc., 1974, 96, 6213).

The mechanism of the protonation of propadiene has proved to
be a particularly fascinating problem. The planar allyl
cation (40), confirmed by *ab initio* calculations (J.A. Pople
et al., *ibid.*, 1973, 95, 6531) to be by far the most stable

$$\underset{\underset{H \quad H}{|\quad|}}{\overset{\overset{H}{|}}{\underset{H}{\bigvee}}}_{+} \qquad (40) \qquad\qquad \underset{\underset{+}{}}{\overset{H_3C}{\diagdown}}C{=}CH_2 \qquad (41)$$

of the possible intermediates, is not directly accessible
from the orthogonal cumulene system, so that addition of HCl,
for example, yields 2-chloropropene not allyl chloride.
Direct protonation in the gas phase, studied by ion cyclotron
resonance (D.H. Aue, W.R. Davidson, and M.T. Bowers, *ibid.*,
1976, 98, 6700) has confirmed that the intermediate is the
vinyl cation (41); calculations suggest that the activation
energy for its conversion into (40) is *ca.* 70 kJ/mol. By
contrast, when 1,3-dimethylallene or more highly alkylated
allenes dissolve in super-acids, the solutions produced
display nmr spectra compatible with planar allylic cations
(C.U. Pittman, Jr., Chem.Comm., 1969, 122).

The hydroboration of propadiene and alkylallenes has received careful attention (H.C. Brown, R. Liotta, and G.W. Kramer, J.Amer.chem.Soc., 1979, $\underline{101}$, 2966). The reaction can be controlled by the use of 9-borabicyclononane (9-BBN), and a pattern of reactivity has emerged. Propadiene is rapidly attacked at both ends, but 1,1-dialkylallenes are exclusively attacked at the unsubstituted π-bond and so give allylboranes; 1,3-dialkyl- and monoalkyl-allenes behave in an intermediate fashion. These results were not reproduced when less hindered dialkylboranes were used.

$$H_2C=C=CR_2 \xrightarrow{R_2^1BH} R_2^1BCH_2CH=CR_2 \xrightarrow[(R\ =\ H)]{R_2^1BH,\ fast}$$

$$(R_2^1BCH_2)_2CH_2 \xrightarrow[NaOH]{H_2O_2} (HOCH_2)_2CH_2$$

$$(R_2^1BH\ =\ 9\text{-BBN})$$

(3) Free-radical additions

Radical addition of hydrogen bromide (E1-A.I. Heiba and W.O. Haag, J.org.Chem., 1966, $\underline{31}$, 3814) and free-radical chlorination (M.-C. Lasne and A. Thuillier, Bull.Soc.chim. Fr., 1974, 249) have both received detailed attention, and the topic has been reviewed (P. Blake, "The Chemistry of Ketenes, Allenes and Related Compounds", ed. S. Patai, Wiley -Interscience, New York, 1980, p. 344). For propadiene there is little difference between the energies of the intermediate vinyl radical formed by terminal attack and the initially non-planar pseudo-allyl radical formed by attack at the central atom; reactions appearing to favour the latter mode may actually involve reversible formation of the vinyl radical. For acyclic allenes, increasing substitution leads to an increased preference for central attack, though the degree of substitution at which the cross over between the two paths occurs depends on the nature of the attacking radical (E. Montaudon, J. Thépénier, and R. Lalande, J.heterocyclic Chem., 1979, $\underline{16}$, 105). ESR studies indicate that tetramethylallene is attacked by RS·, R_3Si·, and R_3Sn· radicals exclusively at the central sp-carbon atom (W.H. Davies, Jr. and J.K. Kochi, Tetrahedron Letters, 1976, 1761) but the equivalence of the methyl groups is incompatible with planarity in the resulting pseudo-allyl radical (42).

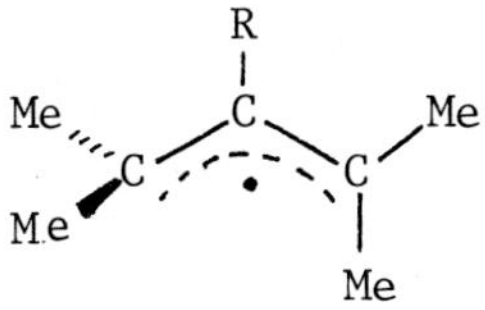

(42)

(4) Oxidation

Whereas unhindered allenes react normally with ozone to give carbon dioxide and carbonyl compounds (P. Kolsaker and B. Teige, Acta chem. Scand., 1970, $\underline{24}$, 2101), ozonolysis of sterically hindered allenes such as 1,1-di- and 1,1,3-tri-t-butylallene gives dioxaspiro[2,2]pentanes (44) consistent with intermediate allene oxides (43) resulting from initial attack at the central carbon (J.K. Crandall *et al.*, J.org. Chem., 1974, $\underline{39}$, 1723) (Scheme 6).

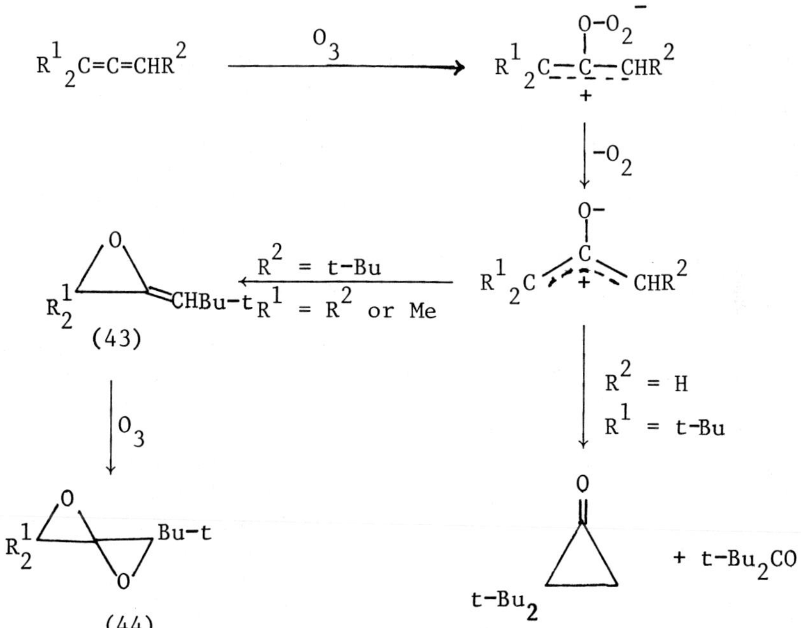

Scheme 6

Allene epoxides have also been implicated in the peracid oxidation of less-hindered alkylallenes (J.K. Crandall, W.H. Machleder, and S.A. Sojka, *ibid.*, 1973, 38, 1149), in which reaction evidence was obtained for valence tautomerism of the epoxides to cyclopropanones. Photo-oxidation of allenes yields carbon dioxide and carbonyl compounds (T. Greibrokk, Tetrahedron Letters, 1973, 1663). Ground state (^3P) oxygen atoms convert alkylallenes into cyclopropanones which, although they decompose readily to CO, olefins, and unsaturated carbonyl compounds, have been trapped as hemi-acetals by adding methanol (J.J. Havel, J.Amer.chem.Soc., 1974, 96, 530). Lead tetra-acetate oxidises (+)-1,3-dimethylallene to mainly (*S*)-(+)-4-acetoxypent-2-yne, suggesting a suprafacial pathway (R.D. Bach, R.N. Brummel, and J.W. Hobulka, J.org.Chem., 1975, 40, 2559).

(5) *Reduction*

The palladium-catalysed semihydrogenation of a series of mainly monosubstituted and 1,1-dialkylated allenes of increasing steric hindrance has been studied, showing that *cis*-hydrogenation occurs mainly at the less hindered π-bond; this technique can be used to determine the absolute configuration of a chiral allene (L. Crombie *et al.*, J.chem. Soc.Perkin I, 1975, 1081, 1090). Other stereospecific reduction techniques reported include the use of Wilkinson's catalyst (M.M. Bhagwat and D. Devaprabhakara, Tetrahedron Letters, 1972, 1391) and the micro-organism *C.Kluyveri* (B. Rambeck and H. Simon, Angew.Chem.internat.Edn., 1974, 13, 609). Selective reduction by optically-active boron hydrides has also been used to isolate one optical isomer of a chiral allene from the racemate (R. Rossi and P. Diversi, Synthesis, 1973, 25; W.R. Moore, H.W. Anderson, and D.S. Clark, J.Amer. chem.Soc., 1973, 95, 835).

$$+ \text{RCH=C(CH}_2\text{R)B(Pin)}_2$$

(6) Photochemical reactions

Only a few photochemical reactions of allenes have been reported. Triplet-sensitized photolysis of propadiene yields a dimer and two trimers, together with products stemming from direct reaction with the sensitizer (H. Gotthardt and G.S. Hammond, Ber., 1975, 108, 657). The isomer ratio of this photo-dimer was different from that obtained by thermal dimerization. Benzene-sensitized photolysis of hexa-1,2-diene gives a range of products most of which can be accounted for by the formation of a biradical by terminal H-transfer to the sp-carbon, but the bicyclohexane seems more likely to arise *via* an allene-cyclopropylidene interconversion (H.R. Ward and E. Karafiath,J.Amer.chem.Soc., 1969, 91, 7475):

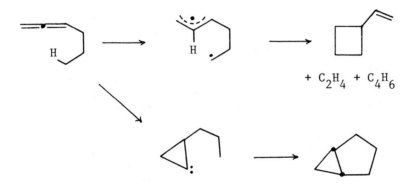

Photochemical addition of propadiene to benzene has also been reported (D. Bryce-Smith, B.E. Foulger, and H. Gilbert, Chem. Comm., 1972, 664).

(7) Cycloaddition reactions

Allene-allene, allene-olefin, and allene-ketene [2+2] cyclo-additions have been reviewed (J.E. Baldwin and R.H. Fleming, Fortsch.Chem.Forsch., 1970, 15, 281 and L. Ghosez and M.J. O'Donnell, "Pericyclic Reactions", Vol. 2, ed. A.P. Marchand and R.E. Lehr, Academic Press, 1977, p. 79). Allene-ketene cycloadditions yield conjugated alkylidenecyclobutanones. Data obtained with optically-active allenes are compatible with a two-step mechanism begun by bonding between the sp-carbon atoms of both components (Scheme 7) (M. Bertrand *et al.*,

Tetrahedron, 1975, <u>31</u>, 849, 857; H.A. Bampfield and P.R. Brook, Chem.Comm., 1974, 171, 172).

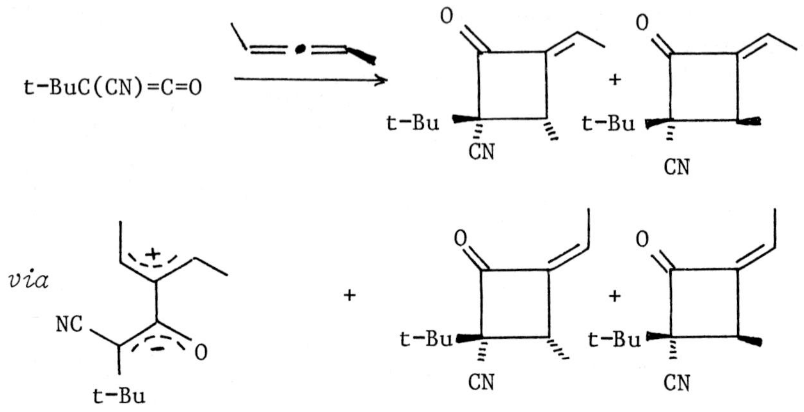

$$t\text{-BuC(CN)=C=O} \longrightarrow$$

Scheme 7

The mechanisms of allene-allene and allene-olefin [2+2] cycloadditions have been hotly debated and the controversy has not yet been fully resolved. Stereospecificity has been observed, for example in the reaction of 1,1-dimethylallene with dimethyl fumarate (E.F. Kiefer, and M.Y. Okamura, J.Amer. chem.Soc., 1968, <u>90</u>, 4187).

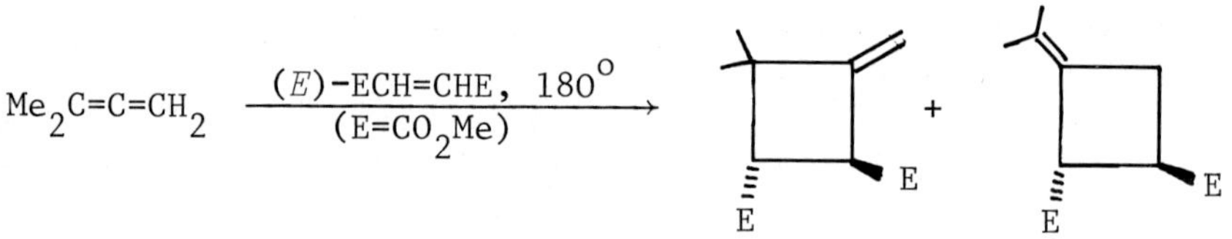

$$Me_2C=C=CH_2 \quad \xrightarrow[\text{(E=CO}_2\text{Me)}]{(E)-ECH=CHE, \ 180^{\circ}}$$

This is capable of interpretation on the basis of a concerted [π2s + π2a] pathway. However, a stepwise but nevertheless orbital symmetry controlled mechanism gained general acceptance following (i) the observation of normal ($k_H/k_D > 1$) intramolecular secondary kinetic isotope effects in the cyclo- addition of acrylonitrile to 1,1-d_2-propadiene (S.H. Dai and W.R. Dolbier, Jr., *ibid.*, 1972, <u>94</u>, 3946) and (ii) the type of stereoselectivity displayed in the cycloaddition of acrylo- nitrile to (R)-(-)-1,3-dimethylallene (J.E. Baldwin and U.V. Roy, Chem.Comm., 1969, 1225). The intermediate was then formulated as an allyl biradical (45), chiral, and forming by a least-hindered suprafacial addition; ring closure was supposed to follow in a disrotatory fashion (Scheme 8).

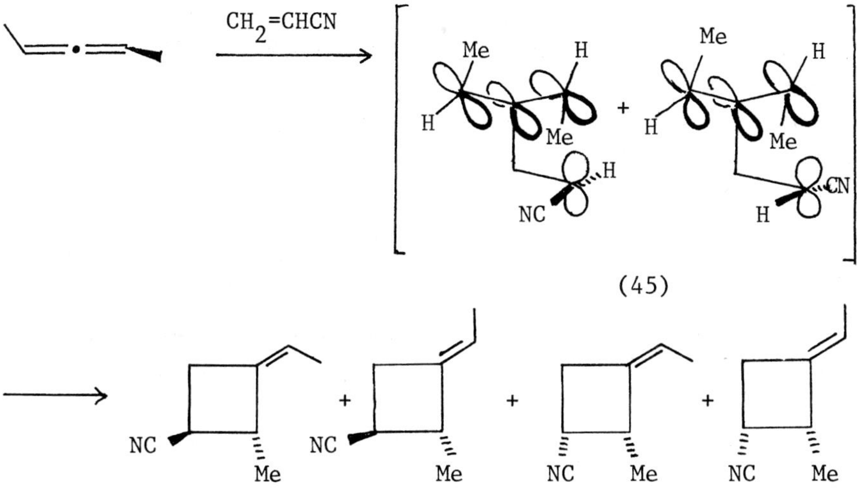

Scheme 8

This result agreed with later work, which suggested a common intermediate bis-allyl diradical in the dimerization of 1,1-dimethylallene and the photochemical or thermal decomposition of the diazabicyclooctadiene (46), reactions which give identical product distributions (T.J. Levek and E.F. Kiefer, J.Amer.chem.Soc., 1976, 98, 1875).

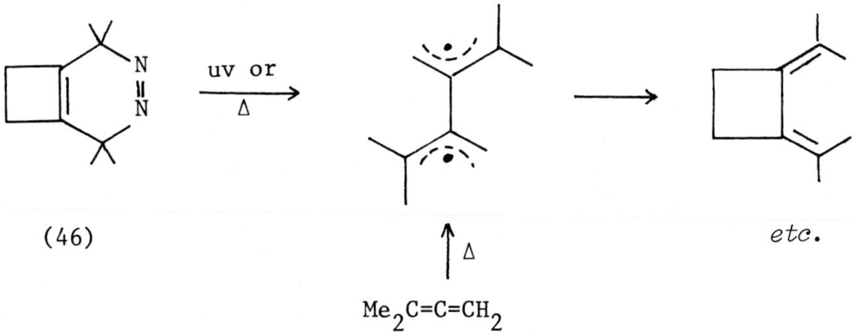

A new proposal for a concerted mechanism, supported by *ab initio* calculations and described as a [π2s + (π2s + π2s)] pathway, has been put forward. This designation is used because it is suggested that the reaction proceeds *via* simultaneous interactions of the non-allenic component with *both* of the orthogonal allenic π-bonds (D.J. Pasto, *ibid.*, 1979, <u>101</u>, 37).

Diels-Alder reactions of allenes with 1,3-dienes are not uncommon; for example, propadiene reacts concertedly with hexachlorocyclopentadiene (Dai and Dolbier, *loc.cit.*), an *inverse* intramolecular secondary kinetic isotope effect being observed with 1,1-d_2-propadiene. The complete regiospecificity of the [4+2] cycloadditions of 1,1-difluoro- and monofluoro-propadiene to various 1,3-dienes has been elegantly shown to confirm HOMO-LUMO control in the concerted process (W.R. Dolbier, Jr., *et al.*, Tetrahedron Letters, 1978, 2231; 1980, 785).

$$
\text{(cyclopentadiene)} \quad \xrightarrow[\text{(X = H or F)}]{H_2C=C=CFX} \quad \text{(bicyclic product)} = CFX
$$

An easy route to alkylidenecyclobutanes is provided by the acid-catalysed cycloaddition of simple allenes with olefins (J.H. Lukas, A.P. Kouwenhoven, and F. Baardman, Angew.Chem. internat.Edn., 1975, <u>14</u>, 709).

MeCH=C=CHMe

\+

MeCH=CHMe

$\xrightarrow{AlCl_3}$

(Me, Me, Me cyclobutylidene product)

Although 1,3-dipolar cycloadditions of various reagents to functionally substituted allenes have been described (e.g., P. Battioni, L. Vo-Quang, and Y. Vo-Quang, Bull.Soc.chim.Fr., 1978, 401 and 415), few instances of analogous cycloadditions to acyclic allenic hydrocarbons have been reported (Dai and Dolbier, *loc.cit.*; J.J. Tufariello, Sk.A. Ali, and H.O. Klingele, J.org.Chem., 1979, <u>44</u>, 4213; and G. Cum *et al.*, J. chem.Soc.Perkin I, 1976, 719) (Scheme 9). No clear pattern of reactivity has emerged.

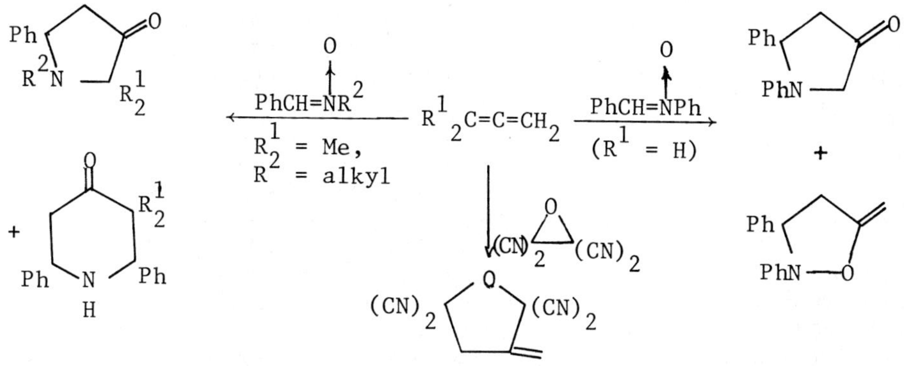

Scheme 9

Carbenes (R.R. Kostikov *et al.*, J.org.Chem.USSR, 1974, 10, 2115, 2339), carbenoids (X. Creary, J.Amer.chem.Soc., 1977, 99, 7632), and nitrenes (E.M. Bingham and J.C. Gilbert, J. org.Chem., 1975, 40, 224) have all been shown to add normally to propadiene and simple alkylallenes. The different regio-selectivities displayed by certain singlet and triplet carbenes in additions to 1,1-dimethylallene have been suggested as a probe for carbene multiplicities (X. Creary, J.Amer.chem.Soc., 1980, 102, 1611).

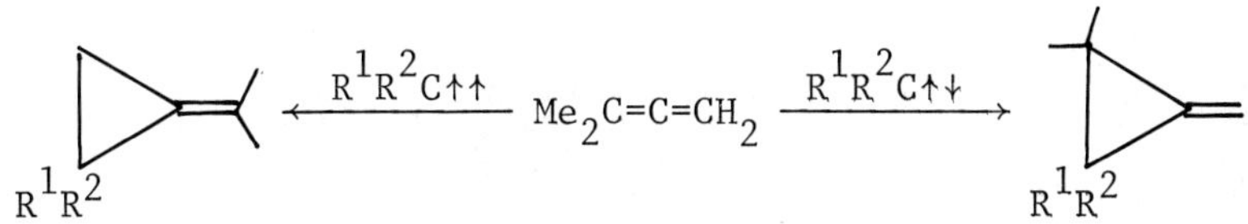

(8) Ene reactions

Alkylallenes undergo ene insertions relatively faster than the correspondingly substituted olefins, due to decreased steric hindrance and the favourable alteration in π-bond structure which occurs during such reactions.

Enophiles so far investigated include sulphonylisocyanates and acrylonitrile (J.C. Martin, P.L. Carter, and J.L. Chitwood, J.org.Chem., 1971, 36, 2225), electron-deficient acetylenes, azoesters, and ketones (D.R. Taylor *et al.*, J. chem.Soc.Perkin I, 1974, 1209, 1884; 1977, 1463; and 1978, 1161), thioarylketones (H. Gotthardt, Tetrahedron Letters, 1971, 2343), and trithiocarbonate S,S-dioxides (J.A. Boerma, N.H. Nilsson, and A. Senning, Tetrahedron, 1974, 30, 2735). Product mixtures may be complicated owing to competing side-reactions such as cycloadditions and rearrangements (Taylor, *loc.cit.*).

(9) Metal catalysed reactions.

Allene reacts with various amines and carbon acids to give bis(butadienyl)derivatives incorporating up to four moles of propadiene: Pd(0) and Rh(I) complexes are suitable catalysts (D.R. Coulson, J.org.Chem., 1973, 38, 1483).

$$(R^3R^4 = CN, CO_2Et, Ac)$$

π-Allyl complexes can be obtained readily from propadiene and, for example, dichlorobis(benzonitrile)palladium (R.G. Schultz, Tetrahedron, 1964, 20, 2809) and, similarly, from tetramethylallene (M.S. Lupin, M.S. Powell, Jr., and B.L. Shaw, J.chem.Soc.(A), 1966, 1687). Carbonylation of such complexes has been reported (J. Tsuji and T. Susuki, Tetrahedron Letters, 1965, 3027):

$$\text{CH}_2=\text{CC1CH}_2\text{CO}_2\text{Et}$$

$$\text{CH}_2=\text{C}=\text{CH}_2$$

Transition metal and Ziegler-Natta catalysts have been used to generate high molecular weight polymers from propadiene (T.F. Rutledge, "Acetylenes and Allenes", Reinhold, New York, 1969, p. 90). Low molecular weight oligomers and co-oligomers have also been prepared from propadiene and butadiene over nickel catalysts (S. Otsuka et al., J.Amer.chem.Soc., 1972, 94, 1037 and J.chem.Soc.Dalton, 1973, 2491; D. Heimbach, H. Selbech and E. Troxler, Angew.Chem.internat.Edn., 1971, 10, 659).

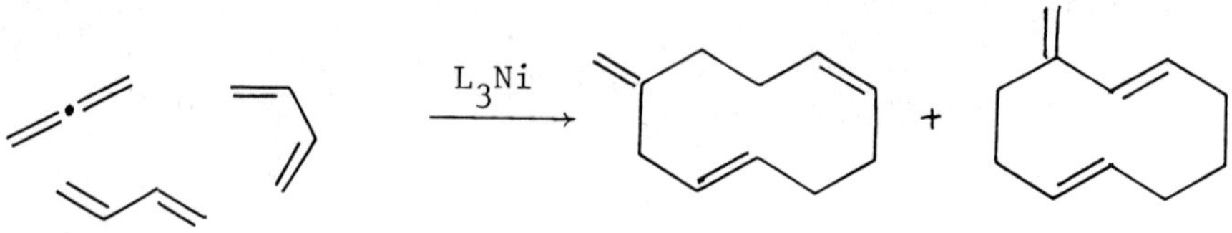

(b) Dienes

This section will concentrate mainly, but not exclusively, upon conjugated dienes. Several useful general texts are available on this topic (G. Pattenden, "Comprehensive Organic Chemistry", ed. D.H.R. Barton, W.D. Ollis and J.F. Stoddart, Pergamon, Oxford, 1979, p. 171; H. von Brachel and U. Bahr in Houben-Weyl's "Methoden der Organischen Chemie", Vol. 5/1c, ed. E. Müller, Thieme-Verlag, Stuttgart, 1970; and M. Cais, "The Chemistry of Alkenes", ed. S. Patai, Wiley-Interscience, London, 1964, ch. 12).

(i) Synthetic methods

(1) Additions to enynes and diynes

Dienes of known stereochemistry are available by selective *syn* partial reduction of diynes (*Z,Z* dienes obtained) or enynes (*E,Z* or *Z,Z* obtained, depending on the geometry of the C=C bond originally present). Suitable reduction procedures are catalytic hydrogenation over Lindlar's catalyst or mono-addition of a hindered or blocked alkylborane, such as thexyl- or catechol-borane, followed by protolysis, (G. Zweifel and N.L. Polston, J.Amer.chem.Soc., 1970, 92, 4068).

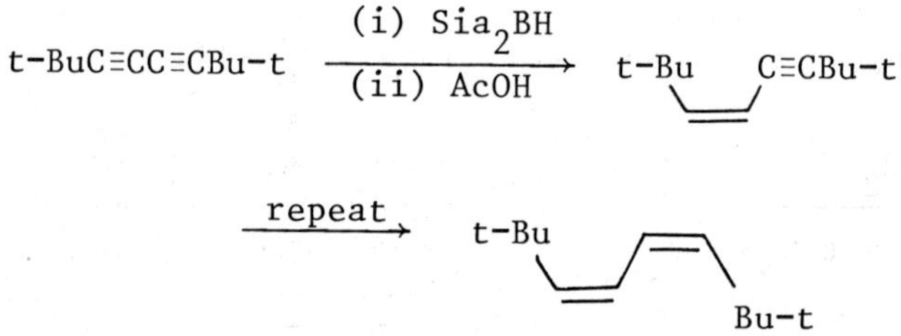

The method may be extended by using the stereospecific addition of organocoppers to the triple bond of enynes, as in a recent synthesis of myrcene (P. Vermeer *et al.*, Tetrahedron

Letters, 1977, 869).

$$HC\equiv CCH=CH_2 \xrightarrow[\text{(ii) } H_3O^+]{\text{(i) } R_2CuMgCl}$$

(R = $Me_2C=CHCH_2CH_2$)

(2) Coupling reactions of vinyl-derivatives

An important group of methods for diene synthesis has emerged following the availability of alkenylboranes by addition of boranes to alkynes. Thus, treatment of dialkenylchloroboranes with methylcopper at low temperature gives pure, symmetrical, E,E-dienes directly (Y. Yamamoto, H. Yatagai, and I. Moritani, J.Amer.chem.Soc., 1975, 97, 5606).

$$EtC\equiv CEt \xrightarrow{H_2BCl} \quad \xrightarrow{MeCu} \quad$$

This method is easily adapted to the stereoselective synthesis of 1,4-dienes (Y. Yamamoto et al., Chem.Comm., 1976, 452).

Symmetrical E,Z-dienes are obtained if alkyne-borane adducts are iodinated, preferably after initially converting them into alkoxyboranes with an amine oxide (G. Zweifel, N.L. Polston, and C.C. Whitney, J.Amer.chem.Soc., 1968, 90, 6243).

$$R^1C\equiv CH \xrightarrow[(R^2 = t-C_6H_{13})]{R^2BH_2} \quad \xrightarrow[\text{(ii) } I_2,OH^-]{\text{(i) } Me_3\overset{+-}{NO}}$$

Alternative routes to *E,E*-dienes are given below.
(i) Oxidative coupling of vinylalanes with copper(I)chloride
(G. Zweifel and R.L. Miller, *ibid.*, 1970, 92, 6678).

$$R^1C{\equiv}CH \xrightarrow{R^2{}_2AlH} \underset{AlR^2{}_2}{\overset{R^1}{\diagup\!\!\!\diagdown}} \xrightarrow{Cu(I)Cl} \overset{R^1}{\diagup\!\!\!\diagdown\!\!\!\diagup}R^1$$

(65-75%)

$(R^1 = Et,Bu;\ R^2 = i\text{-}Bu)$

(ii) Treatment of alkoxyalkenyldialkylborates, obtained from
alkoxides and alkyne-borane adducts, with copper(I)bromide as
its dimethylsulphide complex at 0 °C (J.B. Campbell and H.C.
Brown, J.org.Chem., 1980, 45, 549), a procedure which can
readily be adapted to stereospecific 1,4-diene synthesis by
performing the last step at -15 °C and rapidly adding an
allylic halide (*idem.*, *ibid.*, p. 550).

$$R^1C{\equiv}CR^2 \xrightarrow{R_2BH} \underset{BR_2}{\overset{R^1\quad R^2}{\diagdown\!\!=\!\!\diagup}} \xrightarrow[(ii)\ Cu(I)Br.SMe_2]{(i)\ NaOMe} [Cu\text{-complex}]$$

$$\left\downarrow\begin{array}{l}-15\ ^\circ C\\ R^3Br\end{array}\right.$$

[structure: diene with R^1, R^2, R^2, R^1] $\xleftarrow{\quad 0\ ^\circ C\quad}$ [structure with R^1, R^2, R^3]

$(R^1 = Et,Bu;\ R^2 = H\ or\ R^1;\ R^3 = Me_2C{:}CHCH_2)$

(iii) By oxidative coupling of vinylmercurials using
palladium(II) or nickel(II) halides in polar aprotic
solutions (R.C. Larock, *ibid.*, 1976, 41, 2241), a procedure
which has the advantage that it tolerates other functional
groups.

$$\text{MeC}\equiv\text{CMe} \xrightarrow[\text{NaCl}]{\text{Hg(OAc)}_2}$$

[structure: AcO, Me substituted alkene with Me, HgCl]

$$\xrightarrow[\text{HMPA}]{\text{PdCl}_2,\text{LiCl}}$$

[structure: diene with AcO, Me, Me, Me, Me, OAc]

It should be noted, however, that oxidative coupling of
vinylmercurials by palladium(II) salts has been found to give
head-to-tail, rather than head-to-head, dimers (R.C. Larock
and B. Riefling, *ibid.*, 1978, 43, 1468). Another variation
of oxidative coupling which provides isomerically pure *E,E*-
dienes by head-to-head coupling is the silver ion induced
dimerization of alkenylpentafluorosilyl anions (M. Kumada *et
al.*, Tetrahedron Letters, 1979, 1137).

$$\text{RC}\equiv\text{CH} \xrightarrow[\text{H}_2\text{PtCl}_6]{\text{HSiCl}_3}$$

[structure: R substituted alkene with SiCl$_3$]

$$\xrightarrow{\text{KF}}$$

$$\left[\text{R} \diagup\diagdown \text{SiF}_5 \right]^{-} (\text{K}^{+})_2$$

$$\xrightarrow{\text{AgF,MeCN}}$$

[structure: R, R diene]

Other applications of alkenylboranes to diene synthesis are
given below.
(i) The formation of isomerically pure unsymmetrical *E,E*-
dienes by the reactions between chlorovinylboranes (47) and
terminal alkynes (E.-i. Negishi and T. Yoshida, Chem.Comm.,
1973, 606).

$$\text{R}^1\text{C}\equiv\text{CCl} \xrightarrow[(\text{R}^2 = t\text{-}\text{C}_6\text{H}_{13})]{\text{R}^2\text{BH}_2}$$

[structure: R^1, Cl alkene with BHR^2]

(47)

$$\xrightarrow[\substack{\text{(ii) NaOMe} \\ \text{(iii) PrCO}_2\text{H}}]{\text{(i) R}^3\text{C}\equiv\text{CH}}$$

[structure: R^1, R^2 diene]

(ii) The formation of unsymmetrically substituted E,Z-dienes by the treatment of alkenylalkynylborates with Lewis acids (G. Zweifel and S.J. Backlund, J.organometal.Chem., 1978, 156, 159).
(iii) Palladium(II)-catalysed coupling of dihydroxyalkenyl-boranes to give excellent yields of E,E-dienes (N. Miyaura, K. Yamada, and A. Suzuki, Tetrahedron Letters, 1979, 3437).

An important group of synthetic methods derives from the increased availability of isomerically pure vinylhalides (see for example R.B. Miller and G. McGarvey, J.org.Chem., 1978, 43, 4424 and J.F. Normant et al., Synthesis, 1974, 803). These vinyl halides may be coupled with alkenylalanes in the presence of palladium(0)- or nickel(0)-catalysts (E.-i. Negishi and S. Baba, J.Amer.chem.Soc., 1976, 98, 6729).

$$t\text{-BuC}{\equiv}\text{CH} \xrightarrow{i\text{-Bu}_2\text{AlH}} t\text{-Bu}\diagup\diagdown\diagup^{\text{Al(Bu-i)}_2} \xrightarrow[n\text{-Bu}\diagup\diagdown\diagup_{\text{Hal}}]{5\%\ \text{Pd(PPh}_3)_4}$$

$$t\text{-Bu}\diagup\diagdown\diagup\diagdown_{\text{Bu-n}}$$

(>93%, E,E)

Similar coupling is achieved with alkenylzirconiums, obtained by the hydrozirconation of alkynes (E.-i. Negishi et al., Tetrahedron Letters, 1978, 1027), or with alkenyl Grignard reagents provided that the important Pd(0) catalyst is present (H.P. Dang and G. Linstrumelle, ibid., p. 191).

Certain nickel(0) catalysts will catalyse the direct
oxidative coupling of alkenylhalides, though the yields of
E,E-dienes reported were only moderate (M.F. Semmelhack, P.M.
Helquist and J.D. Gorzynski, J.Amer.chem.Soc., 1972, 94,
9234). The utility of the alkenylzirconium reagents has been
further improved by Negishi's group, who have shown that the
hydrozirconation of an alkyne is accompanied by alkylation
if a trialkylalane is present (E.-i. Negishi *et al.*, *ibid.*,
1978, 100, 2252). The importance of a co-catalyst such as
zinc chloride in the coupling step has been stressed by
Negishi (E.-i. Negishi *et al.*, *ibid.*, p. 2254).

$$n\text{-AmC}{\equiv}\text{CH} \xrightarrow[\substack{(ii)\ \ CH_2{=}CHBr,\ ZnCl_2, \\ Pd(Ph_3P)_4}]{(i)\ \ Me_3Al\text{-}Cp_2ZrClH}$$

(>98% E)

Vinyl halides of known stereochemistry are also ideal
substrates for conversion into dienes *via* preliminary
replacement of halogen by copper, lithium, or aluminium.
Thus, vinyl cuprates can be coupled thermally or by oxidation
(R.B. Banks and H.M. Walborsky, J.Amer.chem.Soc., 1976, 98,
3732, and references cited therein), or used as a source of
dienes by virtue of their clean Michael addition to α,β-
alkynylcarbonyl compounds (F. Näf and P. Degen, Helv., 1971,
54, 1939).

$$\underset{\displaystyle (R = n\text{-Am})}{\text{R}\diagdown\!\!\diagup\text{Hal}} \xrightarrow[\text{Cu(I)iodide}]{n\text{-BuLi}} \Big(\text{R}\diagdown\!\!\diagup\Big)_2\text{CuLi} \xrightarrow{\text{HC}{\equiv}\text{CCO}_2\text{Et}}$$

$$\underset{\substack{R \\ (>95\%\ Z,E)}}{\diagup\!\!\diagdown\!\!\diagup\!\!\diagdown}\text{CO}_2\text{Et}$$

(3) From allylic derivatives

The classical methods of diene synthesis by dehydration,

dehydrohalogenation, etc., of allylic alcohols, halides, and related derivatives, have been reviewed (H. von Brachel and U. Bahr, Houben-Weyl's "Methoden der Organischen Chemie", vol. 5/1c, ed. E. Müller, Thieme-Verlag, Stuttgart, 1970). Milder conditions have occasionally been found for such procedures, as in the dehydration of allylic alcohols using a phosphonium iodide in hexamethylphosphoramide (C.W. Spangler and T.W. Hartford, Synthesis, 1976, 108).

Similarly, enediols may be converted into conjugated dienes in moderate yields *via* corresponding dibromides (H. Yamamoto *et al.*, J.Amer.chem.Soc., 1975, _97_, 3252). Methods for the synthesis of the required enediols from allyl alcohols have also been evaluated (A. Yasuda *et al.*, Bull.chem.Soc.Japan, 1979, _52_, 1752).

$$RCH_2CH=CHCH_2OH \xrightarrow[\text{t-BuO}_2\text{H}]{\text{VO(acac)}_2} RCH=CHCH(OH)CH_2OH \xrightarrow[\text{(ii) zinc}]{\text{(i) } PBr_3,Cu_2Br_2}$$

$$RCH=CHCH=CH_2$$

Dimethylformamide acetals can be used to advantage for the direct conversion of hydroxyacids into dienes (J. Mulzer, U. Kühl and G. Brüntrup, Tetrahedron Letters, 1978, 2953). Equally attractive is the synthesis of terminal 1,3-dienes by the palladium-catalysed elimination of acetic acid or phenol from allylic esters or phenyl ethers (J. Tsuji *et al.*, *ibid.*, 2075).

The influence of this general group of reactions was much enhanced by the advent of the Wittig reaction, which can lead to conjugated dienes in a number of ways. Allylic halides can be converted into corresponding P-ylides for treatment with aldehydes or reactive ketones (von Brachel and Bahr, Houben Weyl, *loc.cit.*, p. 590); the usual problem of lack of stereochemical control remains, but stereochemistry within the allylic portion is preserved.

Similarly, (Z)-α,β-unsaturated aldehydes, when treated with conventional Wittig reagents, yield (Z)-dienes (G. Pattenden and B.C.L. Weedon, J.chem.Soc.(C), 1968, 1984), and if salt-free conditions are used, (Z,Z)-isomers should predominate.

The Horner modification using ylides from allylic diphenylphosphine oxides has been advocated for stereochemical control of diene synthesis (A.H. Davidson and S. Warren, J. chem.Soc. Perkin I, 1976, 639). Factors affecting the phosphinoyl group migration during preparation of the precursors for this new procedure have been discussed (S. Warren et al., ibid., 1977, 550, 1452). The key to successful stereochemical control evidently lies in chromatographic separation of the intermediate diastereoisomeric alcohols (48).

$$R^1CHMeP(O)Ph_2 \xrightarrow[\substack{R^2CHO}]{Bu^nLi} Ph_2P(O)CMeR^1CH(OH)R^2$$

(48)

$$R^1 = \text{(cyclohexenyl)}$$

Sulphoxyl-stabilised (E.R. de Waard *et al.*, Tetrahedron, 1977, 33, 579) and silyl-stabilised (P.W.K. Lau and T.H. Chan, Tetrahedron Letters, 1978, 2383) allylcarbanions have also been used in efficient di- and tri-ene syntheses. There is an excellent account by C.A. Henrick of the stereochemical problems in the Wittig and related procedures used in the synthesis of dienes of natural origin (Tetrahedron, 1977, 33, 1845).

(4) Electrocyclic rearrangements and retrocycloadditions.

Although pyrolytic retrocycloadditions have been suggested as routes to specific 1,3-dienes (e.g., B.M. Trost, S.A. Godleski, and J. Ippen, J.org.Chem., 1978, 43, 4559), the main advantage of such an approach lies in the possibility of removing unwanted (Z)-isomers (B.F. Nesbitt *et al.*, Tetrahedron Letters, 1973, 4669).

The reaction is stereospecific, proceeding suprafacially; for example, *cis*-2,5-dimethyldihydrothiophene dioxide gives, at 100 °C, 99.9% (E,E)-hexa-2,4-diene; and the *trans*-isomer a correspondingly high yield of the (E,Z)-diene (W.L. Mock, J. Amer.chem.Soc., 1975, 97, 3666). In this context, the much more facile elimination of sulphur trioxide from dihydro-oxathiin-2-oxides could be a useful alternative means to the same end (T. Durst *et al.*, *ibid.*, 1974, 96, 935). Cyclo-reversions have also been studied which give 1,4- and 1,5-dienes stereospecifically (J.A. Berson *et al.*, Tetrahedron, 1974, 30, 1639).

The stereospecificity of electrocyclic rearrangements has naturally attracted attention as a means to obtain isomerically pure dienes. One approach features isomerizations of allyl vinyl ethers, the second double bond being introduced at a later stage(R.C. Cookson and R. Gopalan, Chem.Comm., 1978, 608, 924), or allenyl vinyl ethers can be used (M. Bertrand and J. Viala, Tetrahedron Letters, 1978, 2575).

A number of synthetically useful dienes bearing reactive functional groups have been obtained by electrocyclic ring-opening of isomeric cyclobutenes (B.M. Trost and A.J. Bridges, J.Amer.chem.Soc., 1976, 98, 5017; D. Bellŭs and C.D. Weis, Tetrahedron Letters, 1973, 999).

(ii) Physical properties of dienes

Acyclic conjugated dienes are normally in rapid equilibrium between their s-*trans* (49) and s-*cis* (50) conformations, the former predominating in uncrowded butadienes and being increasingly favoured by large terminal substituents. Bulky groups at the 2-position, on the other hand, shift the equilibrium towards the s-*cis* form or, in extreme cases, prevent the diene attaining a planar structure.

(49)

s-*trans*

(50)

s-*cis*

The conformational preferences of butadiene (S. Skaarup, J.E. Boggs and P.N. Skancke, Tetrahedron, 1976, 32, 1179; A.J.P. Devaguet, R.E. Townshend, and W.J. Hehre, J.Amer.chem.Soc., 1976, 98, 4068) and of a large number of its lower homologues and small polyenes (J.C. Tai and N.L. Allinger, *ibid.*, p. 7928) have been evaluated by *ab initio* calculations, and the energy barrier to rotation computed. Infrared analysis of the thermally trapped s-*cis* conformer of butadiene has provided an experimental value of the barrier, 163 kJ/mol (O.L. Chapman *et al.*, *ibid.*, 1979, 101, 3657), and the proportion of s-*cis* at equilibrium at room temperature in various dienes has been estimated by vibrational analysis (D.A.C. Compton, W.O. George, and W.F. Maddams, J.chem.Soc. Perkin II, 1976, 1666); thus there is 3% in butadiene and 11% in isoprene. Proton nmr has been used to probe for the existence of the non-planar or gauche form of hindered butadienes, e.g., with 2-t-butyl-substitution, (A.A. Bothner-By and R.K. Harris, J.Amer.chem. Soc., 1965, 87, 3451), but a more modern approach is to use photoelectron spectroscopy (T.H. Morton *et al.*, *ibid.*, 1976, 98, 4732); thus the diene (51) is twisted by 83–85° out of the plane.

$$\underset{\underset{R}{|}}{\overset{\overset{R}{|}}{Me_2C=C-C=CMe_2}} \quad (51) \qquad (R = i\text{-}Pr)$$

The rate of Diels-Alder cycloadditions, which can only proceed if the diene attains the planar s-*cis* form, has been used in the past to investigate the conformational equilibrium (J. Sauer, Angew.Chem.internat.Edn., 1967, 6, 16). Thus, 2,3-di-t-butylbutadiene, for which a high-yield synthesis has been reported (L. Brandsma *et al.*, Chem.Comm., 1980, 922), fails to react with maleic anhydride although it will undergo other types of cyclizations.

The vertical ionization potentials of the occupied π-molecular orbitals (MO) have been determined by photoelectron spectroscopy for butadiene, isoprene, and small trienes (M. Beez *et al.*, Helv., 1973, 56, 1028), setting experimental values to their MO energies (ψ_1, -11.4 to -12.2 eV, ψ_2, -9.0 eV, for butadiene) for comparison with computed values. The atomic orbital (AO) coefficients in these and the vacant π-MO's have also been calculated (Table 2) (A. Streitwieser, J.I. Brauman, and C.A. Coulson, "Supplemental Tables of Molecular Orbital Calculations", Pergamon, Oxford, 1965). The effect of various types of substituents at the 1- and 2-positions of butadiene on its MO energies and the AO coefficients have been evaluated (K.N. Houk *et al.*, J.Amer.chem.Soc., 1973, 95, 7287): these parameters now have great significance in cyclo-addition reaction theory (K.N. Houk, Topics curr.Chem., 1979, 79, 2; Acc.chem.Res., 1975, 8, 361).

Table 2. Atomic orbital coefficients (c_n) and orbital energies of the π-molecular orbitals (MO's) of buta-1,3-diene

MO	c_1	c_2	c_3	c_4	energy in eV
ψ_4*	0.371	- 0.60	0.60	- 0.371	3.3
ψ_3*	0.60	- 0.371	- 0.371	0.60	1.0
ψ_2	0.60	0.371	- 0.371	- 0.60	- 9.1
ψ_1	0.371	0.60	0.60	0.371	- 11.4

Ultraviolet spectroscopy played an historic role in the determination of the structures of dienes and polyenes, being formalised for this purpose in the Woodward Rules. Early work in this area, reviewed by Cais ("The Chemistry of Alkenes", ed. S. Patai, Wiley-Interscience, London, 1964, ch. 12) has been up-dated (R. Grinter and T.L. Threlfall, Tetrahedron, 1979, 35, 1543). The determination of *E/Z* ratios and related structural features is usually achieved by nmr nowadays (J.K. Saunders and J.W. Eaton, Determin.org.Struct.Phys.Methods, 1976, 6, 271). The application of the nuclear Overhauser effect to the structural assignment of complex dienes has been reported (J. de Bruijn *et al.*, Tetrahedron, 1973, <u>29</u>, 1541) but has not been used very extensively.

(iii) Chemical reactions of dienes

The most important developments of the last two decades in the chemical reactions of conjugated dienes are undoubtedly in the areas of pericyclic reactions, diene photochemistry, and metal-catalysed processes.

(1) Thermal rearrangements

For acyclic hydrocarbons these are essentially limited to the Cope reaction and related sigmatropic shifts, and to reversible electrocyclic ring-forming rearrangements. All such processes are governed by the restraints of orbital symmetry (see also Section 2d-vii), which for electrocyclic rearrangements lead to the satisfyingly correct prediction of the variation of stereochemistry with the number of π-electrons in the open-chain isomer (Table 3). Thus the allowed stereochemistry for closure of a 1,3-diene is conrotation (con), whereas that for a 1,3,5-triene is dis-rotation (dis). The former process displays a C_2-axis of symmetry, the latter a plane of symmetry; the stereochemistry which is observed in photochemical isomerizations is exactly the reverse of that of the thermal processes of the molecules in their ground states.

Table 3. Stereochemistry predicted by orbital symmetry rules for electrocyclic rearrangements (R.B. Woodward and R. Hoffmann, Angew.Chem.internat.Edn., 1969, 8, 781).

Number of π-electrons in acyclic polyene (n = integer)	Stereochemistry (C = con; D = dis)	
	Thermal	Photochemical
2	D	C
4	C	D
6	D	C
4n	C	D
4n+2	D	C

For uncongested dienes the ring-opened isomer is the one favoured thermodynamically, *i.e.*, the reaction is more useful for diene synthesis (Section 3b-i). Typical examples are shown (R.E.K. Winter, Tetrahedron Letters, 1965, 1207; R. Criegee *et al.*, Ber., 1965, 98, 2339).

(R = D, Cl or Me)

For trienes, the cyclohexadiene isomer is more stable under normal circumstances, and ring closure of the triene often occurs smoothly at 130 °C, [e.g. for (*E,Z,E*)-octa-2,4,6-triene, E.N. Marvell, G. Caple, and B. Schatz, Tetrahedron Letters, 1965, 385]. The latter reaction is historically important, since it was the difference in the stereochemistry

of thermal and photochemical cyclization of precalciferol,
essentially a 1,3,5-triene, first noted in 1961 (E. Havinga
and J.L.M.A. Schlatmann, Tetrahedron, 1961, 16, 146), which
eventually led to the formulation of the so-called orbital
symmetry rules (R.B. Woodward, "Aromaticity", Chem.Soc.spec.
Publ., 1967, No. 21, p. 217).

Pentadienyl and longer chain dienes can undergo, in addition
to electrocyclic rearrangements, thermal sigmatropic shifts,
most readily of hydrogen although alkyl and other groups can
migrate. 1,4-Dienes are thermally stable in the absence of
catalysts, and 1,5-dienes are capable of sigmatropic
isomerizations but not electrocyclic ring closure. It is one
of the notable triumphs of the theoretical advances of the
last fifteen years that diverse reactions of this type can be
explained on the same basis. For an explanation of the under-
lying theories of sigmatropic shifts, specialist texts should
be consulted (T.L. Gilchrist and R.C. Storr, "Organic
Reactions and Orbital Symmetry", 2nd Edn., CUP, Cambridge, 1979,
ch. 7; for sigmatropic H shifts see C.W. Spangler, Chem.Rev.,
1976, 76, 187; for Cope rearrangements, see S.J. Rhoads and
N.R. Raulins, Org.Reactions, 1975, 22, 1). The kinetics of
such processes have been treated separately (H.M. Frey and R.
Walsh, Chem.Rev., 1969, 69, 103; R.H. DeWolfe, "Comprehensive
Chemical Kinetics", Vol. 9, ed. C.H. Bamford and C.F.H.
Tipper, Elsevier, London, 1973, ch. 5).

If the migrating group is hydrogen and it shifts to a
terminus j atoms (the count includes the atom it was
originally attached to) the process is termed a [1,j] shift.
If the σ-bonds broken and made lie on the same side of the
π-system, the process is termed suprafacial, the reverse
shift from one face to the other being antarafacial. The
selection rules for 'allowed' thermal sigmatropic shifts of
hydrogen in neutral polyenes are shown in Table 4. They are
exactly reversed for photochemical shifts, and for those
shifts of, for example, alkyl groups in which the migrating
link atom undergoes inversion of configuration (rare in
acyclic polyenes). The related Cope rearrangement involves
migration of a σ-bond but without atom migration: the same
type of terminology leads to its designation as a [3,3]
sigmatropic shift, and its stereochemistry is best understood
on the basis of a chair-like conformation for the transition
state, corresponding boat-like transition states being some
25 kJ/mol less accessible (R.K. Hill, and N.W. Gilman, Chem.

Table 4. Selection rules for hydrogen sigmatropic shifts in
          neutral polyenes (Gilchrist and Storr, *loc.cit.*).

Number of π electrons in the polyene (n = integer)	Type of shift	Allowed stereochemistry A = antarafacial S = suprafacial
2	[1,3]	A (not observed)
4	[1,5]	S
6	[1,7]	A
4n	[1,(4n+1)]	S
4n+2	[1,(4n+3)]	A

Comm., 1967, 619). Some representative examples are shown
below.

[1,5] suprafacial shifts

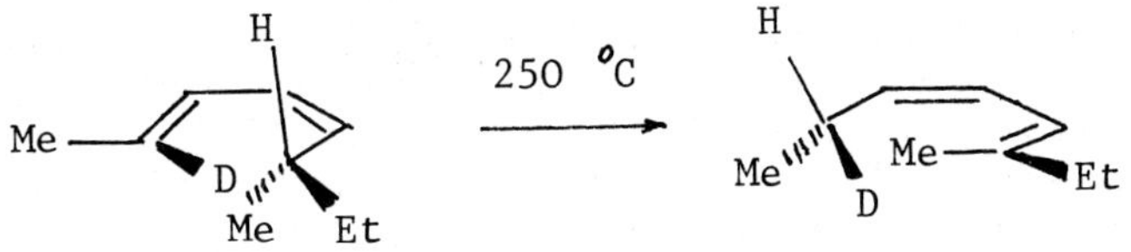

(W.R. Roth, J. König, and W. Stein, Ber., 1970, 103, 426).

[1,7] antarafacial shifts

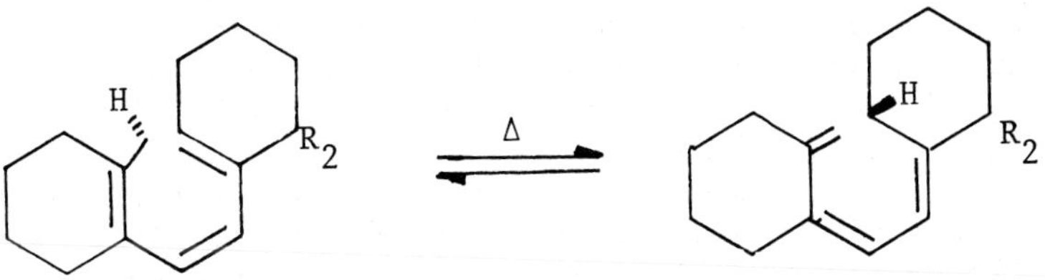

(E. Havinga and J.L.M.A. Schlatmann, Tetrahedron, 1961, 16,
146).

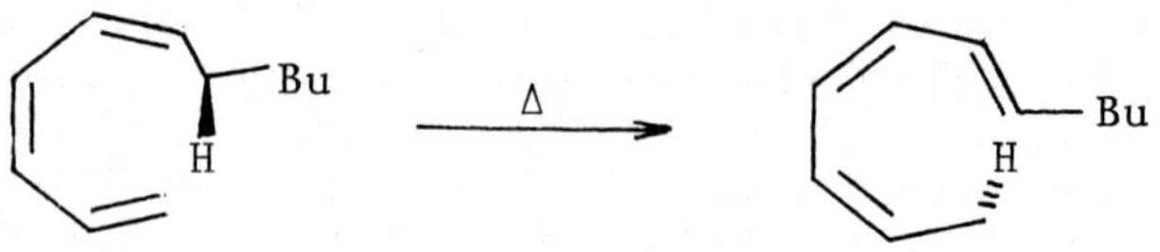

(G. Ohloff *et al.*, Helv.,1975, <u>58</u>, 1016).

[3,3] Cope sigmatropic shifts

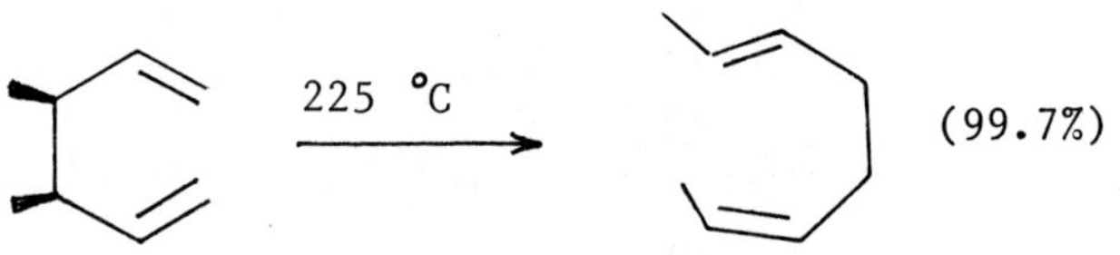

(99.7%)

(W. von E. Doering and W.R. Roth, Angew.Chem.internat.Edn., 1963, <u>2</u>, 115).

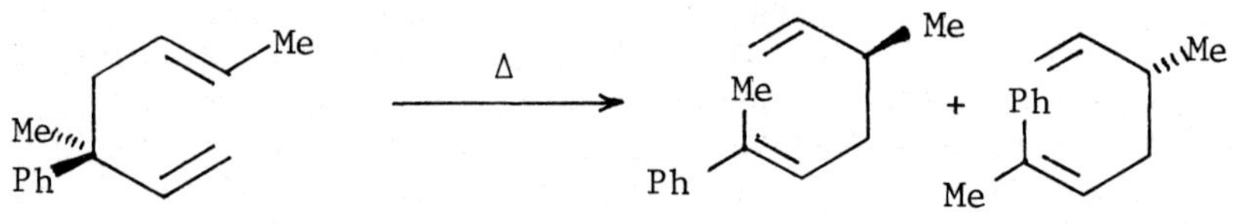

(R)   (S)(87%)   (R)(13%)

(Hill and Gilman, *loc.cit.*)

(2) Thermal cyclizations

The general features of the Diels-Alder reaction have been known for many years (2nd Edn., Section 2b-iii). For more detailed accounts of the early work, additional reviews are now available (J. Sauer, Angew.Chem.internat.Edn., 1966, <u>5</u>, 211; 1967, <u>6</u>, 16; various authors in "1,4 Cycloaddition Reactions", ed. J. Hamer, Academic Press, New York, 1967).

Kinetic data have also been reviewed (P. Beltrame, "Comprehensive Chemical Kinetics", Vol. 9, ed. C.H. Bamford and C.F.H. Tipper, Elsevier, London, 1973, ch. 2). In modern terminology the reaction is viewed as a thermally allowed [π2s + π4s] process (R.B. Woodward and R. Hoffmann, Angew.Chem.internat.Edn., 1969, 8, 781; T.L. Gilchrist and R.C. Storr, "Organic Reactions and Orbital Symmetry", 2nd Edn., CUP, Cambridge, 1979, chapters 4-5) proceeding *via* a transition state (52) in which both of the new σ-bonds are being formed concurrently; detailed analysis of this transition state has been achieved by *ab initio* calculations (W.J. Hehre *et al.*, J.Amer.chem.Soc., 1976, 98, 2190).

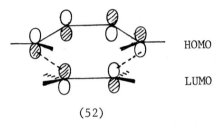

HOMO

LUMO

(52)

Experimental evidence for the correctness of this mechanism is overwhelming, including stereospecificity, small inverse secondary kinetic isotope effects (D.E. Van Sickle and J.O. Rodin, *ibid.*, 1964, 86, 3091), activation parameters ($\Delta S^{\ddagger}$ in the range -125 to -170 J/deg/mol: Beltrame, *loc.cit.*), and low solvent effect. However, well-known features of the reaction have proved less easily rationalised; these include *endo*-specificity, relative rates of reaction of particular dienes with various dienophiles, and, most significant of all for synthetic organic chemists, regioselectivities with unsymmetrically substituted reactants. Although arguments based on the correlation of orbital symmetry can be advanced for the interpretation of some of these phenomena (Woodward and Hoffmann, *loc.cit.*), frontier orbital interaction theory has been much more successfully predictive (B. Fleming, "Frontier Orbitals and Organic Chemical Reactions", Wiley, London, 1976).

In shortened form this theory states that for hydrocarbon dienes (electron rich) the decisive interactions are between diene HOMO and dienophile LUMO as shown in (52): the closer these are in energy, the lower the activation barrier. This explains why electron-deficient dienophiles such as

acrylonitrile and maleic anhydride are so much more reactive
than ethylene or styrene (conjugation and withdrawal lowers
the energy of the LUMO). Regioselectivity is determined
on the basis that the relative orientation of the dominant
HOMO-LUMO pair is controlled by those termini with large AO
coefficients, which interact favourably. Thus, those
orbitals interact most which overlap best and are closest in
energy (K.N. Houk, Acc.chem.Res., 1975, 8, 361). It has
therefore become important to establish the HOMO and LUMO
energies and their AO coefficients in common dienes and
dienophiles, for which task computational methods and photoelectron spectroscopy have been used extensively (K.N. Houk,
J.Amer.chem.Soc., 1973, 95, 4092).

Frontier orbital theory used in this way is able to explain
the usual pattern of orientation in Diels-Alder reactions
with unsymmetrical dienes and dienophiles, which is for
'*ortho*' adducts to be favoured from 1-substituted dienes and
'*para*' adducts to be favoured when 2-substituted dienes are
used, as long as both diene and dienophile are not electron-
rich (Scheme 10, D = donor, W = electron-withdrawing group).

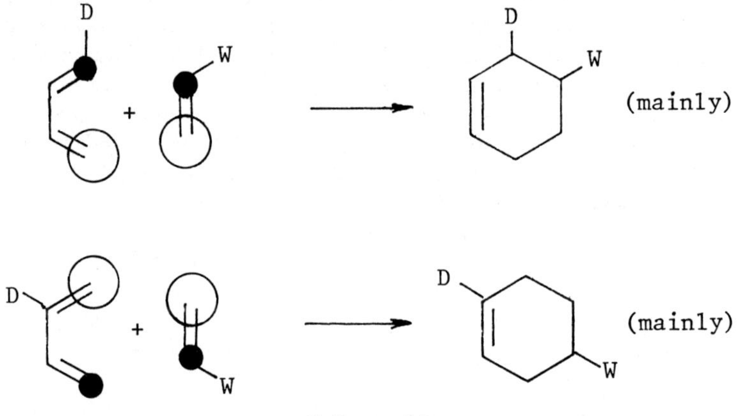

HOMO   LUMO   Scheme 10

The theoretical prediction that '*meta*' adducts would be
formed in the unusual circumstances that both reagents were
electron-rich was recently verified by experiment (I. Fleming,
F.L. Gianni, and T. Mah, Tetrahedron Letters, 1976, 881).

MeC≡CH

(65%)

The role played by secondary orbital interactions in determining orientation in Diels-Alder reactions is debated (P.V. Alson, R.M. Ottenbrite, and T. Cohen, J.org.Chem., 1978, 43, 1864; I. Fleming *et al.*, Tetrahedron Letters, 1978, 1313). Its part in establishing the preference for *endo* adducts is widely accepted (Woodward and Hoffmann, *loc.cit.*), though even in that case it is not the only explanation (W.C. Herndon and L.H. Hall, *ibid.*, 1967, 3095; Theoret.chim.Acta, 1967, 7, 4).

The importance of the Diels-Alder reaction in synthesis is growing, partly as a result of the availability of dienes bearing latently useful functionality, e.g. (53), (54), and (55) (M.J. Carter and I. Fleming, Chem.Comm., 1976, 679; I. Fleming and A. Percival, *ibid.*, p. 681; M.E. Jung and C.A. McCombs, Tetrahedron Letters, 1976, 2935; B.M. Trost and A.J. Bridges, J.Amer.chem.Soc., 1976, 98, 5017), and also through the use of novel dienophiles such as the highly reactive nitroethylene (S. Ranganathan, D. Ranganathan, and A.K. Mehrota, *ibid.*, 1974, 96, 5261). Intramolecular Diels-

SiMe$_3$

(53)

Me$_3$SiO

(54)

MeO

PhS

(55)

Alder reactions (review: R.G. Carlson, Ann.Reports medic. Chem., 1974, 9, 270) have also been used elegantly in the synthesis of, for example, the cedrane skeleton (56) (G. Fráter, Tetrahedron Letters, 1976, 4517).

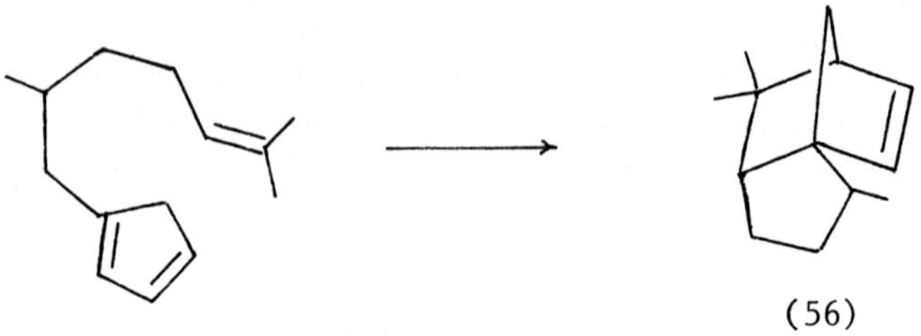

(56)

Acid-catalysed Diels-Alder reactions are useful not only because of their relatively milder conditions which lead to rate ratios of 10^5 at 25 °C in favour of the catalysed process (T. Inukai and T. Kojima, J.org.Chem., 1967, 32, 872), but also because addition of a Lewis acid may change completely the orientation of addition (Z. Stojonac et al., Canad.J.Chem., 1975, 616, 619). Even when this is not the case, acid catalysis may lead to a higher regiospecificity.

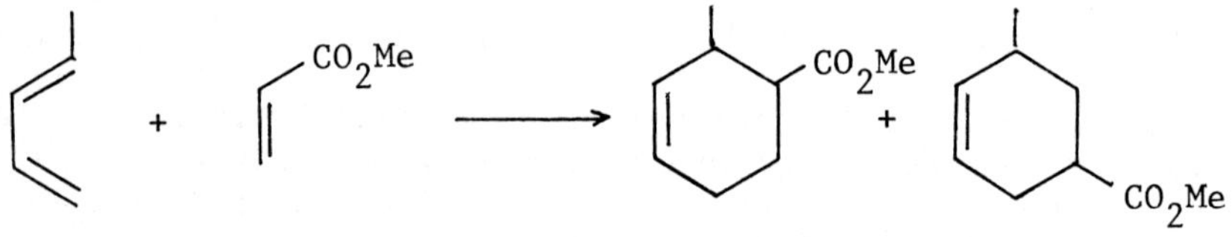

no   AlCl$_3$;   90 : 10

with   AlCl$_3$;   98 : 2

These effects have also been successfully rationalised by the frontier orbital treatment (Fleming, loc.cit., p. 161).

The possibility that thermal dimerization of dienes such as butadiene and piperylene proceed non-concertedly via bis-allyl radicals has been thoroughly investigated using isotopically labelled dienes, and the concerted mechanism vindicated (J.H. Berson et al., J.Amer.chem.Soc., 1976, 98, 5937).

(3) Photochemical Reactions

The photochemistry of acyclic dienes is very complicated

but considerable advances have been made in understanding it. Direct irradiation of the equilibrium mixture of s-*cis* and s-*trans* conformers in dilute inert solutions produces mainly cyclobutenes by symmetry-allowed electrocyclic rearrangement of the excited singlet state. Bicyclobutanes may also arise, their structure suggesting that they stem mainly from the s-*trans* form, e.g. (R. Srinivasan and F.I. Sonntag, J.Amer.chem. Soc., 1965, 87, 3778):

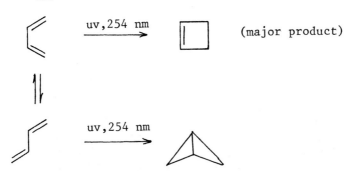

uv,254 nm   (major product)

uv,254 nm

The products are thermodynamically unstable relative to the open-chain isomer, but are essentially transparent to u.v. radiation at 254 nm and so survive. The expected disrotatory stereospecificity of the cyclobutene formation has been demonstrated using (*E,E*)-hexa-2,4-diene (J. Saltiel, L. Metts, and M. Wrighton, *ibid.*, 1970, 92, 3227; J. Srinivasan, *ibid.*, 1968, 90, 4498).

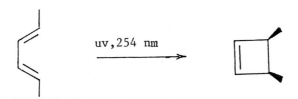

uv,254 nm

Mercury-sensitized photolysis of conjugated dienes leads to cyclopropenes (S. Boué and R. Srinivasan, Mol.Photochem., 1972, 4, 93), but for many dienes this is accompanied by dehydrogenation to complex mixtures of allenes, alkynes, and their photodecomposition products (R. Srinivasan and S. Boué,

Tetrahedron Letters, 1970, 203). Such reactions seem most likely to proceed *via* intermediate biradicals; the formation of ethers by photolysis in alcohols could be viewed as evidence for this (J.A. Barltrop and H.E. Browning, Chem. Comm., 1968, 1481).

At increased concentration, unsensitized photolysis in the liquid phase produces dimers, presumably resulting from bimolecular collisions of the vibrationally excited ground state, since the excited singlet state would be expected to have too short a lifetime for efficient bimolecular processes (J.A. Barltrop and J.D. Coyle, "Excited States in Organic Chemistry", Wiley, London, 1975; Srinivasan and Sonntag, *loc. cit.*).

(50%) + (30%)
(*cis* + *trans*)

Photosensitized dimerization of alkylbutadienes leads to a variety of dimers *via* triplet diradicals (G.S. Hammond, N.J. Turro, and R.S.H. Liu, J.org.Chem., 1963, 28, 3297; J.Amer. chem.Soc., 1965, 87, 3406). The results suggest that the s-*cis* conformer requires less energy for excitation to the triplet state than the s-*trans* form, and that *cis*-triplets give more vinylcyclohexene dimers whereas *trans*-triplets give mainly divinylcyclobutanes (Scheme 11). Thus the proportion

of cyclobutane dimers formed falls to a minimum of *ca.* 35–
40% at triplet sensitizer energies of E$_T$ *ca.* 210 kJ/mol.
1,4-Cyclooctadiene is one of the products, but is believed

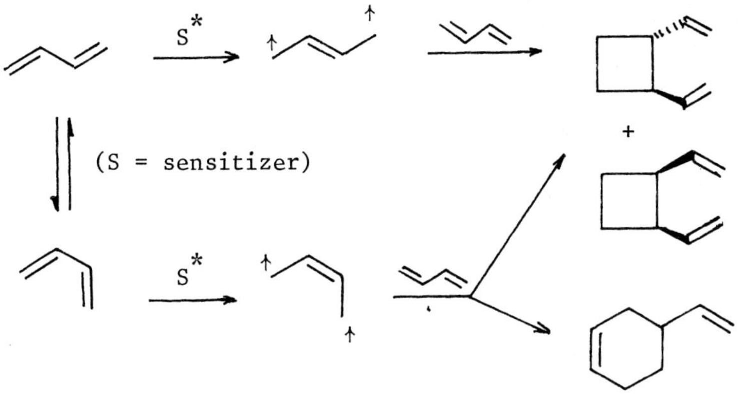

(S = sensitizer)

Scheme 11

to arise by sigmatropic rearrangement of *cis*-divinylcyclo-
butane. For dienes such as penta-1,3-diene and hexa-2,4-
diene the situation is further complicated by the occurrence,
during dimerization, of photosensitized geometrical
isomerization (J. Saltiel *et al.*, *ibid.*, 1971, **93**, 5302), so
that the number of possible isomers of the dimer becomes
dauntingly large. Sensitized photochemical co-dimerizations
of alkadienes with alkenes (W.L. Dilling and J.C. Little,
ibid., 1967, **89**, 2741) and with carbonyl compounds (R.R.
Hautala, K. Dawes, and N.J. Turro, Tetrahedron Letters,
1972, 1229), and photo-oxidation by singlet oxygen to give
1,4-peroxides (K. Kondo and M. Matsumoto, Chem.Comm., 1972,
1332) have also been reported.

Non-conjugated dienes also give a confusing variety of
cyclic and acyclic products by several different intramolecular photochemical processes; such reactions can
provide useful routes to bicyclic compounds. Important
classes of reaction include the di-π-methane rearrangement
of 1,4-dienes (S.S. Hixson, P.S. Mariano, and H.E. Zimmerman,
Chem.Rev., 1973, **73**, 531):

mercury-sensitized cyclization of 1,4- and 1,5-dienes *via* triplet diradicals, e.g. (R. Srinivasan and K.H. Carlough, J.Amer.chem.Soc., 1967, 89, 4932),

and sigmatropic shifts (R.C. Cookson, Quart.Rev., 1968, 22, 423).

In myrcene and related acyclic trienes the two main types of photochemical reaction which occur on direct photolysis are intramolecular [2+2] cycloaddition and the electrocyclic rearrangement (P.A. Leermakers and F.C. James, J.org.Chem., 1967, 32, 2898), whereas mercury-photosensitized photolysis gives an entirely different type of product (57) (R.H.S. Liu and G.S. Hammond, J.Amer.chem.Soc., 1967, 89, 4936).

(57)

There are several reviews of this fascinating area of chemistry (R. Srinivasan, Adv.Photochem., 1966, 4, 113; G. Kaupp, Angew.Chem.internat.Edn., 1978, 17, 150; W.L. Dilling, Chem.Rev., 1969, 19, 845; J.D. Coyle, Quart.Rev., 1974, 3, 329).

Photochemical substitutive chlorination of isoprene has been achieved: the main product is 2-chloromethylbutadiene, which was isolated by ingenious use of the reversible sulphur dioxide addition (see Section 3b-i/4) (F. Borg-Visse, F. Dawans, and E. Maréchal, Synthesis, 1979, 817).

(4) Reactions of dienes catalysed by metals and their complexes

Enormous strides have been made in the selective catalysis of diene reactions by metals such as nickel and palladium and their derivatives. The variety of possible conversions is exemplified for butadiene in Schemes 12 to 14 and specialist reviews are available giving a more comprehensive survey (R. Baker, Chem.Rev., 1973, 73, 487; P. Heimbach, **Angew.Chem.internat.Edn.**, 1973, 12, 975; R.F. Heck, "Organo-transition Metal Chemistry", Academic Press, New York, 1974, ch. 6; P.W. Jolly and G. Wilke, "Organic Chemistry of Nickel", Vol. 2, Academic Press, New York, 1975, ch. 3 and 4; G.P. Chiusoli and G. Salerno, Adv.organometal.Chem., 1979, 17, 195; A.C.L. Su, *ibid.*, p. 269; J. Tsuji, *ibid.*, p. 141; J. Tsuji, "Organic Synthesis with Palladium Compounds", **Springer-Verlag**, Berlin, 1980 and Topics curr.Chem., 1980, 91, 29; see also Section 2d-x).

Iron carbonyls react with 1,3-dienes to produce stable iron tricarbonyl complexes, (58), which are useful for the protection of diene units during synthesis. The masked diene is deprotected using ferric chloride (for an example see D.H.R. Barton *et al.*, J.chem.Soc. Perkin I, 1976, 821, 829). Alternatively, such iron tricarbonyl complexes can be

Scheme 12

Cyclo-oligomerizations of butadiene catalysed
by nickel derivatives

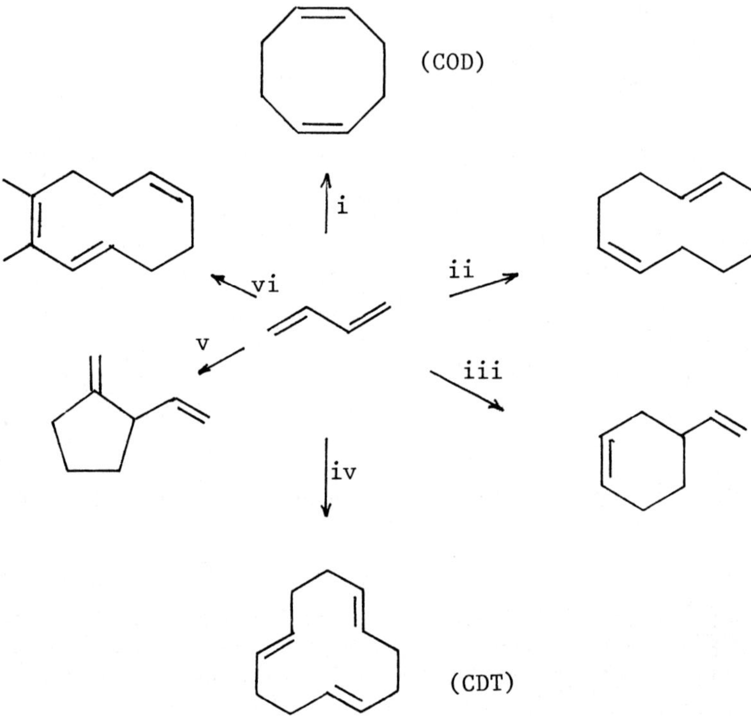

Reagents. (i) Ni(COD)$_2$P(OAr)$_3$ (G. Wilke *et al.*, Ann., 1969, 727, 161); (ii) C$_2$H$_4$, Ni(CDT) (P. Heimbach and G. Wilke, *ibid.*, p. 183); (iii) Ni(O), P(c-C$_6$H$_{11}$)$_3$ (Wilke *et al.*, *loc. cit.*); (iv) Ni(COD)$_2$ (G. Wilke *et al.*, Makromol.Chem., 1963, 69, 18); (v) NiBr$_2$(PBu$_3$)$_2$, n-BuLi, MeOH-PhH (J. Kiji, K. Masui and J. Furukawa, Bull.chem.Soc.Japan, 1971, 44, 1956); (vi) MeC≡CMe, Ni(COD)$_2$PPh$_3$, (W. Brenner, P. Heimbach, and G. Wilke, Ann. 1969, 727, 194).

Scheme 13
Conversions of butadiene catalysed by palladium
salts and complexes

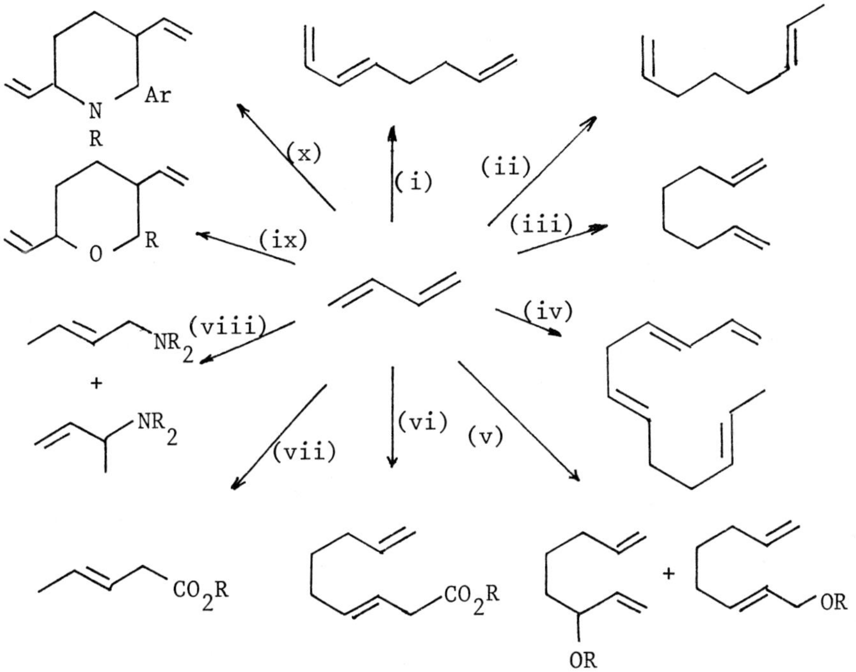

Reagents. (i) Pd(PPh₃)₂-maleic anhydride, i-PrOH (S. Takahashi, T. Shibano, and N. Hagihara, Bull.chem.Soc.Japan, 1968, 41, 454; (ii) Pd(OAc)₂, HCO₂H, Et₃N, (S. Gardner and D. Wright, Tetrahedron Letters, 1972, 163); (iii) PdCl₂(PPh₃)₂, i-PrONa, then HCO₂H (R. Ugo et al., J.organometal.Chem., 1973, 55, 405; (iv) π-C₃H₅Pd(OAc), PhH (D. Medema and R. van Helden, Rec.trav. Chim., 1971, 90, 324; (v) ROH, various catalysts (R=H,Ac,Me,Ph, SiMe₃) (see J. Tsuji "Organic Synthesis with Palladium Compounds", Springer-Verlag, Berlin, 1980); (vi) Pd(acac)₂(PPh₃)₂, CO, ROH (W.E. Billups, W.E. Walker, and T.C. Shields, Chem.Comm., 1971, 1067); (vii) PdBr₂(Bu₃P)₂, CO, MeOH (S. Brewis and P.R. Hughes, ibid., 1965, 157; (viii) Pd(OAc)₂(Ph₂PCH₂CH₂PPh₂) (K. Takahashi, A. Miyake, and G. Hata, Bull.chem.Soc.Japan, 1972, 45, 1183); (ix) [π-C₃H₅PdCl], PPh₃, NaOPh, RCHO (K. Ohno, T. Mitsuyasu, and J. Tsuji, Tetrahedron Letters, 1971, 67); (x) Pd(NO₃)₂(PPh₃)₂, ArCH=NR (J. Kiji et al., Chem.Comm., 1974, 506).

112

Scheme 14
Conversions of butadiene catalysed by metals
other than palladium or nickel

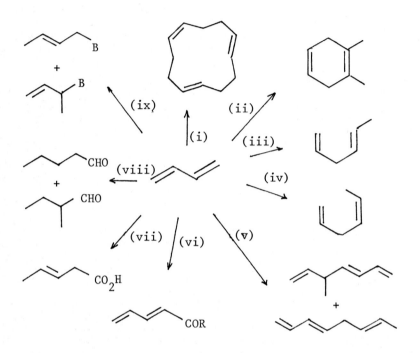

Reagents. (i) TiCl₄, 5Et₂AlCl (G. Wilke *et al.*, Makromol.
Chemie, 1963, <u>69</u>, 18); (ii) MeC≡CMe, Fe(COD)₂ (A. Carbonaro,
A. Greco, and G. Dall'Asta, J.org.Chem., 1968, <u>33</u>, 3948);
(iii) C₂H₄, RhCl₃, EtOH (T. Alderson, E.L. Jenner, and R.V.
Lindsey, J.Amer.chem.Soc., 1965, <u>87</u>, 5638); (iv) C₂H₄,
+ CoCl₂, Et₂AlCl, Ph₂PCH₂CH₂PPh₂ or FeCl₃, Et₃Al,
Ph₂PCH₂CH₂PPh₂ (A.C.L. Su, Adv.organometal.Chem., 1979, <u>17</u>,
269); (v) CoCl₂, Et₃Al (S. Tanaka, K. Mabuchi, and N.
Shimazaki, J.org.Chem., 1964, <u>29</u>, 1626); (vi) NaCo(CO)₄,
(C₆H₁₁)₂NH, RCl, CO (R.F. Heck, J.Amer.chem.Soc., 1963, <u>85</u>,
3381, 3383); (vii) CO, H₂O, Co₂(CO)₈, pyridine (N.S. Imyanitov
and D.M. Rudkovskii, Zh.prikl.Khim., 1968, <u>41</u>, 172); (viii)
Co₂(CO)₈, H₂, CO (H. Adkins and J.L.R. Williams, J.org.Chem.,
1952, <u>17</u>, 980); (ix) RhCl₃, BH (B = EtO, Me₂N, etc.) (K.C.
Dewhirst, *ibid.*, 1967, <u>32</u>, 1297).

protonated stereoselectively to give π-allyliron carbonyl
cations (59) or (60), which react with reactive nucleophiles
to give products of overall 1,4-addition to the diene (T.H.
Whitesides, R.W. Arhart, and R.W. Slaven, J.Amer.chem.Soc.,
1973, 95, 5792) (Scheme 15). Iron carbonyl complexes have
been reviewed (R.B. King, "Organic Chemistry of Iron", Vol. 1,
ed., E.A. Koerner von Gustorf, F.-W. Grevels, and I. Fischler,
Academic Press, New York, 1978, p. 525).

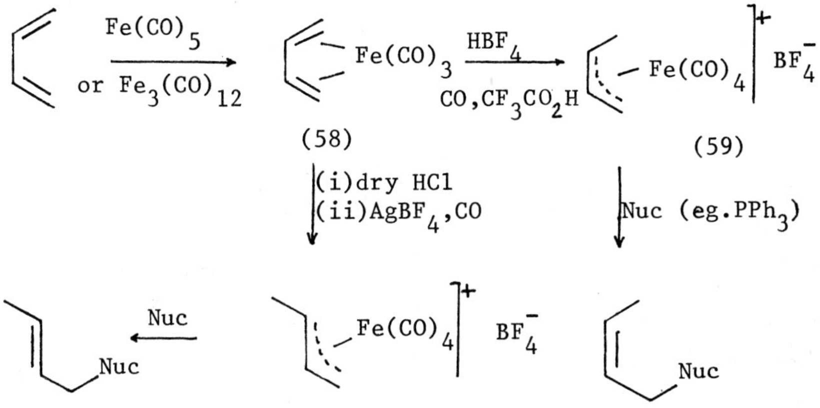

(58)

(59)

(i)dry HCl
(ii)AgBF$_4$,CO

Nuc (eg.PPh$_3$)

(60)

Scheme 15

(5) Polymerization and Copolymerization

Techniques for the polymerization and copolymerization of
butadiene and isoprene include the use of:
(i) organolithium catalysts (review: A.F. Halasa *et al.*, Adv.
organometal. Chem., 1980, 18, 35) which convert butadiene
mainly into 1,4-polybutadiene (61) in non-polar solvents,
but to 1,2-polybutadiene (62) in solvents such as tetrahydrofuran; the 1,4-polymer is a mixture of *cis* and *trans* units;
(ii) Alfin catalysts (such as a mixture of allylsodium, sodium
chloride, and sodium isopropoxide) (review: R. Newberg, H.
Greenberg, and T. Sato, Rubber.Chem.Technol., 1970, 43, 333),
which give mainly very high-molecular weight all-*trans*-1,4-
polybutadiene;
(iii) Ziegler-Natta catalysts (reviews: Ph. Teyssié *et al.*,
"Coordination Polymerization", ed., J.C.W. Chien, Academic
Press, London, 1975, p. 327; and various authors in "Ziegler-
Natta Catalysts and Polymerizations", ed. J. Boor, Academic

Press, New York, 1979), which can be made highly selective for most types of polybutadiene.

Butadiene polymerization has been reviewed in a general survey (D.H. Richards, Quart.Rev., 1977, $\underline{6}$, 235).

$$-\text{[}CH_2CH=CHCH_2\text{]}_{\underline{n}} \qquad\qquad -\text{[}CH\text{---}CH_2\text{]}_{\underline{n}}$$
$$\qquad\qquad\qquad\qquad\qquad\qquad\qquad\qquad | $$
$$\qquad\qquad\qquad\qquad\qquad\qquad\qquad CH=CH_2$$

(61)   (62)

4. Acetylenes or Alkynes

Important books on the chemistry of acetylenes include L. Brandsma, "Preparative Acetylene Chemistry", Elsevier, Amsterdam, 1971; T.F. Rutledge, "Acetylenic Compounds", and "Acetylenes and Allenes", Reinhold, New York, 1968 and 1969; and various authors in "Chemistry of Acetylenes", ed. H.G. Viehe, Marcel Dekker, New York, 1969, and in "The Chemistry of the Carbon-Carbon Triple Bond", ed. S. Patai, Wiley-Interscience, Chichester, 1978, together with monographs on acetylene itself "Acetylene, its Properties and Uses", S.A. Miller, E. Benn, London, 1965-66 and on naturally-occurring acetylenes, F. Bohlmann, T. Burkhardt, and C. Zdero, "Naturally-Occurring Acetylenes", Academic Press, London, 1973.

(a) Preparation of acetylenes

In addition to the texts cited above (especially Brandsma, *loc.cit.*) there are several reviews containing accounts of the available synthetic routes to acetylenes (V. Jäger and H.G. Viehe, in Houben-Weyl's "Methoden der Organischen Chemie", Vol. 5/2a, 4th Edn., ed. E. Müller, Thieme-Verlag, Stuttgart, 1977, p. 1; G. Pattenden, "Comprehensive Organic Chemistry", Vol. 1, ed. D.H.R. Barton, W.D. Ollis and J.F. Stoddart, Pergamon, Oxford, 1979, p. 171; C.A. Buehler and D.E. Pearson, "Survey of Organic Synthesis", Wiley-Interscience, New York, 1970, p. 150).

(i) Dehydrohalogenation methods

Routes to the precursors for these methods, usually geminal dihalides ($R^1CH_2CX_2R^2$) or halogenoalkenes $R^1CH=CXR^2$), have

been improved. Geminal dihalides suitable for conversion
into alk-1-ynes are best made by the aluminium trichloride-
catalysed additions of a branched alkyl chloride to vinyl
chloride (J.H. van Boom *et al.*, Rec.trav.Chim., 1965, 84,
31).

$$R^1R^2R^3CCl \xrightarrow[\text{AlCl}_3]{\text{CH}_2\text{=CHCl}} R^1R^2R^3CCH_2CHCl_2$$

or by the coupling of a primary alkyl halide with lithio-
dichloromethane (J. Villieras, P. Perriot, and J. Normant,
Synthesis, 1979, 502). The most effective bases to promote
the subsequent elimination to alk-1-ynes are potassium t-
butoxide either dissolved in dimethylsulphoxide (DMSO)
(P.J. Kocienski, J.org.Chem., 1974, 39, 3285) or as the
solid activated by a catalytic amount of a crown ether (E.V.
Dehmlow and M. Lissel, Ann., 1980, 1), sodamide in liquid
ammonia (Brandsma, *loc.cit.*), n-butyl lithium (Villieras
et al., *loc.cit.*), or sodium hydride in DMSO (J. Klein and
E. Gurfinkel, Tetrahedron, 1970, 26, 2127) or hexamethylphosphoramide (P. Caubère, Bull.Soc.chim.France, 1966, 1293).
Phase transfer catalysts (A. Gorgues and A. Le Coq, *ibid.*,
1976, 125) and crown ethers (F. Naso and L. Ronzini, J.chem.
Soc. Perkin I, 1974, 340) may also be used as promoters of
less powerful bases.

1,1-Dihalogenoalkenes are also suitable for use as
precursors of alk-1-ynes; they may be obtained from
aldehydes by a modified Wittig reaction (J. Villieras, B.
Perriot, and J.F. Normant, Synthesis, 1975, 458; E.J.
Corey and P.L. Fuchs, Tetrahedron Letters, 1972, 3769).

$$RCHO \xrightarrow[\text{or LiCCl}_2\text{P(O)(OEt)}_2]{\text{Ph}_3\text{P, CBr}_4\text{, Zn}} RCH=CX_2 \xrightarrow[\text{(ii) H}_3\text{O}^+]{\text{(i) n-BuLi, THF}} RC\equiv CH$$

Alk-2-ynes may be obtained in such eliminations if the base
selected is sufficiently strong to set up the prototropic
equilibrium with the thermodynamically favoured isomer.

$$RCH_2CHBrCH_2Br \xrightarrow{-2HBr} RCH_2C\equiv CH \xrightarrow[\text{DMSO, heat}]{\text{NaNH}_2} RC\equiv CCH_3$$

Such rearrangements are especially facile when an aryl migrating group is available (D.F. Bender, T. Thippeswamy, and W.L. Rellahan, J.org.Chem., 1970, __35__, 939).

Base-promoted eliminations from olefins have been reviewed (G. Köbrich, Angew.Chem.Internat.Edn., 1965, __4__, 49).

(ii) Other eliminations

Several alternative routes from carbonyl compounds to alkynes have been developed. β-Ketoesters may be converted in two steps into alkynoic esters *via* 2-pyrazolones; conditions are mild and yields excellent (E.C. Taylor, R.L. Robey, and A. McKillop, *ibid.*, 1972, __11__, 48).

$$RCOCH_2CO_2Et \xrightarrow{N_2H_4 \cdot H_2O} \qquad \xrightarrow[MeOH]{Tl(NO_3)_2} RC\equiv CCO_2Me$$

β-Hydroxyesters, themselves obtainable by the Reformatsky reaction, yield N-nitroso-oxazolidinones which decompose when treated with alkali (M.S. Newman and S.J. Gromelski, J.org. Chem., 1972, __37__, 3220) or, better, butylamine (H.P. Hogan and J. Seehafer, *ibid.*, 4446).

$$R^1R^2CO \longrightarrow R^1R^2C(OH)CH_2CO_2Et \longrightarrow$$

$$\longrightarrow R^1C\equiv CR^2$$

A rather similar approach involves β-ketosulphones, which are sensitized to electron-transfer initiated elimination by first converting them to enol phosphate esters (P.A. Bartlett, F.R. Green III, and E.H. Rose, J.Amer.chem.Soc., 1978, __100__, 4852). Alkynes with one additional carbon atom can also be derived from ketones by treatment with trimethylsilyl- or phosphinyl-diazomethane and butyllithium (E.W. Colvin and B.J. Hamill, Chem.Comm., 1973, 151).

$$R^1R^2CO \xrightarrow[\text{n-BuLi, THF}]{(MeO)_2P(O)CHN_2} R^1C{\equiv}CR^2$$

This procedure has also been applied successfully to aldehydes (J.C. Gilbert and U. Weerasooriya, J.org.Chem., 1979, 44, 4497). Ketones may be converted to alkynes without increase in chain length *via* enol triflates. Very pure t-butylacetylene is best prepared in this way (R.J. Hargrove and P.J. Stang, J.org.Chem., 1974, 39, 581).

$$t\text{-BuCOCH}_3 \xrightarrow{(T_fO)_2O} t\text{-BuC(OT}_f)=CH_2 \xrightarrow{pyr.} t\text{-BuC}{\equiv}CH$$

Eschenmoser's method for the preparation of alkynes is based on elimination from alkoxytosylhydrazones; it is usually used for the synthesis of steroidal alkynes (A. Eschenmoser *et al.*, Helv., 1967, 50, 2101) but is not restricted to this area of application.

$$\underset{R^3 \quad R^4}{\overset{R^1CO \quad R^2}{\diagdown\text{C}\diagup}} \xrightarrow[\substack{CH_2Cl_2, \text{ AcOH} \\ 0-20\ ^\circ C}]{TsNHNH_2} R^1C{\equiv}CR^2 \quad + \quad R^3R^4CO$$

2,4-Dinitrobenzenesulphonylhydrazine has been shown to offer some advantages over tosylhydrazine in this sequence (E.J. Corey and H.S. Sachdev, J.org.Chem., 1975, 40, 579). A further refinement recommends the use of β-thioalkylsulphonyl hydrazones (S. Kano, T. Yokomatsu, and S. Shibuya, *ibid.*, 1978, 43, 4366).

Vicinal diketones may affectively be reduced to the corresponding acetylenes by sequences involving oxidation of their bis-hydrazones; suitable oxidants are (i) mercuric oxide (A.C. Cope, D.S. Smith, and R.J. Cotter, Org.Synth., 1963, Coll.Vol. 4, 377), (ii) lead tetraacetate (A. Krebs and H. Kimling, Angew.Chem.internat.Edn., 1971, 10, 509) or (iii) oxygen in the presence of Cu(II) salts and pyridine (J. Tsuji, H. Takahashi, and T. Kajimoto, Tetrahedron Letters, 1973, 4573).

$$R^1CO.COR^2 \xrightarrow{N_2H_4 \cdot H_2O} R^1C(=NNH_2)C(=NNH_2)R^2 \xrightarrow{[O]} R^1C{\equiv}CR^2$$

When this approach fails, an alternative route involving the synthesis of intermediate vinyl bis(trimethylsilyl)ethers may be effective (D.P. Bauer and R.S. Macomber, J.org.Chem., 1976, 41, 2640). Overall yields for the procedure are 30-35%.

$$R^1CO.COR^2 \xrightarrow[Me_3SiCl]{EtMgBr} Me_3SiOCR^1{=}CR^2OSiMe_3$$

(i) MeLi or KH
(ii) CS$_2$, MeI

$$R^1C{\equiv}CR^2 \xleftarrow{(EtO)_3P}$$

(iii) Via metal acetylides

There has been a **resurgence** of interest in the synthesis of alkynes *via* metal acetylides, long recognized as an important route to internal alkynes. Improvements upon older procedures began with the use of sodium acetylide in polar aprotic solvents such as hexamethylphosphoramide (HMPA), and with the **realisation** that substrates containing other leaving groups besides halides, such as tosylates, could be used (J.F. Normant, Bull.Soc.chim.Fr., 1965, 859).

$$HC{\equiv}CH \xrightarrow{Na, THF} HC{\equiv}CNa \xrightarrow[HMPA, 0{-}50\,^{\circ}C]{n{-}BuOTs} HC{\equiv}CBu{-}n$$

Techniques are available for obtaining amine-free lithium acetylide at low temperature, which prevents the unwanted disproportionation to insoluble Li$_2$C$_2$ (M.M. Midland, J.org. Chem., 1975, 40, 2250; W. Beckmann *et al.*, Synthesis, 1975, 423).

$$HC{\equiv}CH \xrightarrow[THF, 0\,^{\circ}C]{n{-}BuLi} HC{\equiv}CLi \xrightarrow[HMPA]{Br(CH_2)_{12}Br} HC{\equiv}C(CH_2)_{12}C{\equiv}CH$$

Alternatively, monolithium acetylide may be conveniently handled as its complex with ethylenediamine (EDA), which can be isolated as a solid stable at up to 45 $^{\circ}$C, thereby avoiding the necessity for liquid ammonia to be used as solvent (W. Novis Smith and O.F. Beumel, *ibid.*, 1974, 441).

A further modification which has made available a very large number of terminal and non-terminal, branched and straight-chain alkynes and their derivatives, evolved from the observation that the dianions of propyne, but-1-yne, and other alkynes containing the $CH_2C{\equiv}CH$ residue, may be alkylated either selectively at C_3 or, in a stepwise fashion, at both C_1 and C_3 (F. Scheinmann *et al.*, J.chem.Soc., Perkin I, 1979, 1218; 1976, 1609; Synthesis, 1976, 321).

$$R^1CH_2C{\equiv}CH \xrightarrow[\text{TMEDA } (R^1 = H)]{\text{n-BuLi, hexane}} [R^1\bar{C}HC{\equiv}C^-]\ 2Li^+$$

$$R^1CHR^2C{\equiv}CR^3 \xleftarrow[\text{DMSO or HMPA}]{R^3Br} R^1CHR^2C{\equiv}CLi \qquad R^1CHMeC{\equiv}CMe$$

with R^2Br, hexane (downward) and Me_2SO_4 (to the right)

The power of this approach will be appreciated when it is seen that (a) other electrophiles such as carbonyl compounds can be used to trap the anionic centres and (b) the branched non-terminal alkynes formed can be converted into isomeric terminal alkynes by means of the acetylene zipper (see Section 4a-iv), and so re-routed through the di-alkylation procedure. Propargyl dianions have also been generated from terminal allenes (L. Brandsma and E. Mugge, Rec.trav.Chim., 1973, *92*, 628). Even higher poly-lithiated forms have been reported (W. Priestner, R. West, and T.L. Chwang, J.Amer. chem.Soc., 1976, *98*, 8413).

$$CH_3C{\equiv}CH \xrightarrow{4BuLi} C_3Li_4 \xrightarrow{Et_2SO_4} Et_3CC{\equiv}CEt$$

$$+$$

$$Et_2C{=}C{=}CEt_2$$

The dipotassium salts of propynylarenes have been used in a similar procedure (N.M. Libman, P.A. Brestkin, and S.G. Kuznetsov, J.org.Chem.USSR, 1976, *12*, 2480).

Lithium trialkylalkynylborates (63), generated by the interaction of a trialkylborane with a lithium alkynide, react with iodine (H.C. Brown *et al.*, J.Amer.chem.Soc., 1973, 95, 3080) to give iodovinylboranes, convertible into alkynes under mild conditions.

$$R^1_3B \xrightarrow{R^2C\equiv CLi} [\ R^1_3BC\equiv CR^2\ ]^- Li^+ \xrightarrow[-78\ ^{\circ}C]{I_2} R^1C\equiv CR^2$$

$$(63)$$

Methylsulphinyl chloride may replace iodine in this procedure (M. Naruse, K. Utimoto, and H. Nozaki, Tetrahedron, 1974, 30, 2159).

One problem which arises in the above reactions is that organometallic reagents such as alkali-metal acetylides can cause elimination rather than the desired substitution when tertiary, or even in some instances secondary, alkyl halides are used as substrates. This difficulty has been overcome by the use of alkynylalanes (E.-i. Negishi and S. Baba, J.Amer. chem.Soc., 1975, 97, 7385).

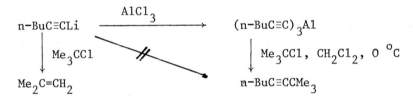

An alternative strategy is to introduce the branched alkyl group first, as illustrated by the reaction of 1-bromo-3,4,4-trimethylhex-1-yne with tri-isobutylaluminium in the presence of a nickel catalyst (G. Giacomelli and L. Lardicci, Tetrahedron Letters, 1978, 2831).

$$Me_3CCHMeC\equiv CBr \xrightarrow[Ni(mesal)_2]{i-Bu_3Al} Me_3CCHMeC\equiv CBu-i$$

(mesal = bis-N-methylsalicylaldimine)

A useful three-carbon homologation of alkynes is made possible by the reaction of tetra-alkylaluminates with bromopropadiene in the presence of Cu(I) salts (F. Sato, H. Kodama, and M. Sato, Chem.Letters, 1978, 789):

$$RC \equiv CH \xrightarrow{\text{H}_2, \text{ cat.}} RCH=CH_2 \xrightarrow[\text{TiCl}_4]{\text{LiAlH}_4} (RCH_2CH_2)_4 AlLi$$

$$CH_2=C=CHBr, \; Cu(I)Br$$

$$RCH_2CH_2CH_2C \equiv CH$$

(iv) Via isomerization of other acetylenes

Undoubtedly the most important reagent to appear in this context is the so-called acetylene zipper, potassium amino-propylamide (KAPA) (C.A. Brown and A. Yamashita, J.Amer.chem. Soc., 1975, 97, 891), which in diaminopropane solution promotes the multipositional migration of the $C \equiv C$ bond, to such an extent that it occurs 10^5 times faster than in the presence of, for example, potassium t-butoxide in DMSO. Under the usual conditions for its use, KAPA leads exclusively to terminal alkynes; it should be noted, however, that the triple bond cannot migrate past a branch in the chain or a functional group such as a carbonyl group. The reagent is best generated from potassium amide and the diamine in ammonia (H. Hommes and L. Brandsma, Rec.trav.Chim., 1977, 160).

$$n\text{-}BuC \equiv CCH_2OH \xrightarrow[80\ ^{\circ}C, \; 1 \; h]{H_2N(CH_2)_3NHK} HC \equiv C(CH_2)_5OH \quad (80\%)$$

Base-catalysed isomerizations of acetylenes by more conventional techniques have been reviewed (R.J. Bushby, Quart.Reviews, 1970, 24, 585).

(b) Structure and physical properties of acetylenes

The structure of the $C \equiv C$ bond and related theoretical treatments have been discussed at length (J. Dale, "Chemistry of Acetylenes", ed. H.G. Viehe, Marcel Dekker, New York, 1969, ch. 1; M. Simonetta and A. Gavezzotti, "The Chemistry of the Carbon-Carbon Triple Bond", ed. S. Patai, Wiley-Interscience, Chichester, 1978, ch. 1; J.L. Hencher, *ibid.*, ch. 2). Electron diffraction measurements (W. Zeil, J. Haase, and

M. Dakkouri, Disc.Faraday Soc., 1969, <u>47</u>, 149) and high
resolution infrared and Raman spectroscopy (A. Baldacci *et
al.*, J.mol.Spec., 1976, <u>59</u>, 116) indicate that the C≡C bond
length in simple alkynes lies in the range 1.203 ± 0.001 Å;
X-ray diffraction has given slightly lower values.

The characteristic infrared vibrational frequencies of
alkynes have been assigned (L.J. Bellamy, "Advances in Infra-
red Group Frequencies", Methuen, London, 1968). Ionization
potentials (IP's) of a number of terminal and non-terminal
alkynes have been measured by UV photoelectron spectroscopy
(PES) (G. Mouvier *et al.*, J.electron Spec.Rel.Phenomena, 1975,
<u>7</u>, 55) and the values of the first vertical IP related to the
sum of the inductive effects of the alkyl groups.
Correlations of such PES data with the predictions by
molecular orbital theory of electronic energy levels have
been evaluated (F. Brogli *et al.*, Helv., 1975, <u>58</u>, 2620).

Proton nmr shifts for ethynyl, ≡CH, protons in various
environments lie in the range δ 1.7 - 3.0 (L.M. Jackmann and
S. Sternhell, "Applications of NMR in Organic Chemistry", 2nd
Edn., Pergamon, Oxford, 1969). Coupling constants across the
triple bond in alkynes lie in the range 2.0-2.7 Hz (for
acetylene the value is 9.1 Hz). Values of ^{13}C nmr shifts for
the carbon atoms along the chain leading to an acetylenic
bond, [e.g. in (64), J.-M. Kornprobst and J.-P. Doucet,
J.chim.Phys., 1974, <u>71</u>, 1129] are valuable aids in locating
the position of the triple bond in an alkyne.

$$CH_3-CH_2-CH_2-CH_2-CH_2-C≡CH$$

22.5		28.9		87.6
14.1	31.1		20.9	69.3

(64)

These and other physical methods useful in the identification
of alkynes have been reviewed (K.A. Connors, "The Chemistry
of the Carbon-carbon Triple Bond", ed. Patai, *loc.cit.*, ch.
5).

(c) Reactions of acetylenes

(i) Oxidation

Selective oxidation of the methylene groups adjacent to a

triple bond in an internal alkyne can be achieved by means of
the chromium(VI)oxide-pyridine complex (W.G. Dauben, M.
Lorber, and D.S. Fullerton, J.org.Chem., 1969, $\underline{34}$, 3587).

$$\text{n-PrC}\equiv\text{CCH}_2\text{Et} \xrightarrow[\text{CH}_2\text{Cl}_2, \ 20 \ ^\circ\text{C}]{\text{CrO}_3 \cdot 2\text{pyr}} \text{n-PrC}\equiv\text{CCOEt}$$

The mechanism of ozonolysis of non-terminal alkynes is
generally agreed to be the Criegee mechanism shown in Scheme
16 (S. Jackson and L.A. Hull, $ibid.$, 1976, $\underline{41}$, 3340). If
ozonolysis is carried out in ethyl acetate containing tetracyanoethylene (1 mol/mol) at −78 $^\circ$C, enhanced yields of α-
diketones are obtained (N.C. Yang and J. Libman, J.org.Chem.,
1974, $\underline{39}$, 1782), whereas addition of pinacolone increases
the proportion of carboxylic acids formed.

Scheme 16

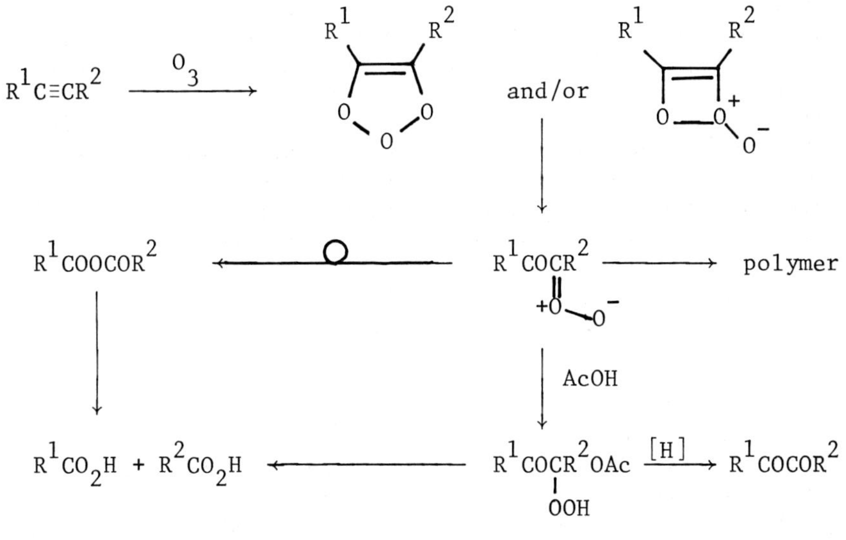

The intermediate adduct formed by low-temperature ozonation
of but-2-yne has been shown to epoxidize olefins at sub-
ambient temperature (R.E. Keay and G.A. Hamilton, J.Amer.chem.
Soc., 1976, $\underline{98}$, 6578). Ozonolysis of terminal alkynes in
methanol is a practical method for preparing α-ketoacids (S.
Cacchi, L. Cagliotti, and P. Zappelli, J.org.Chem., 1973, $\underline{38}$,

3653).

Di-t-butylacetylene is converted by peracids into a complex mixture of cyclic and acyclic products, rationalised on the basis of ketene formation by a Wolff–like rearrangement (J. Ciabattoni *et al.*, J.Amer.chem.Soc., 1970, <u>92</u>, 3826).

$$t\text{-BuC}{\equiv}\text{CBu-t} \xrightarrow{\ RCO_3H\ } t\text{-Bu}\underset{O}{\triangle}\text{Bu-t} \rightleftharpoons t\text{-BuCO-}\overset{..}{\text{C}}\text{-Bu-t}$$

$$t\text{-Bu}_2\text{C=C=O} \ + \ t\text{-BuCO}\underset{}{\triangledown}\text{Me}_2 \ + \ t\text{-BuCOCMe=CMe}_2$$

$$\Big\downarrow RCO_3H$$

$$t\text{-BuCO}\underset{\text{Me}\ \ O}{\overset{}{\bowtie}}\text{Me}_2$$

Ruthenium(IV)oxide, generated *in situ*, is an effective ambient temperature oxidant for the triple bond in alkynes (H. Gopal and A.J. Gordon, Tetrahedron Letters, 1971, 2941). The use of phase transfer reagents in the potassium permanganate oxidation of internal alkynes enables vicinal diones to be isolated (D.G. Lee and V.S. Chang, Synthesis, 1978, 462).

(ii) Reduction

A variety of methods exist for the semi-reduction of a carbon-carbon triple bond to give an alkene. *cis*-Isomers result when a non-terminal alkyne is hydrogenated in the presence of:

(a) Lindlar's catalyst to which quinoline or a sulphide has been added (E.N. Marvel and T. Li, Synthesis, 1973, 457);
(b) P-2 nickel, obtained by borohydride reduction of Ni(II) acetate (C.A. Brown and V.K. Ahuja, J.org.Chem., 1973, <u>38</u>, 2226);
(c) cationic rhodium complexes (R.R. Schrock and J.A. Osborn, J.Amer.chem.Soc., 1976, <u>98</u>, 2143).

$$R^1C{\equiv}CR^2 \xrightarrow{\ [H]\ } \begin{array}{c} R^1 \diagdown \qquad \diagup R^2 \\ C{=}C \\ H \diagup \qquad \diagdown H \end{array}$$

The same product geometry is achieved in two-step processes based on initial *syn*-additions of:

(a) dialkylboranes (G. Zweifel, Intra-Science Chem.Rept., 1973, 7, 181);
(b) bis-cyclopentadienylzirconium hydrochloride (J. Schwartz and J.A. Labinger, Angew.Chem.internat.Edn., 1976, 15, 333);
(c) di-isobutylaluminium hydride (G. Zweifel and R.B. Steele, J.Amer.chem.Soc., 1967, 89, 5085).

All three of these procedures require protolysis at the last step, which enables selective monodeuteriation to be achieved.

$$R^1C{\equiv}CR^2 \xrightarrow{\ MH\ } \begin{array}{c} R^1 \diagdown \qquad \diagup R^2 \\ C{=}C \\ M \diagup \qquad \diagdown H \end{array} \xrightarrow{\ D^+\ } \begin{array}{c} R^1 \diagdown \qquad \diagup R^2 \\ C{=}C \\ D \diagup \qquad \diagdown H \end{array}$$

$$M = R_2B, \ R_2Al, \ or \ \pi Cp_2ZrCl$$

Stereospecific *trans* reduction of alkynes is usually achieved by dissolving metals in liquid ammonia, but an alternative is the treatment of the adducts obtained from alkynes and dialkylorganoboranes with silver nitrate in alkali (K. Avasthi, S.S. Ghosh, and D. Devaprabhakara, Tetrahedron Letters, 1976, 4871).

Partial catalytic reduction of alkynes has been reviewed (H. Gutmann and H. Lindlar in "The Chemistry of Acetylenes", ed. H.G. Viehe, Marcel Dekker, New York, 1969, p. 355).

(iii) Electrophilic addition

This topic has been reviewed in several texts, notably by G.H. Schmid in "The Chemistry of the Carbon-Carbon Triple Bond", ed. S. Patai, Wiley-Interscience, Chichester, 1978, ch. 8; R.C. Fahey, Topics in Stereochemistry, 1968, 3, 237; and by V. Jäger and H.G. Viehe in Houben-Weyl, "Methoden der Organischen Chemie", Vol. 5/2a, 4th Edn., ed. E. Müller, Thieme-Verlag, Stuttgart, 1977, p. 687).

(1) Halogenation

Fluorine reacts with alkynes at low temperature to give tetrafluorides (R.F. Merritt, J.org.Chem., 1967, 32, 4124). The same result is achieved more controllably by the use of xenon difluoride (M. Zupan and A. Pollak, *ibid.*, 1974, 39, 2646), but intermediate difluoroalkenes cannot be isolated in either case.

$$R^1C{\equiv}CR^2 \quad \xrightarrow[\text{or } XeF_2]{F_2, \ CCl_3F} \quad R^1CF_2CF_2R^2$$

Dichlorides can be obtained by the addition of chlorine to alkynes either in a polar solvent or in the presence of metal halides such as Cu(II) chloride. The reaction is first order in both reagents in the absence of any catalyst, and electron release in the alkyne accelerates the rate of addition, in keeping with an Ad_E2 mechanism *via* an open-chain intermediate cation (65). Capture of the intermediate by a nucleophilic

$$R^1C{\equiv}CR^2 \quad \xrightarrow{Cl_2, AcOH} \quad R^1CCl{=}\overset{+}{C}R^2 \quad \longrightarrow \quad R^1CCl{=}CClR^2$$

$$Cl^- \qquad\qquad\qquad cis > trans$$

$$(65)$$

solvent yields *anti*-adducts, suggesting that (65) is a tight ion-pair in keeping with theoretical predictions (V. Lucchini, G. Modena, and I.G. Ciszmadia, Gazz.Chim.Ital., 1975, 105, 675). Molybdenum(V) chloride in dichloromethane and antimony(V) chloride are just two of the proposed reagents for controlled *syn* addition of chlorine to alkynes, but yields are not high (J.S. Filippo, Jr., A.F. Sowinski, and L.J. Romano, J.Amer.chem.Soc., 1975, 97, 1599; S. Uemura, A. Onoe, and M. Okano, Chem.Comm., 1976, 145). The opposite stereochemistry is observed when a combination of Cu(II)-chloride and lithium chloride is used to promote chlorination (S. Uemura *et al.*, J.chem.Soc. Perkin I, 1977, 676):

$$\underset{Cl}{\overset{Ar}{>}}C{=}C\underset{R}{\overset{Cl}{<}} \quad \xleftarrow{\underset{LiCl}{CuCl_2}} \quad ArC{\equiv}CR \quad \xrightarrow[CH_2Cl_2]{MoCl_5} \quad \underset{Cl}{\overset{Ar}{>}}C{=}C\underset{Cl}{\overset{R}{<}}$$

Iodobenzene dichloride has been recommended for the controlled *anti*-chlorination of internal alkynes (A. Debon, S. Masson, and A. Thuillier, Bull.Soc.chim.Fr., 1975, 2493).

Bromination has been shown by kinetic studies to follow a clean second-order rate law at low concentration (less than 3×10^{-4} M in Br_2) and in the absence of added metal bromides, but in contrast to chlorination an intermediate cyclic bromonium ion is proposed to account for the predominantly *anti*-stereochemistry of addition (G.H. Schmid, A. Modro, and K. Yates, J.org.Chem., 1980, 45, 665).

$$R^1C{\equiv}CR^2 \xrightarrow[\text{AcOH}]{Br_2} \underset{+}{\overset{R^1}{\underset{Br}{\diagdown}}C{=}C\overset{R^2}{\diagup}} \longrightarrow \overset{R^1}{\underset{Br}{\diagdown}}C{=}C\underset{Br}{\overset{R^2}{\diagup}}$$

This scheme has been thought to be misleadingly simple, and refinements have been proposed, including bromide ion catalysis (J.-M. Kornprobst, X.-Q. Huynh, and J.E. Dubois, J.chim.Phys., 1974, 71, 1126) and a pre-formed π-complex (G.A. Olah and T.R. Hochswender, Jr., J.Amer.chem.Soc., 1974, 96, 3574).

Closely related mechanisms have been proposed for the addition of iodine (R.A. Hollins and M.P.A. Campos, J.org. Chem., 1979, 44, 3931), arenesulphenyl chlorides (A. Modro, G.H. Schmid and K. Yates, *ibid.*, p. 4221), areneselenyl chlorides (G.H. Schmid and D.G. Garratt, Tetrahedron Letters, 1975, 3991) and mercury(II) salts (S. Uemura *et al.*, Chem. Comm., 1975, 548), though not always with completely adequate experimental support.

Addition of F-X, where X=Cl, Br, or I, to symmetrical alkynes has been achieved using *N*-halogenosuccinimides in polyhydrogen fluoride-pyridine-tetramethylenesulphone (G.A. Olah, M. Nojima, and I. Kerekes, Synthesis, 1973, 780); *anti*-addition was claimed. Interaction of iodomethane with xenon difluoride produces a reagent which brings about the *syn*-addition of I-F to non-terminal alkynes (M. Zupan, Synthesis, 1976, 473).

(2) Additions of H-X (where X=OH, halide, etc.)

The addition of hydrogen chloride to alkynes has been shown (R.C. Fahey, M.T. Payne and D.-J. Lee, J.org.Chem., 1974, $\underline{39}$, 1124) to proceed by an Ad_E3 mechanism, although an Ad_E2 mechanism can compete, especially in arylalkynes and alk-1-ynes.

$$EtC\equiv CEt \; \xrightleftharpoons{HCl} \; [\; EtC\overset{\underset{HCl}{\uparrow}}{\text{=}}CEt \;] \; \xrightarrow{Cl^-} $$

$$\left[\begin{array}{c} H\text{----}Cl \\ \vdots \\ EtC\text{===}\overset{\vdots}{C}Et \\ \vert \\ \overset{\vert}{C}l \end{array}\right]^- \xrightarrow{-Cl^-} \underset{Cl}{\overset{Et}{>}}C=C\underset{Et}{\overset{H}{<}}$$

Triethylammonium hydrogen dichloride is perhaps the most effective reagent for exclusively *syn*-addition of HCl to alkynes, again by an Ad_E3 mechanism (J. Cousseau and L. Gouin, J.chem.Soc. Perkin I, 1977, 1797). Rearrangements frequently occur, especially when a suitably branched centre lies close to the incipient vinyl carbocation (K. Griesbaum and Z. Rehman, J.Amer.chem.Soc., 1970, $\underline{92}$, 1417).

$$Me_3C\equiv CH \; \xrightarrow[\text{no solvent}]{HCl} \; [Me_3C\overset{+}{C}=CH_2] \; \xrightarrow[\text{Me shift}]{\circlearrowright} \; Me_2\overset{+}{C}CMe=CH_2$$

$$\big\downarrow \begin{array}{c} Cl^- \\ HCl \end{array} \qquad\qquad \big\downarrow \begin{array}{c} Cl^- \\ HCl \end{array}$$

$$Me_3CCCl=CH_2 \qquad\qquad Me_2CClCClMe_2$$

$$+ \; Me_3CCCl_2CH_3 \qquad\qquad + \; Me_2CClCHMeCH_2Cl$$

syn-Adducts predominate in additions to arylalkynes, their amount depending upon the bulk of the alkyl-group (F. Marcuzzi and G. Melloni, *ibid.*, 1976, $\underline{98}$, 3295).

In the liquid phase, treatment of but-2-yne with dry HCl causes cyclodimerization and trimerization as well as addition; this can produce an extraordinary variety of products (H. Schneider and K. Griesbaum, J.org.Chem., 1979, 44, 3316), a number of which arise by the capture of an intermediate vinylcation by further molecules of alkyne. Similar, but less complex, mixtures arise by the liquid phase hydrobromination of terminal alkynes (K. Griesbaum *et al.*, Ann., 1979, 1137). Careful studies have established that additions of trifluoroacetic acid to dialkylacetylenes give non-stereospecific addition, except with but-2-yne from which mainly the Z-isomer is formed by *anti*-addition (P.E. Peterson and J.E. Duddey, J.Amer.chem.Soc., 1966, 88, 4990; P.H. Summerville and P.von R. Schleyer, *ibid.*, 1974, 96, 1110). The unusual behaviour of but-2-yne towards protic reagents is again revealed in its greater reactivity than other alkynes towards fluorosulphonic acid at low temperature, its greater tendency towards *anti*-addition, and the detectable formation of cyclobutenylcation (66) (G.A. Olah and R.J. Spear, *ibid.*, 1975, 97, 1845).

Reviews of the important metal- and acid-catalysed hydration of alkynes have appeared (P.J. Stang, Prog.phys. Org.Chem., 1972, 10, 205; G. Modena and U. Tonellato, Adv. phys.Org.Chem., 1971, 9, 185; Ya.A. Dorfman and N.A.

Karazhanova, Russ.J.phys.Chem., 1975, $\underline{49}$, 751).

(3) Hydrometalation

Ionic addition of the B–H and of the Al–H bond to alkynes leads to the important and synthetically useful alkenyl-boranes and alkenyl-alanes. Such additions appear to proceed in an *anti*-Markownikov sense, because the hydrogen is relatively negative, the boron or aluminium atom constituting the electrophile. These reactions are best understood as four-centre bimolecular additions, which explains the *syn*-stereospecificity:

$$R^1C\equiv CH$$

$$\delta- \quad \delta+$$
$$H-MR_2$$

$$(M = B \text{ or } Al)$$

In such additions to non-terminal acetylenes the metal becomes regioselectively attached to the less hindered terminus.

(97%)   (85%)

Although diborane yields tris(alkenyl)boranes $(RCH=CH)_3B$, selective mono-addition can be achieved by the use of dialkylboranes such as dicyclohexyl-, disiamyl-, or catechyl-borane, or with the highly hindered thexylborane $(Me_2CHCMe_2BH_2)$.

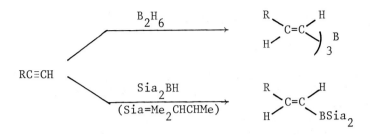

A few examples of the utility of these reactions in synthesis follow, Scheme 17. Excellent reviews abound (G. Zweifel, Intra-Science Chem.Rept., 1973, 7, 181; C.F. Lane and G.W. Kabalka, Tetrahedron, 1976, 32, 981; H.C. Brown "Organic Syntheses *via* Boranes", Wiley, New York, 1975; P.F. Hudrlik and A.M. Hudrlik in "The Chemistry of the Carbon-Carbon Triple Bond", ed. S. Patai, Wiley-Interscience, Chichester, 1978, ch. 7).

(4) C-C Bond-forming electrophilic additions

These are of two types, namely, addition of metal alkyls such as organocuprates and additions initiated by carbenium ions.

Organocopper(I) reagents (review J.F. Normant in "New Applications of Organometallic Reagents in Organic Synthesis", ed. D. Seyferth, Elsevier, Amsterdam, 1976, p. 215) such as RCu, MgI$_2$ add readily to terminal alkynes in a regio- and *syn*-stereo-specific manner. The resulting alkenylcoppers are susceptible to attack, without loss of stereochemical integrity, by a variety of electrophiles (N.J. LaLima and A.B. Levy, J.org.Chem., 1978, 43, 1279; A. Marfat, P.R. McGuirk, and P. Helquist, Tetrahedron Letters, 1978, 1363, 2465; and H.J.T. Bos *et al.*, Rec.trav.Chim., 1976, 95, 299, 304) (Scheme 18).

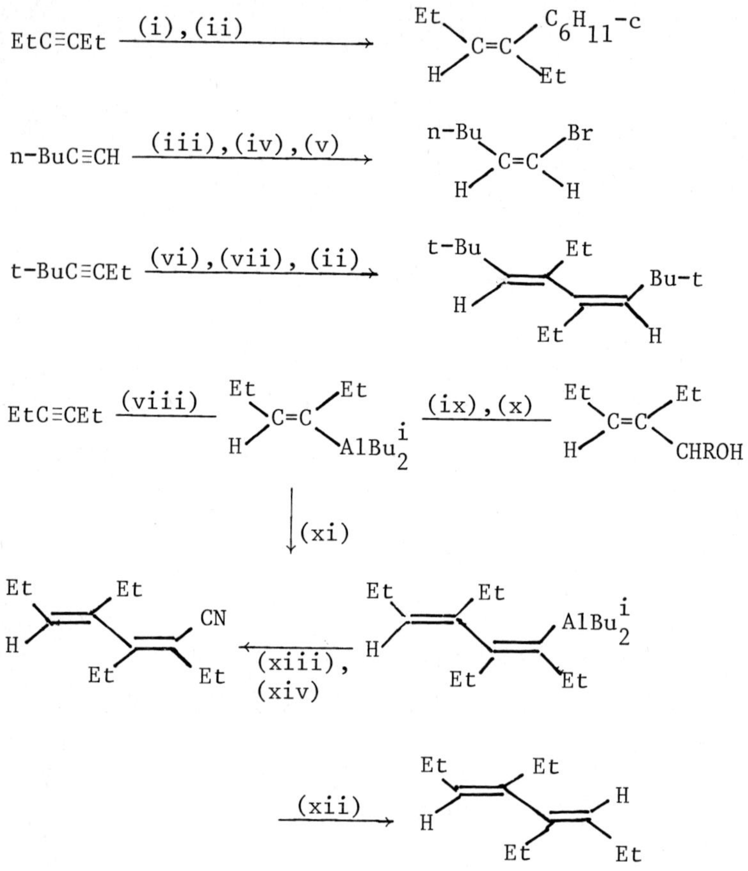

$EtC\equiv CEt$ $\xrightarrow{\text{(i),(ii)}}$

$$\underset{H}{\overset{Et}{>}}C=C\underset{Et}{\overset{C_6H_{11}\text{-}c}{<}}$$

$n\text{-}BuC\equiv CH$ $\xrightarrow{\text{(iii),(iv),(v)}}$

$$\underset{H}{\overset{n\text{-}Bu}{>}}C=C\underset{H}{\overset{Br}{<}}$$

$t\text{-}BuC\equiv CEt$ $\xrightarrow{\text{(vi),(vii),(ii)}}$

$EtC\equiv CEt$ $\xrightarrow{\text{(viii)}}$ $\underset{H}{\overset{Et}{>}}C=C\underset{AlBu^i_2}{\overset{Et}{<}}$ $\xrightarrow{\text{(ix),(x)}}$ $\underset{H}{\overset{Et}{>}}C=C\underset{CHROH}{\overset{Et}{<}}$

$\downarrow$ (xi)

$\xleftarrow{\text{(xiii),} \atop \text{(xiv)}}$

$\xrightarrow{\text{(xii)}}$

Reagents: (i) $(c\text{-}C_6H_{11})_2BH$; (ii) I_2, NaOH; (iii) Sia_2BH; (iv) Br_2; (v) H_2O; (vi) $Me_2CHCMe_2BH_2$; (vii) $Me_3\overset{+}{N}\text{-}\overset{-}{O}$ (viii) $i\text{-}Bu_2AlH$; (ix) MeLi; (x) RCHO; (xi) $EtC\equiv CEt$; (xii) H_3O^+; (xiii) $n\text{-}BuLi$; (xiv) $(CN)_2$.

Scheme 17

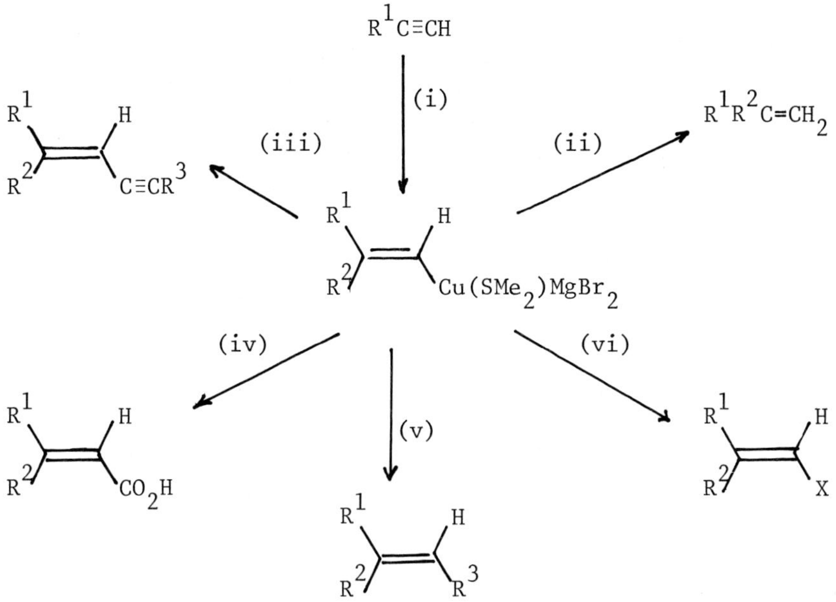

$$R^1C\equiv CH$$

(i)

(iii)

(ii)

$$R^1R^2C=CH_2$$

R¹ H

C=C

R² C≡CR³

$$R^1 \quad H$$
$$C=C$$
$$R^2 \quad Cu(SMe_2)MgBr_2$$

(iv)

(vi)

(v)

R¹ H

C=C

R² CO₂H

R¹ H

C=C

R² X

R¹ H

C=C

R² R³

Reagents: (i) $R^2Cu.MgBr_2$, Me_2S; (ii) acid;
(iii) $R^3C\equiv CHal$, EDA; (iv) CO_2, acid; (v) R^3X; (vi) X_2
(X = Cl or Br).

Scheme 18

Terminal alkynes usually undergo metal–hydrogen exchange with lithium dialkylcuprates, but this can be prevented either by suitable choice of solvent or by the addition of lithium bromide to the reaction mixture in tetrahydrofuran (H. Westmijze, H. Kleijn, and P. Vermeer, Tetrahedron Letters, 1977, 2023). Internal alkynes (without the assistance of an activating functional group) are not readily attacked by either lithium dialkylcuprates or organocopper/ magnesium halide complexes. However, if a suitable leaving group is attached to the propargylic carbon, allenic hydro- carbons are formed (see Section 3a-i/2).

Acetylene and alkynes can be used to trap carbenium ions generated either from alcohols and concentrated sulphuric acid or alkyl or acyl halides and a Lewis acid, preferably a

zinc halide. In the former case, aldehydes or ketones are
formed; in the second situation, mainly *anti*-addition occurs
(H. Martens, F. Janssens, and G. Hoornaert, Tetrahedron,
1975, 31, 177; P.T. Lansbury *et al.*, J.Amer.chem.Soc., 1975,
97, 394).

$$R^1C{\equiv}CH \xrightarrow[\text{conc. } H_2SO_4]{R^2OH} R^1CH_2COR^2$$

$$R^1C{\equiv}CR^2 \longrightarrow \begin{array}{l} \xrightarrow[\text{ZnX}_2]{R^3X} \quad \begin{array}{c} R^1 \\ \diagdown \\ X \end{array} C{=}C \begin{array}{c} R^3 \\ \diagup \\ R^2 \end{array} \\[2em] \xrightarrow[\text{AlCl}_3]{R^3COCl} \quad \begin{array}{c} R^1 \\ \diagdown \\ Cl \end{array} C{=}C \begin{array}{c} COR^3 \\ \diagup \\ R^2 \end{array} \end{array}$$

(iv) Nucleophilic additions

Reactions which occur by uncatalysed nucleophilic addition
pathways to non-activated alkynes are rare. Principal
examples of reagents which successfully accomplish this mode
of addition are lithium aluminium hydride (E.F. Magoon and
L.H. Slaugh, Tetrahedron, 1967, 23, 4509), alkoxides (E.S.
Rothman *et al.*, J.Amer.oil Chem.Soc., 1972, 49, 376) and
S-nucleophiles.

$$MeC{\equiv}CEt \xrightarrow[\text{THF, heat}]{LiAlH_4} \textit{trans}\text{-pent-2-ene}$$

$$i\text{-BuC}{\equiv}CH \xrightarrow[225\ ^{\circ}C]{t\text{-BuOK, } t\text{-BuOH}} i\text{-BuC(OBu-t)}{=}CH_2 \quad (51\%)$$

$$+$$

$$i\text{-BuCH}{=}CHOBu\text{-t} \quad (9\%)$$

Nucleophilic additions to alkynes have been reviewed
(J.I. Dickstein and S.I. Miller, "The Chemistry of the
Carbon-Carbon Triple Bond", ed. S. Patai, Wiley-Interscience,
Chichester, 1978, ch. 19).

(v) Non-photochemical free-radical additions

Few significant advances have been made since the last major review (M. Julia in "The Chemistry of Acetylenes", ed. H.G. Viehe, Marcel Dekker, New York, 1969, ch. 5). Free-radical addition of benzenesulphonyl chloride gives β-chlorovinylsulphones, the stereochemistry of which may be controlled by a suitable choice of solvent (Y. Amiel, J.org. Chem., 1971, 36, 3691, 3697).

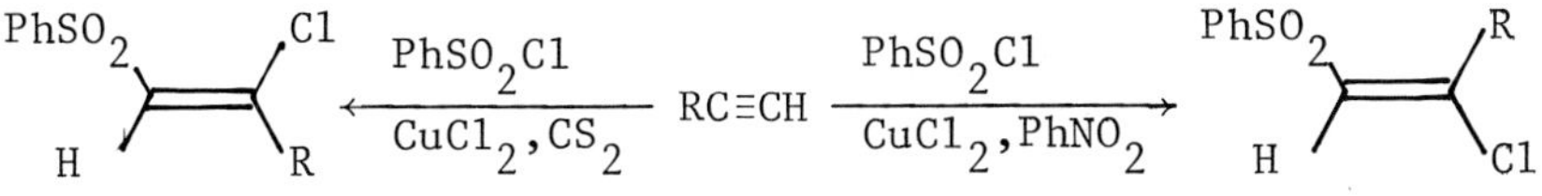

Free-radical additions of trimethyltin hydride to conjugated enynes and of thiols to monosubstituted alkynes have been reported to proceed without unexpected results (M.L. Poutsma and P.A. Ibarbia, J.Amer.chem.Soc., 1973, 95, 6000; R. Mantione and J.F. Normant, Bull.Soc.chim.Fr., 1973, 2261).

(vi) Photochemical reactions

A review of this topic is available (J.D. Coyle, "The Chemistry of the Carbon-Carbon Triple Bond", *loc.cit.*, ch. 12). Whereas photofragmentation and phototrimerization of acetylene both occur on irradiation below 200 nm (H. Okabe, J.chem.Phys., 1975, 62, 2782):

$$HC{\equiv}CH \xrightarrow{124/147\ nm} C + H_2$$

$$HC{\equiv}CH \xrightarrow{185\ nm} C_6H_6 + HC{\equiv}CC{\equiv}CH + H_2 + C_2H_4$$

propargylic cleavage of longer chain alkynes can occur on irradiation at longer wavelength (A. Galli, P. Harteck, and R.R. Reeves, J.phys.Chem., 1967, 71, 2719):

$$CH_3C{\equiv}CH \xrightarrow[-H\cdot]{250\ nm} \dot{C}H_2C{\equiv}CH \longrightarrow (HC{\equiv}CCH_2)_2$$

Photoadditions to alkynes are of two types, those that occur *via* π-π* or Rydberg photoexcitation of the alkyne (for

136

example the photoaddition of isopropanol, G. Büchi and S.H.
Feairheller, J.org.Chem., 1969, 34, 609) and those that do
not. In the latter category are the photoadditions of
hydrogen bromide and tetrahalogenomethanes, which proceed
predictably but with much simultaneous polymerization.
Photoaddition of aldehydes (J.S. Bradshaw, R.D. Knudsen,
and W.W. Parish, *ibid.*, 1975, 40, 529) and formamide (G.
Friedman and A. Komen, Tetrahedron Letters, 1968, 3357) give
mainly 1:2 adducts.

$$RC{\equiv}CR \xrightarrow[\text{Ph}_2\text{CO, uv}]{\text{H}_2\text{NCHO}} RCH(CONH_2)CH(CONH_2)R$$

Interesting results have been obtained by the photoaddition
of α,β-unsaturated carbonyl compounds to alkynes, whereby a
useful variety of cycloadducts can be prepared (P. de Mayo
and M.C. Usselman, Anal.Quim., 1972, 68, 779; W. Hartmann,
Ber., 1971, 104, 2864; and J.A. Barltrop and D. Giles, J.
chem.Soc.(C), 1969, 105).

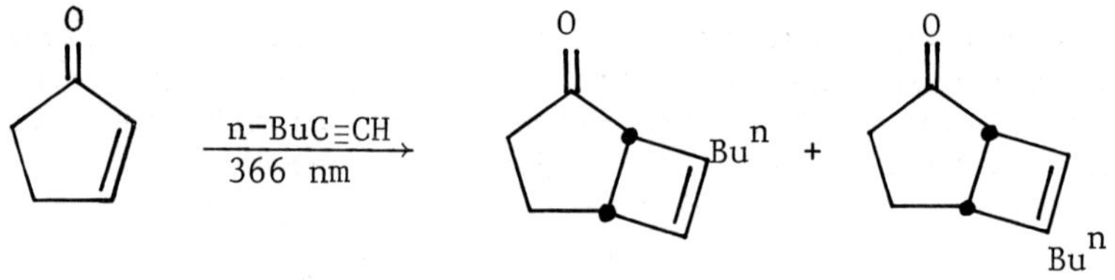

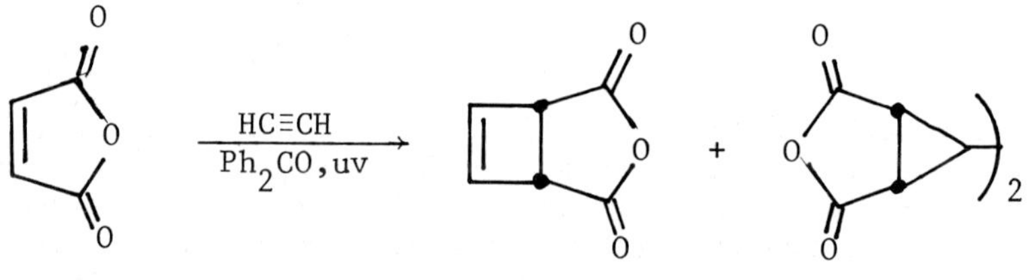

(vii) Cycloadditions

Although electron-deficient acetylenes, notably esters of
propiolic acid and acetylenedicarboxylic acid, are much more
versatile reagents for cycloadditions than simple alkynes,
examples can be found of most types of cycloadditions.

(1) Carbene additions

Diphenylcarbene reacts with both terminal and non-terminal alkynes to give cyclopropenes, usually admixed with other products such as indenes (M.E. Hendrick, W.J. Baron, and M. Jones, Jr., J.Amer.chem.Soc., 1971, 93, 1554), e.g.

$$RC{\equiv}CH \xrightarrow{\ Ph_2C:\ } \quad \underset{R}{\overset{Ph_2}{\triangle}} \quad + \quad \text{(indene, Ph, R)}$$

Even triflate-based vinylidenecarbenes react normally, yielding alkylidenecyclopropanes from, for example, but-2-yne (P.J. Stang and M.G. Mangum, *ibid.*, 1975, 97, 3854).

(2) [2 + 2] cycloadditions

Lewis acids catalyse the otherwise symmetry-forbidden four electron suprafacial-suprafacial cycloaddition of alkynes to alkenes (J.H. Lukas, F. Baardman, and A.P. Kouwenhoven, Angew.Chem.internat.Edn., 1976, 15, 369).

$$Me_2C{=}CMe_2 \xrightarrow[\text{EtAlCl}_2]{\ RC{\equiv}CH\ } \quad \underset{Me_2}{\overset{R\qquad Me_2}{\square}}$$

Under such reaction conditions electron-deficient acetylenes would give ene-insertions (Section 2d-viii). A further complication is the possibility of Lewis-acid catalysed cyclotrimerization; anhydrous aluminium chloride, for example, converts but-2-yne into various cyclic oligomers depending on the precise conditions (Section 4c-viii).

(3) Diels-Alder cycloadditions

Acetylene reacts as a dienophile at high temperatures and pressures with dienes such as cyclopentadiene (M.A. Pryanishnikova *et al.*, Izvest.Akad.Nauk, SSR, Ser.Khim., 1967, 1127) and anthracene (J.C. Muller and J. Vergne, Compt. rend., 1966, 263, 1452). Reverse electron demand Diels-Alder reactions occur with alkynes under less drastic conditions. Examples are the reaction between hex-1-yne and hexachlorocyclopentadiene (E.T. McBee, J.D. Idol, and C.W. Roberts, J.Amer.chem.Soc., 1955, 77, 6674) and between t-butylacetylene and tri-t-butylcyclopentadienone (C. Hoogzand and W. Hubel, Tetrahedron Letters, 1961, 637), e.g.

(4) 1,3-Dipolar cycloadditions

As in the Diels-Alder reaction, simple alkynes are much less reactive dipolarophiles than electron-deficient acetylenes, and so are less commonly used. Nevertheless, dipolar cycloadditions to alkynes provide valuable routes to selected heterocyclic systems; dipolar reagents which have been successfully used include nitrilimines (R. Huisgen et al., Ber., 1967, 100, 1580), nitrile oxides (G. Stagno D'Alcontres and G. Lo Vecchio, Gazz.chim.Ital., 1960, 90, 1239), diazoalkanes (M.E. Hendrick, W.J. Baron, and M. Jones, Jr., J.Amer.chem.Soc., 1971, 93, 1554), nitrile ylides (K. Burger and J. Fehn, Ber., 1972, 105, 3814), and sydnones (R. Huisgen, H. Gotthardt, and R. Grashey, ibid., 1968, 101, 536), amongst others. The regiospecificity of the nitrile oxide and nitrilimine cycloadditions to propyne agree with predictions based on frontier orbital treatments (J. Bastide, N. El Ghandour, and O. Henri-Rousseau, Bull.Soc.chim.France, 1973, 2290, 2294; ibid., 1974, 1037).

The application of frontier orbital theory to this type of cycloaddition has been reviewed (J. Bastide *et al.*, *ibid.*, 1973, 2555, 2871). The whole gamut of alkyne cycloadditions has been thoroughly reviewed (see especially J. Bastide and O. Henri-Rousseau, "The Chemistry of the Carbon-Carbon Triple Bond", ed. S. Patai, Wiley-Interscience, Chichester, 1978, ch. 11).

(viii) Cyclo-oligomerization and polymerization

The well-known trimerization of acetylenes to benzene derivatives often gives rise to mixtures containing rings of other sizes, hence the utility of the Reppe procedures, namely, trimerization to benzene over nickelcarbonyl-phosphine catalysts and tetramerization to cyclo-octatetraene over nickel(II) cyanide catalyst (2nd Edition, Vol. IA, p. 451). Such industrially useful cyclo-oligomerizations of alkynes have been reviewed (W. Reppe, N.V. Kutepow, and A. Magin, Angew.Chem.internat.Edn., 1969, 8, 727; P.M. Maitlis, Pure Appl.Chem., 1972, 30, 427; P.W. Jolly and G. Wilke, "Organic Chemistry of Nickel", Vol. 2, Academic Press, New York, 1975, p. 94). Mechanisms for such ring-forming catalytic interactions between alkynes and transition metals are very complex (P.M. Maitlis, Acc.chem.Res., 1976, 9, 93) and beyond the scope of this chapter. The unexpectedness of such processes is well illustrated by the isolation of pentamethylcyclopentadienyl-hexamethylbenzene chromium(II) chloride as the initial product of cyclization of but-2-yne over chromium(III) chloride (H. Benn, G. Wilke, and D. Henneberg, Angew.Chem.internat.Edn., 1973, 12, 1001): each C_5-ring has formed from two-and-a-half butyne units!

When but-2-yne is treated with a larger than catalytic proportion (50% molar) of anhydrous aluminium chloride, a dimer of tetramethylcyclobutadiene is obtained as major product (H.M. Rosenberg and E.C. Eimutis, Canad.J.Chem., 1967, 45, 2263). If a lower, catalytic (*ca.* 5% molar) amount

is used, 60-70% of hexamethyl Dewar benzene is obtained,
which isomerizes in refluxing benzene to hexamethylbenzene
(W. Schäfer and H. Hellmann, Angew.Chem.internat.Edn., 1967,
6, 518).

The presence of a catalyst in such processes is not always
essential; t-butyl-fluoroacetylene trimerizes spontaneously
to 1,2,3-tri-t-butyl-trifluorobenzene. The isomer obtained
can only be formed if the isomeric Dewar benzene intervenes,
a deduction proved by isolation of the intermediate (H.G.
Viehe, *ibid.*, 1965, **4**, 746). 1,2,3-Trisubstituted benzenes
are not usually formed in such processes, however; the
predominant isomers from both non-catalytic and catalytic
trimerizations of unsymmetrical alkynes are 1,2,4- and
1,3,5-trialkylbenzenes, suggesting that other pathways
normally dominate (R. Fuks and H.G. Viehe, "Chemistry of
Acetylenes", ed. H.G. Viehe, Marcel Dekker, New York, 1969,
ch. 8).

Promising synthetic procedures have been developed for the
conversion of α,ω-diynes into annelated benzenes by catalytic
cotrimerization with a third alkyne unit, frequently bis-
trimethylsilylacetylene; a suitable catalyst is π-cyclo-
pentadienylcobaltdicarbonyl (K.P.C. Vollhardt, Acc.chem.Res.,
1977, **10**, 1), e.g.,

$$(CH_2)_2 \begin{array}{c} C\equiv CMe \\ \\ C\equiv CMe \end{array} \xrightarrow[\text{CpCo(CO)}_2]{RC\equiv CR} \text{(benzocyclobutene: Me, R, R, Me)}$$

Spectacularly complex intramolecular trimerizations have been achieved over Ziegler–Natta catalytic systems such as titanium(IV)chloride/tri–isobutyl–aluminium (A.J. Hubert, J.chem.Soc.(C), 1967, 6, 11,13).

Linear polymerization of acetylene, alkynes, and ethynylarenes, especially over Ziegler catalysts (e.g. H. Shirakawa and S. Ikeda, Polymer J., 1971, 2, (2), 231), have been extensively investigated, but to date no material sufficiently desirable for its large-scale manufacture to be undertaken has emerged. The products obtained, however, display interesting properties connected with their highly delocalised electronic structure, such as semi-conduction and pressure-sensitive resistivity (H.A. Pohl and R.P. Chartoff, J.poly.Science A, 1964, 2, 2787; A.G. MacDiarmid et al., J.Amer.chem.Soc., 1978, 100, 1013). Reviews of these researches are available (T.F. Rutledge, "Acetylenes and Allenes", Reinhold, New York, 1969, ch. 5; E.M. Smolin and D.S. Hoffenberg, Encycl.Polymer Sci.Technol., 1964, 1, 46; V.V. Pen'kovskii, Russ.chem.Rev., 1964, 33, 532).

(ix) Transition metal complexes

For details of the formation, structure, and reactions of the vast number of complexes of alkynes with transition metals, the reader is referred to the many excellent specialist reviews, a representative selection of which are provided here (for general reviews, see A.C. Hopkinson, "The Chemistry of the Carbon-Carbon Triple Bond", ed. S. Patai, Wiley-Interscience, Chichester, 1978, ch. 4, and G.W. Parshall, "Homogeneous Catalysis", Wiley-Interscience, New York, 1980, ch. 8).

Olefin and acetylene complexes of the nickel triad: J.H.

Nelson and H.B. Jonassen, Co-ordination chem.Rev., 1971, <u>6</u>, 27; olefin and acetylene complexes of nickel: P.W. Jolly and G. Wilke, "Organic Chemistry of Nickel", Vol. 2, Academic Press, New York, 1975; olefin and acetylene complexes of platinum and palladium: F.R. Hartley, Chem. Rev., 1969, <u>69</u>, 799 and P.M. Maitlis, "Organic Chemistry of Palladium", Vol. 1, Academic Press, New York, 1971, ch. 3; palladium-catalysed reactions of alkenes, dienes, and alkynes: P.M. Henry, Adv.organometal.Chem., 1975, <u>13</u>, 363; palladium-catalysed cyclotrimerizations of alkynes: P.M. Maitlis, Pure appl.Chem., 1973, <u>33</u>, 498; alkyne complexes of cobalt: R.S. Dickson and P.J. Fraser, Adv.organometal.Chem., 1974, <u>12</u>, 217; acetylene and allene complexes of transition metals, S. Otsuka and A. Nakamura, *ibid.*, 1976, <u>14</u>, 245; transition metal complexes of alkenes and alkynes, their stability and structure; L.D. Pettit and D.S. Barnes, Topics curr.Chem., 1972, <u>28</u>, 85; organometallic derivatives from metal carbonyls and acetylenic compounds; W. Hübel, "Organic Synthesis *via* Metal Carbonyls", Vol. 1, ed. I. Wender and P. Pino, Wiley-Interscience, New York, 1968, p. 273; carbonylation of acetylenes, P. Pino and G. Braca, *ibid.*, Vol. 2, 1977, p. 419; cyclopolymerization of acetylenes, C. Hoogzand and W. Hübel, *ibid.*, p. 343, (see also Section 2d-x).

Propargylic metallation, which has been discussed earlier with reference to its synthetic applications (Section 4a-iii), has been reviewed (J. Klein, "The Chemistry of the Carbon-Carbon Triple Bond", ed. Patai, *loc.cit.*, ch. 9).

5. Hydrocarbons with more than two double-bond equivalents

The following list of books and reviews may serve as an introduction to the main types of compounds in this class.

(a) Conjugated enynes

W.R. Franke and H. Neunhoeffer in Houben-Weyl's "Methoden der Organischen Chemie", Vol. 5/1d, 4th Edn., ed. E. Müller, Thieme-Verlag, Stuttgart, 1972, p. 609; G. Pattenden, "Comprehensive Organic Chemistry", eds. D.H.R. Barton, W.D. Ollis and J.F. Stoddart, Pergamon, Oxford, 1979, Vol. 1, p. 205.

(b) Conjugated diynes and polyacetylenes

U. Niedballa in Houben-Weyl's "Methoden der Organischen Chemie", *loc.cit.*, Vol. 5/2a, p. 913; J.D. Bu'lock, Prog. org.Chem., 1964, 6, 86; P. Cadiot and W. Chodkiewicz (ch. 9) and F. Bohlmann (ch. 14) in "Chemistry of Acetylenes", ed. H.G. Viehe, Marcel Dekker, New York, 1969; F. Bohlmann, T. Burkhardt, and C. Zdero, "Naturally Occurring Acetylenes", Academic, New York, 1973; W.D. Huntsman (ch. 13) and E.R.H. Jones and V. Thaller (ch. 14) in "The Chemistry of the Carbon-Carbon Triple Bond", ed. S. Patai, Wiley-Interscience, Chichester, 1978.

(c) Open-chain polyenes

K. Reppe in Houben-Weyl's "Methoden der Organischen Chemie", *loc.cit.*, Vol. 5/1d, p. 1; W.S. Johnson, Angew. Chem.internat.Edn., 1976, 15, 9; various authors in "Carotenoids", ed. O. Isler, Birkhauser Verlag, Basel, 1971; and various authors in Pure appl.Chem., 1976, 47, 97ff and 1979, 51, 435ff.

(d) Conjugated ene-allenes

I.Z. Egenburg, Russ.chem.Rev., 1978, 47, 470.

(e) Cumulenes

H. Fischer, "The Chemistry of Alkenes", ed. S. Patai, Wiley-Interscience, London, 1964, Vol. 1, p. 1025; M. Murray in Houben-Weyl's "Methoden der Organischen Chemie", *loc.cit.*, Vol. 5/2a, p. 693; W.D. Roth and H.-D. Exner, Ber., 1976, 109, 1158; J.L. Ripoll and A. Thuillier, Tetrahedron, 1977, 33, 1333.

Chapter 18

TRIHYDRIC ALCOHOLS, THEIR ANALOGUES AND DERIVATIVES AND
THEIR OXIDATION PRODUCTS: TRIHYDRIC ALCOHOLS TO TRIKETONES

B.J.COFFIN

Glycerol, 1,2,3-trihydroxypropane, propan-1,2,3-triol.
    Since the commencement of the manufacture on a commercial
scale of glycerol from propylene in 1948, the formation of
glycerol has been based on several routes with improvements
over the past few years.
1. By the oxidation of allyl alcohol with aqueous
permanganate to glycidol, followed by hydrolysis (J.Berthoux,
Fr. Pat., 1,548,678; H.P.Liao *et al.*, *ibid.*, 1,509,278).
2. By the hydroxylation of allyl alcohol with aqueous sodium
hypochlorite in the preence of osmium tetroxide (F.M.C.Corp.,
Neth. Pat., 74 11,150). This method is also used in the
formation of other polyols. A 9.2% tungstic catalyst can also
be used (Z.Maciejewski *et al.*, Pol. Pat., 57,650).
3. From hydrogen and carbon monoxide using a rhodium
catalyst formed *in situ* from dicarbonyl(2,4-pentanedionato)
rhodium in the presence of a solvent such as pyridine or
morpholine (E.P.Walker and R.L.Proelt, Ger. Pat., 2,426,411).
4. By the hydrolysis of glycidyl alcohol with carbon
dioxide at 6-12 atmosphers at 200°C in the presence of a
base catalyst (N.N.Lebedev *et al.*. U.S.S.R. Pat., 322,973).
5. Glycerol labelled with ^{14}C in the β - position can be
prepared from β - ^{14}C labelled acrylic acid by bromination
followed by hydrolysis and reduction (J.P. Guermont,M. Herbert
and L. Pichat, Bull. Inform. Sci. Tech. (Paris) 1967, 118, 32).

Successive bromination, esterification and hydrolysis gives glycerol labelled with carbon-14.

$1-^{14}C$-Glycerol has been synthesised from glycerol-3-phosphate $-1-^{14}C$ by enzymic dephosphorylation (P.F.Blackmore et $al.$, J. label.Compd., 1972,8, 71).

Glycerol can be estimated by titration with sodium periodate (P.S.Verma and K.C.Grover, Vijnana Parished Anusandh ars Patrika, 1971, 14, 13), or by colorimetry of its periodic oxidation product, formaldehyde, by colour development by the addition of hydrochloric acid, 3-methyl-2-benzo-thiazolinone, hydrazine and ferric chloride (M.Pays et $al.$, Ann. Pharm. Fr., 1967, 25, 29).

When glycerol is heated with calcium oxide, a mixture of metal glyceroxides is formed, or if glycerol is heated with metal acetates at $100^{\circ}C$ crystalline metal glycerolato-complexes are obtained (K.Fujii and W.Kondo, Z. Anorg. Allg. Chem., 1968, 359, 296). When various ratios of glycerol and calcium oxide are used the resultant mixture contains CaHL, $Ca(H_2L)_2$, $Ca_3(H_2L)_6.2H_3L$ and $Ca(H_2L)_2.4H_3L$ where H_3L represents glycerol. At different temperatures various equilibria exist.

$$Ca(H_2L).4H_3L \underset{}{\overset{18^{\circ}C}{\rightleftharpoons}} Ca_3(H_2L).2H_3L$$

$$Ca_3(H_2L).2H_3L \overset{44^{\circ}C}{\longrightarrow} Ca(H_2L)_2$$

X-ray analysis shows that there are both linear calcium-oxygen bonds and calcium-oxygen-calcium rings.

Glycerol, on treatment with hydrogen chloride in acetic acid at $100^{\circ}C$ followed by sodium periodate oxidation, gives chloracetaldehyde. Thus, glycerol-2-^{14}C can be converted into chloracetaldehyde-1-^{14}C (J.C.Greenfield and N.J.Leonard, J. labelled Compd., 1976,12, 545).

Glycerol reacts with formaldehyde and aniline at pH 4 and $75^{\circ}C$ to give:

$$CH_2OCH_2NHPh$$
$$|$$
$$CHOCH_2NHPh$$
$$|$$
$$CH_2OCH_2NHPh$$

Glycerol can be converted into its per-O-diethyl borylated by reaction with activated diethylborane (W.K.Dahlhoff and R.Koester, Ann. Chem., 1975, 10, 1914). These derivatives can be converted either thermally or by a BH catalysed reaction into their corresponding dioxaboracycloalkanes, dioxaboralanes and spiro-compounds.

Similarly, butane-1,2,4- and hexane-1,2,6-triols can be borylated.

$n = 1, m = 2$
$n = 1, m = 4$

Pentane-1,2,5-triol reacts with boric acid in an acid medium to provide a cyclic derivative (A.Vegorere, E.Svarcs and A.Levins, Laty. PSR Zinat. Akad. Vestis, Kim. Ser., 1973, 5, 549) which polymerises in the presence of phosphorus pentoxide.

A number of trioxaboratricycloundecanes can be made by the reaction of glycidol with 2-aminoethanol and tributyl borate

(J.H.Ludwig and K.J.Witsken, U.S.Pat., 3,755,388).

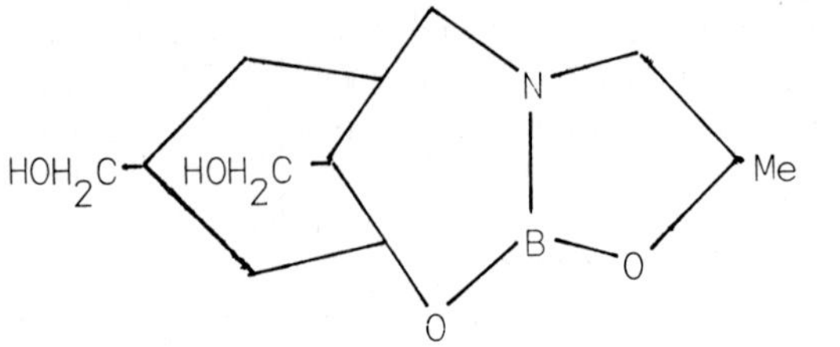

The reaction between glycerol and trimethylphosphite gives 2,6,7-trioxa-1-phosphabicyclo [2.2.1] heptane (D.B.Denney and V.Sander, Tet. Letters, 1966, 40, 4935). Some of the reactions of the latter are shown below (N.A.Makarova *et al.*, Dokl. Akad. Nauk. SSSR., 1973, 219, 1331).

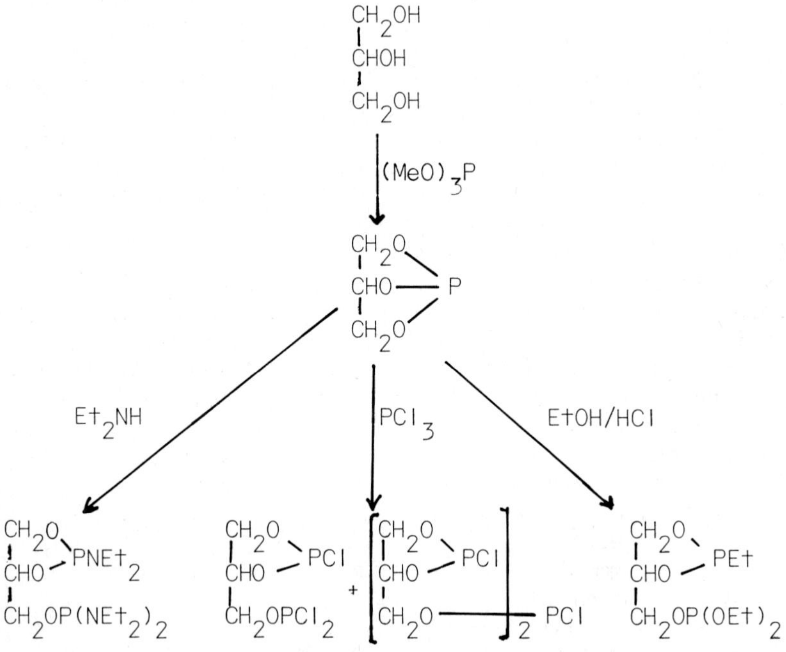

The cyclocondensation of l-chloro-2,3-dihydroxypropane with triethyl phosphite also leads to thr formation of a substituted dioxaphospholane (N.A.Makorova *et al.*, Izv. Akad. Nauk. SSSR SSSR Ser. Khim., 1974, 7, 1637). This, with oxygen in benzene, gives 2-ethoxy-C-chloromethyl-1,3,2-dioxaphospholane.

The reaction between glycerol and thionyl chloride in methanol leads to the formation of the *cis* and the *trans* isomer of 2-hydroxy-propylene-1,3-sulphite and of 3-hydroxy-propylene-1,2-sulphite (A.T.Rasschaert, Ind. Chim. Belge, 1967, 32, 115; H.F. van Woerden, Rec. Trav. Chem., 1967, 86, 601). In contrast the reaction between glycerol and thionyl chloride in carbon tetrachloride leads to the formation of the *cis* and the *trans* isomer of 2-chloropropylene-1,3-sulphite and of 3-chloropropylene-1,2-sulphite.

The analysis of radioactively labelled glycerol-3-phosphate is achieved by its cleavage with periodic acid to give formaldehyde, derived from carbon I (isolated as its dimedone derivative) and phosphoglycolaldehyde. The latter is dephosphorylated with alkaline phosphatase to glycolaldehyde which is then cleaved with periodic acid to formaldehyde from carbon -2 and formic acid from carbon-3 (P.Rauschenbach and P.Lamprecht, Z. physiol. Chem. 1965, 346, 290).

Diacyl-DL- and L-α-glycerol phosphates are prepared by the direct acylation of glycerol phosphate with alkanoic acid anhydrides in the **presence** of the appropriate tetraethyl ammonium salt. Only the phosphatidic acids are made with no cyclic phosphates being formed. Thus, DL-α-glycerol phosphate monotetraethylammonium salt is added to a solution of tetra-ethylammonium palmitate in methanol to give an oily product. This is refluxed with palmitic anhydride in carbon tetrachloride to give dipalmityl-DL-α-glycerol phosphate.

Prostaglandins with increased stability in the blood-stream are made by the action of glycerol or glycerol-3-phosphate compounds with either an unsaturated prostanoic acid with protected hydroxy groups, or with an unsaturated prostanoic

acid in the presence of a phospholipase and cofactors
(M.Gordon and J.A.Weisbach, U.S.Pat., 3,632,627). Thus,
$9\alpha,11\alpha,15\alpha$,-tris(tert-butoxycarbonyloxy)-13-prostenoic acid
and 1,3-bis(tert-butoxycarbonyl)glyceride are reacted in
solution containing glycerol. The removal of the tert-butoxycarbonyl groups with hydrogen chloride in benzene gives the
prostenoyl glyceride.

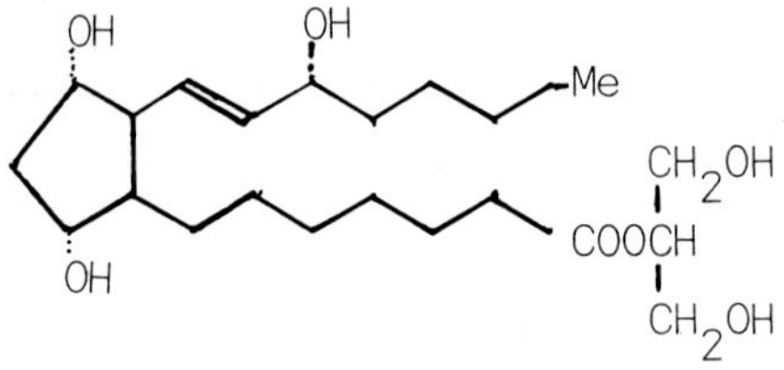

Phosphatidylalkanolamine derivatives are made from 1,2-
isopropylidene-sn-glycerol by its reaction firstly with
phosphorus oxychloride and then with N-(2-hydroxyethyl)
phthalimide to give a mixture of 1,2-isopropylidene-sn-
glycerol-3-(2-phthalimidoethyl hydrogen phosphate) and
sn-glycerol-3-(2-phthalimidoethyl hydrogen phosphate)
(S.Rakhit, U.S.Pat., 3,577,446). The latter compound on
reaction with 8,12,15-octadecatrienoyl chloride in dimethylformamide gives sn-glycerol-1,2-di(8,12,15-octadecatrienoate)-
3-(2-aminoethyl hydrogen phosphate).

Cephalin, α-(β,γ-dipalmitoyl)glyceryl aminoethylphospho
(P→N)glycine may be synthesised by the following scheme
(M.K.Pretova et $al.$, Zh. org Khim., 1969, 5, 1883).

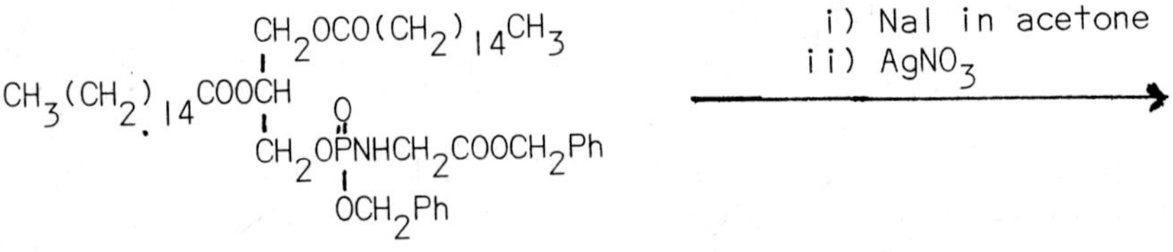

i) NaI in acetone
ii) $AgNO_3$

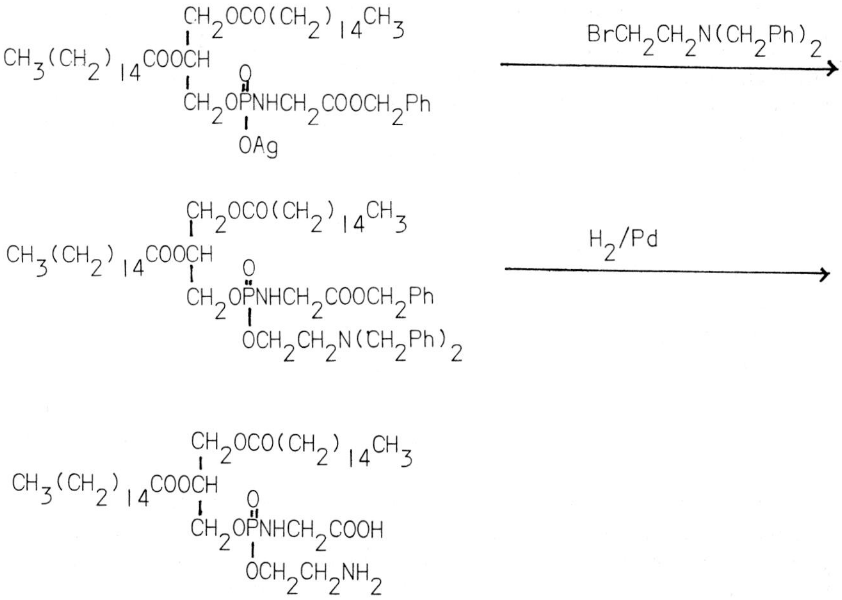

$$CH_3(CH_2)_{14}COOCH \begin{array}{c} CH_2OCO(CH_2)_{14}CH_3 \\ | \\ | \\ CH_2O\overset{O}{\overset{||}{P}}NHCH_2COOCH_2Ph \\ | \\ OAg \end{array}$$

$$\xrightarrow{\quad BrCH_2CH_2N(CH_2Ph)_2 \quad}$$

$$CH_3(CH_2)_{14}COOCH \begin{array}{c} CH_2OCO(CH_2)_{14}CH_3 \\ | \\ | \\ CH_2O\overset{O}{\overset{||}{P}}NHCH_2COOCH_2Ph \\ | \\ OCH_2CH_2N(CH_2Ph)_2 \end{array}$$

$$\xrightarrow{\quad H_2/Pd \quad}$$

$$CH_3(CH_2)_{14}COOCH \begin{array}{c} CH_2OCO(CH_2)_{14}CH_3 \\ | \\ | \\ CH_2O\overset{O}{\overset{||}{P}}NHCH_2COOH \\ | \\ OCH_2CH_2NH_2 \end{array}$$

$(\overset{+}{-})$-|-Glyceryl phenyl-3-aminopropylphosphate may be formed from $(\overset{+}{-})$-2,3-bis(palmitoyloxy)propyl bromide by the following scheme (T.Moramatsu and I.Hara, Bull. soc. Chim. Fr., 1971, $\underline{9}$, 3335).

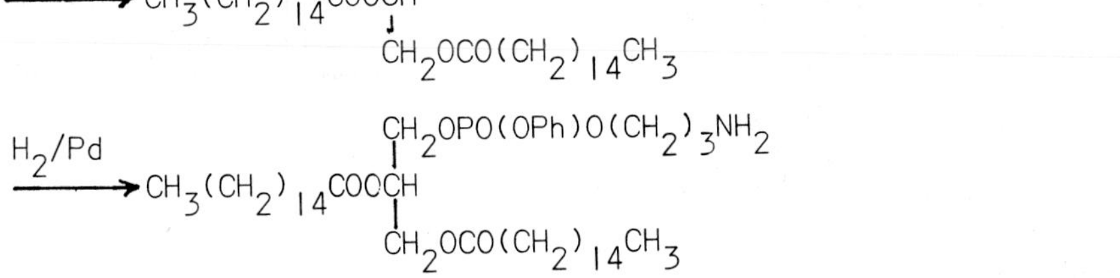

$$CH_3(CH_2)_{14}COOCH \begin{array}{c} CH_2Br \\ | \\ | \\ CH_2OCO(CH_2)_{14}CH_3 \end{array} \quad + \quad PhOP{=}O \begin{array}{c} OAg \\ | \\ | \\ O(CH_2)_3NHCOOCH_2Ph \end{array}$$

$$\xrightarrow{\quad} CH_3(CH_2)_{14}COOCH \begin{array}{c} CH_2OPO(OPh)O(CH_2)_3NHCOOCH_2Ph \\ | \\ | \\ CH_2OCO(CH_2)_{14}CH_3 \end{array}$$

$$\xrightarrow{H_2/Pd} CH_3(CH_2)_{14}COOCH \begin{array}{c} CH_2OPO(OPh)O(CH_2)_3NH_2 \\ | \\ | \\ CH_2OCO(CH_2)_{14}CH_3 \end{array}$$

The reaction of 2-(dimethylamino)-4-nitrophenyl phosphate with alcohols containing primary and secondary hydroxyl groups leads to the selective phosphorylation of the primary groups.

Thus, glycerol is phosphorylated to glycerol-3-phosphate (Y.Taguchi and Y.Mushika, Chem. pharm. Bull. Japan, 1975, 23, 1586).

Glyceraldehyde, 2,3-dihydroxypropan-1-al, $HOCH_2CH(OH)CHO$
D-Glyceraldehyde- 3,3-d_2-3-phosphate can be prepared from 2-O-benzyl-D-arabinose as indicated below (G.R.Gray and R. Barker, Carbohydrate Res., 1971, 20, 31).

Mass spectrometry shows that glyceraldehyde is dimeric in the vapour state. It fragments by isomerisation of the molecular ion *via* 4-membered transition states (E.F.H.Brittain *et al.*, J. chem. Soc. (B)., 1971, 2414).

Oxidation of glyceraldehyde with molecular oxygen at $90^{\circ}C$ proceeds *via* the intermediates, glycolaldehyde and methyl-glyoxal, to a mixture of glyceric acid, acetic acid, formic acid and carbon dioxide (L.M.Andronov and Z.K.Maizus, Izv.

Akad. Nauk. SSSR Ser. Khim., 1967, 519).

The aldolisation of D-glyceraldehyde is catalysed by organic and inorganic bases to yield a mixture of D-fructose, D-sorbose and DL-dendroketose (C.D.Gutsche *et al.*, J. Amer. chem. Soc., 1965, **89**, 1235).

$$
\begin{array}{l}
CH_2OH \\
| \\
CO \\
| \\
CHOH \\
| \\
HOCCH_2OH \\
| \\
CH_2OH
\end{array}
$$

DL–dendroketose

Oxidation of glyceraldehyde with silver carbonate on Celite in methanol at 35°C gives methyl glycerate (S.Morgenlie, Acta. chem. Scand., 1973, **27**, 3009).

In acidic media glyceraldehyde undergoes dehydration to acrolein where the significant reaction step is the enolisation reaction (D.Fleury and P.Souchey, Compt. rend., 1968,266c, 1035).

D-Glyceraldehyde is used for asymmetric syntheses. Thus, methyl (-)-2-methoxy-2-phenylpropionate can be made from D-glyceraldehyde (R.Meric *et al.*, Bull. Soc. chim. Fr., 1973, **1**, 327). It can also be used as the starting material in a number of syntheses. Thus, D-glyceraldehyde reacts with tert-butylamine in an atmosphere of hydrogen in the presence of a palladium on charcoal catalyst as follows (L.M.Weinsock, R.J. Tull and D.M.Mulvey, Ger . Pat., 4,031,125).

$$
\begin{array}{l}
CHO \\
| \\
CHOH \\
| \\
CH_2OH
\end{array}
\xrightarrow{Me_3CNH_2,\ H_2/PdC}
\begin{array}{l}
CH_2NHCMe_3 \\
| \\
CHOH \\
| \\
CH_2OH
\end{array}
$$

The treatment of (±)-glyceraldehyde with 4-amino-3-hydrazino-5-mercapto-1,2,4-triazole gives the corresponding 1,2,3,4-tetrahydro-6-mercapto-s-triazole 4,3-b -s-tetrazine derivative (E.Humares *et al.*, Carbohydrate Res., 1976, *46*, 284)

Glyceraldehyde undergoes a cycloaddition reaction with 2-(aminomethyl)ethyleneimine to give a diazobicyclohexane (S.Hillers *et al.*, Khim. Geterotsckl. Soeden., 1975, *3*, 425).

D-α-Lysolecithin, (β-trimethylammonio)-ethyl-3-(palmitoyl-oxy)-2-hydroxypropylphosphate is a useful immunological adjuivant and 2,3-di-O-benzylglyceraldehyde is an important intermediate in its synthsis outlined below (K.Thomal, Ger. Pat., 2,033,358).

$$\begin{array}{c}\text{CH}_2\text{OCH}_2\text{Ph} \\ | \\ \text{CHOCH}_2\text{Ph} \\ | \\ \text{CHO}\end{array} \xrightarrow{\text{LiAlH}_4} \begin{array}{c}\text{CH}_2\text{OCH}_2\text{Ph} \\ | \\ \text{CHOCH}_2\text{Ph} \\ | \\ \text{CH}_2\text{OH}\end{array} \longrightarrow \begin{array}{c}\text{CH}_2\text{OCH}_2\text{Ph} \\ | \\ \text{CHOCH}_2\text{Ph} \\ | \\ \text{CH}_2\text{OCO}(\text{CH}_2)_{14}\text{CH}_3\end{array}$$

$$\xrightarrow{\text{H}_2/\text{Pd}} \begin{array}{c}\text{CH}_2\text{OH} \\ | \\ \text{CHOCH}_2\text{Ph} \\ | \\ \text{CH}_2\text{OCO}(\text{CH}_2)_{14}\text{CH}_3\end{array} \xrightarrow[\text{ii)Me}_3\text{N}]{\text{i)Br(CH}_2)_2\overset{\overset{\text{O}}{\|}}{\text{OPCl}}} \begin{array}{c}\text{CH}_2\text{OCH}_2\text{CH}_2\overset{\overset{\text{O}}{\|}}{\text{OP}}\text{NMe}_2 \\ | \\ \text{CHOCH}_2\text{Ph} \\ | \\ \text{CH}_2\text{OCO}(\text{CH}_2)_{14}\text{CH}_3\end{array}$$

$$\xrightarrow{\text{H}_2/\text{Pd}} \begin{array}{c}\text{CH}_2\text{OCH}_2\text{CH}_2\overset{\overset{\text{O}}{\|}}{\underset{\text{OH}}{\text{OP}}}\text{NMe}_2 \\ | \\ \text{CHOH} \\ | \\ \text{CH}_2\text{OCO}(\text{CH}_2)_{14}\text{CH}_3\end{array}$$

Glyceric acid, 2,3-dihydroxypropanoic acid, $\text{HOCH}_2\text{CHOHCOOH}$
Glyceric acid is synthesised by the reaction of disodium
hydrogen phosphate at pH 9 with epoxypropanoic acid (N.V.
Stamicarbon, Neth. Pat., 67 04 361). High specific activity
glyceric acid -3-phosphate-1-^{14}C is made by means of ribulose
diphosphate carboxylase reaction for the preparation of
dihydroxyacetone phosphate-1-^{14}C and fructose-^{14}C with the
aid of the following enzymes: 3-phosphoglyceric acid kinase,
glyceraldehyde phosphate dehydrogenase and triosephosphate
isomerase (E.Sturani, Communicate Eur. Energ. At-EURATOM
Rapp. 1968, EUR-4044i 19). Alternatively, glyceric acid-
3-phosphate-1-^{14}C can be synthesised by the action of D-
erythro-pentulose-diphosphate carboxylase in the prescence
of D-erythro-pentulose-1,5-diphosphate and carbon dioxide-
[14]C, followed by the reversal of the glycolytic reactions
leading from D-glucose (*ibid.,* J. labelled Compd., 1969, 5,
47).
The photolysis of aqueous disodium α-D-glucoso-6-

phosphate with radiation at 254nm. gives a mixture of 6-phosphogluconate, arabinose-5-phosphate, phosphoglycolylate and phosphoglycerate (C.Triantaphylides and M.Halmann, J. chem. Soc. Perkin II, 1975, $\underline{1}$, 34).

Sodium glycerophosphate can be determined colorimetrically by its decomposition with a nitric acid/hydrogen peroxide mixture, followed by reaction with ammonium molybdate solution containing potassium antimonyl tartrate and ascorbic acid which gives a coloured solution, λ_{max} 640-670nm. (A.Cholewicki, Bull. Vet. Inst. Pulawy., 1971, $\underline{15}$, 100).

Ethyl glycerate reacts with paraformaldehyde in benzene in the presence of toluene-p-sulphonic acid to give a high yield of 4-(ethoxycarbonyl)-1,3-dioxolane (A.B.Perstorp, Brit. Pat., 1,272,509).

Ethyl glycerate reacts quantitatively with long-chained aldehydes under acidic conditions to provide both the *cis*- and the *trans*-2-alkyl-4-ethoxycarbonyl-1,3-dioxolane which can be separated chromatographically (Y.Wedmid and W.J. Baumann, J. org. Chem., 1977,$\underline{42}$, 3624).

D-Glyceric acid can be used as a source for D-glyceric acid p-vinylanilide-2,3-*O*-(p-vinylboronates) (G.Wulff *et al.*, Ber., 1974, $\underline{107}$, 3364).

Glycerol esters of inorganic acids.

Convenient methods have been developed for the synthesis of ^{14}C-labelled 1-chloro and 1-bromo-2,3-dihydroxypropanes from ^{14}C-glycerol (A.R.Jones, J. labelled Compd., 1973, 9, 697). These compounds can then be converted by reaction with sodium iodide into the corresponding iodo-compounds.

The product of the reaction of 1-chloro-2,3-dihydroxy-propane with ethyl diazoacetate in the presence of boron trifluoride-ether, followed by hydrolysis provides a useful calcium complexing reagent (G.Borggrefe, Ger. Pat., 2,142,207).

With glyoxal in the presence of phosphorus pentoxide l-chloro-2,3-dihydroxypropane yields a substituted dioxanediol which is claimed as a photohardening agent (N.Yamamoto, Jap. Pat., 74 39,998).

l-Chloro-2,3-dihydroxypropane undergoes a cyclocondensation reaction with formaldehyde to give 5-chloromethyl-1,3-dioxolane (S.B.Richter and J.Krenzer, U.S. Pat., 4,012,222).

The esterification of l-chloro-2,3-dihydroxypropane with substituted benzoic acids in dimethylformamide solution containing barium carbonate gives glycerol benzoate esters, $RC_6H_4COOCH_2CHOHCH_2OH$, (R 2-Cl, 2-I, 3-NO_2) (M.Mihalic, Acta. pharm. Jugoslav., 1975, 24, 1). This method allows the synthesis of glaphenin [R = 2-(7-chloro-4-quinolinyl-amino)] which exhibits analgesic and anti-inflammatory activity. Benzoylation can be achieved by its reaction with trichlorophenylmethane to give a mixture of products presumably *via* the intermediate, $PhCCl_2OCH_2CHOHCH_2Cl$ (G.K.Chizh *et al.*, Zh. org. Khim., 1972, 8, 698).

However the reaction of 1,3-dichloro-2-hydroxypropane provides only one product, 2-benzoyl-1,3-dichloropropane.

$$\begin{array}{c} CH_2Cl \\ | \\ CHOH \\ | \\ CH_2Cl \end{array} \xrightarrow{PhCCl_3} \begin{array}{c} CH_2Cl \\ | \\ CHOCOPh \\ | \\ CH_2Cl \end{array}$$

1,3-Dichloro-2-hydroxypropane is oxidised with potassium chromate to dichloroacetone (K.E.Mtsumoto and J.H-H. Chan, Ger. Pat., 2,703,370).

$$\begin{array}{c} CH_2Cl \\ | \\ CHOH \\ | \\ CH_2Cl \end{array} \xrightarrow{KCrO_4} \begin{array}{c} CH_2Cl \\ | \\ C=O \\ | \\ CH_2Cl \end{array}$$

2,3-Dichloropropanal, $ClCH_2CHClCHO$, is obtained by the addition of chlorine to acrolein in the absence of any solvent (I.V.Andreeva, Zh. org. Khim., 1975, 11, 954).

With cyclohexene in the presence of aluminium chloride 2,3-dichloropropanal gives a mixture of two compounds (Z.I.Sadykh-Zade et al., Uch. Zap.-Minist. Vyssh. Sredn. Spets. Obraz. Az. SSSR Ser. Khim. Nauk., 1975, 1, 42).

Formylaziridines:

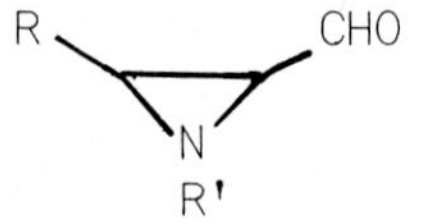

where R = H, Me and R^1 = $CHMe_2$, CMe_3 and X = 0, can be made by treating RCHBrCHBrCHO with R^1NH_2, followed by hydrolysis with acid (L.Wartski and E.A.Sierra, Compt. rend., 1974, <u>279C</u>,149).
   Chloroepoxides with the general formula

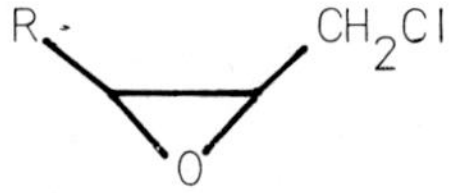

can be made by the reaction of 2,3-dichloropropanal with the appropriate Grignard reagent (T.Shigeo *et al.*, Kagaku Kyokai Shi., 1971,<u>29</u>, 530). The hydrolysis of 2,3-epoxy-1-chloropropane at molar ratios to water of 1:1 to 10:1 at 90-110°C gives 9 monomeric, heptameric, partially and completely cyclic and open-chain products (L.F.Glukahen'ka *et al.*, Zh. Prikl. Khim. (Leningrad), 1974, <u>47</u>, 629).

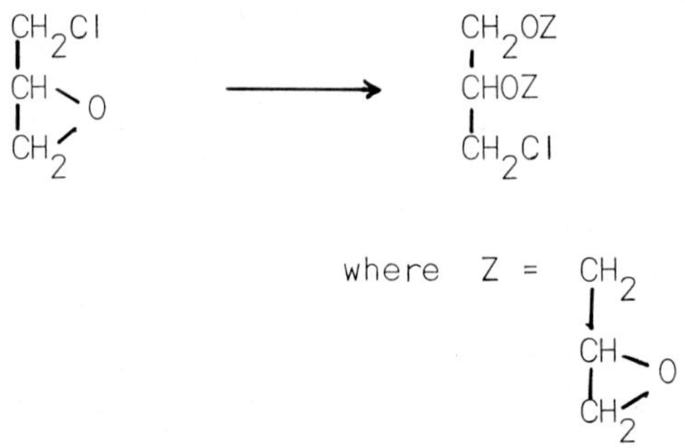

where Z =

and

When reacted with myristic alcohol in the presence of boron trifluoride-ether, 2,3-epoxy-1-chloropropane provides the monoglyceride, 1-tetradecyloxy-3-chloro-2-hydroxypropane (J.Berecoechea and J.Anatol, Bull. soc. Chim. Fr., 1967, 2160). The reaction of 1-chloro-3-deuterio-2,3-epoxypropane with dimethyl sulphide in aqueous sodium perchlorate gives 1,4-dioxan-2,5-ylene bis(monodeuteriomethylene) bis(dimethyl sulphonium) perchlorate.

(S.Fujisaki, Nippon Kagaku Kaishi, 1975, $\underline{2}$, 400). When 2,3-epoxy-1-chloropropane is heated with ethyl salicylate, benzodioxepinone is formed (J.Gilbert *et al.*, Bull. soc. Chim. Fr., 1975,(12 pt.2), 277). Benzoxathiepinones can be derived from thiosalicylic acid derivatives.

The reaction of $Me_4\overset{+}{N}OPO(OCMe_3)_2$ in $(MeOCH_2)_2$ with 2,3-epoxy-1-bromopropane gives the corresponding derivative

which can be hydrolysed with trifluoroacetic acid to

(M.Kluba and A.Zwierak, Roscz. Chim. 1974, $\underline{48}$, 1603).

Nitrogen analogues of glycerol.

3-Amino-1,2-dihydroxypropane, $NH_2CH_2CHOHCH_2OH$, reacts with phosgene to provide 5-hydroxymethyl-2-oxazolidinone (G.Tsuchihashi *et al.*, Jap. Pat., 76 16,661).

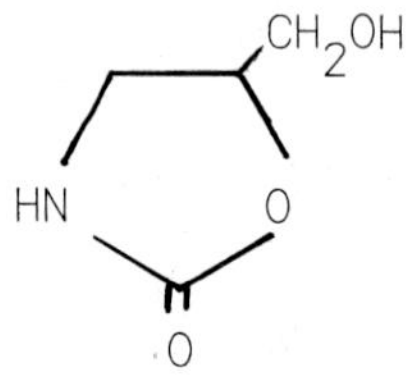

2,6-Dioxa-10-azatricyclo $[5.2.1.0^{4,10}]$ decanes can be made by the following reaction scheme (H.G.Broadbent, W.J.Burnham and R.M.Sheeley, J. heterocycl. Chem., 1976, 13, 337).

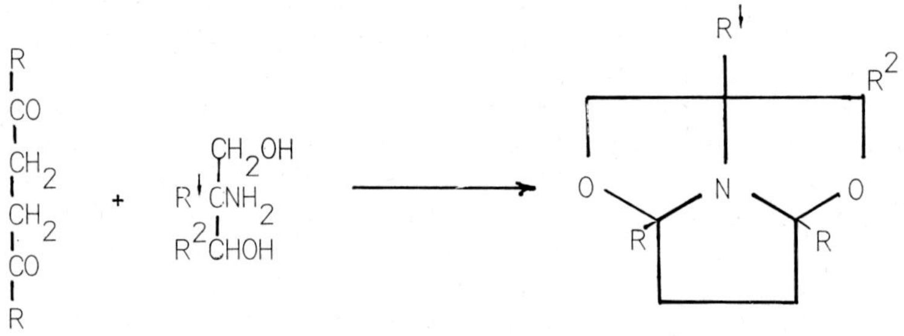

(R = H, Me, Et; R^1 = H, alkyl; R^2 = p-$NO_2C_6H_4$).

1,2-Diamino-3-chloropropane, $NH_2CH_2NH_2CHCH_2Cl$, dihydrochloride m.p. 220-223° (R.Paul, R.P.Williams and E.Cohen, J. org. Chem., 1975, 40, 1653). It can be synthesised by the following methods.
  a) From 2,3-dibromopropanol by fusion with potassium phthalimide, followed by hydrolysis (E.Philippi, R.Seka and L.Ableindinga, Ann. Chem., 1923, 433, 94; J.Stancek and J.Prusova, Chem. Listy., 1952, 46, 491).

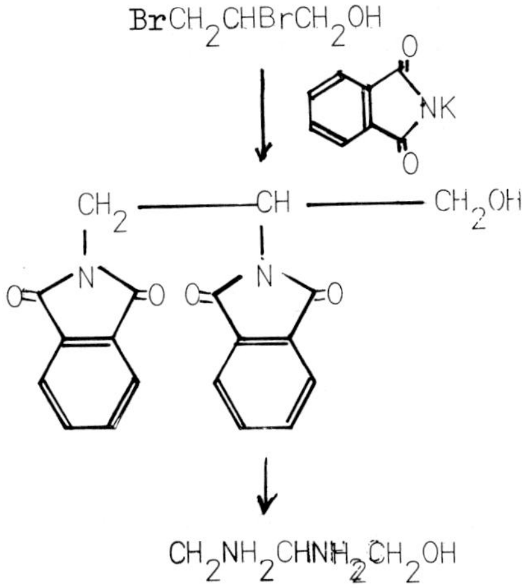

$$\text{Br}CH_2CHBrCH_2OH$$

$$CH_2NH_2CHNH_2CH_2OH$$

This method of synthesis may not, however, be unambiguous.

    b) From ethyl acetamidocyanoacetate, $MeCONH(CN)CHCOOEt$, by reduction with sodium borohydride followed by hydrogenation in the presence of a Raney nickel catalyst in acetic anhydride. This provides ethyl 2,3-bis-(acetamido)-propionate, m.p. 132-133°. This can be hydrolysed to give 2,3-diaminopropan-1-ol dihydrochloride, m.p. 162-163°. Selective hydrolysis is not possible. Carbobenzoxylation of this compound (M.Bergmann and L.Zervas, Ber., 1932, 65, 1192) provides N,N'-dicarbobenzoxy-2,3-diaminopropan-1-ol, m.p. 109-110°. By use of triphenylphosphine-carbon tetra-chloride this alcohol can be converted into N,N'-dicarbo-benzoxy-1,2-diamino-3-chloropropane, m.p. 120-121°; dihydrobromide, m.p. 210-212°.

1,3-Diamino-2-chloropropane, $NH_2CH_2CHClCH_2NH_2$, can be made from 1,3-dichloropropan-2-ol by fusion with potassium phthalimide to give 1,3-bis(phthalimido)-propan-2-ol which, with phosphorus pentachloride followed by hydrolysis, gives 1,3-diamino-2-chloropropane dihydrochloride, m.p. 216° (S.Gabriel and W.Michels, Ber., 1892, 25, 3056; S.Gabriel, *ibid.*, 1889, 22, 224).

2-Amino-3-chloropropan-1-ol, $ClCH_2CHNH_2CH_2OH$, can be prepared from methyl 2-phenyl-2-oxazoline-4-carboxylate according to the following reaction scheme (M.Bergmann and A.Mickely,

Z. physiol. Chem., 1949, 14, 887; R.Paul, R.P.Williams and E.Cohen, J. org. Chem., 1975, 40, 1653; E.M.Fry, *ibid.*, 1949, 14, 887).

1-Amino-3-chloro-propan-2-ol, $NH_2CH_2CHOHCH_2CI$, is obtained as a mixture of its two enantiomers by a number of methods, all of which give poor yields (S.Gabriel and H.Ohle, Ber., 1917, 50, 819; H.Roth, Arch. Pharm., 1959, 292, 76; E.Charbuliez *et al.*, Helv., 1960, 43, 1158). The reaction of 2,3-epoxy-1-chloropropane with concentrated ammonia and benzaldehyde gives (±)-1-benzalimino-3-chloropropan-2-ol which, on hydrolysis, gives (±)-1-amino-3-chloropropan-2-ol (hydrochloride, m.p. 101-102°) (H.E.Carter and P.K.Bhattacharyya, J. Amer. chem. Soc., 1953, 75, 2503). This can be resolved by using dibenzoyl tartaric acid or preferably (+)-10-camphorsulphonic acid when the pure (+)-1-amino-3-chloropropan-2-ol (+)-10-camphorsulphonate, $[\alpha]_D^{25}$ +5.2, m.p. 118-119° and (-)-1-amino-3-chloropropan-2-ol hydrochloride $[\alpha]_D^{25}$ -5.23 and m.p. 144-146° are obtained. Using (+)-1-amino-3-chloropropan-2-ol hydrochloride and acetone by the general method of J.H.Billing and A.C.Diesing (J. org. Chem., 1957. 22, 1068) gives (+)-1-chloro-3-isopropylamino-2-propanol hydrochloride with $[\alpha]_D^{25}$ +33.6 and m.p. 105-107°.

Sulphur analogues of glycerol and other triols.

Many simple thiol, dithiol and hydroxy-thiol compounds have attracted interest because some have shown significant anti-tubercular activity.

1,3-Dimercaptopropan-2-ol, $HSCH_2CHOHCH_2SH$, b.p. 52-54°/0.15mm., is made from 3-chloro-1,2-epoxypropane with sodium hydrogen sulphide solution (E.P.Adams *et al.*, J. chem. Soc., 1960, 2649). It reacts with acetone in the presence of strong acid to give 1,3-isopropylidenedithiopropan-2-ol (5-hydroxy-2,2-dimethyl-1,3-dithian) b.p. 117-118°/11mm. This product can be oxidised to the disulphone with hydrogen peroxide.

CH_2SH
|
CHOH   →(Me_2CO)   HOCH(CH_2S / CH_2S)CMe_2   →(H_2O_2)   HOCH(CH_2SO_2 / CH_2SO_2)CMe_2
|
CH_2SH

On treatment with hydrochloric acid followed by the reaction of the first formed product with sodium hydrogen carbonate 1,3-dimercaptopropan-2-ol gives 3-mercaptopropylene sulphide (b.p. 62-64°/22mm.) from which acyl and alkoxycarbonyl derivatives of 3-mercaptopropylene sulphide can be prepared by treatment with either carboxylic acid chlorides or chloroformic esters in dry ether at 0-4°C, or carboxylic acids in toluene containing triethylamine.

CH_2SH
|
CHOH   →(HCl)   CH_2SH | CH⟍SH | CH_2   →(HCl)   CH_2SH | CHSH | CH_2Cl   →(aq. NaHCO_3)   CH_2SH | CH⟍S | CH_2

        →(RCOCl)   CH_2SH | CHCl | CH_2SCOR   →(aq. NaHCO_3)   CH_2⟍S | CH | CH_2SCOR

1,3-Dimercapto-2-methyl-propan-2-ol, $HSCH_2C(Me)OHCH_2SH$,
b.p. $54°/0.13mm$., can be prepared from 3-bromo-2-methyl
propylene oxide and sodium hydrogen sulphide solution, or
by reduction of 1,3-dibenzylthiopropan-2-ol with sodium and
ethanol in liquid ammonia (E.P.Adams *et al.*,*loc. cit.*).
1,3-Dibenzylthiopropan-2-ol is formed by the reaction of
2,3-dibromo-2-methyl-propan-1-ol with the sodium derivative
of benzyl thiol *via* an epoxide intermediate.

With benzaldehyde 1,3-dimercapto-2-methyl-propan-2-ol gives
1,3-benzylidenedithio-2-methyl-propan-2-ol (5-hydroxy-5-
methyl-2-phenyl-1,3-dithian), m.p. $145-146°$. It may be
converted into 1-mercaptomethyl-1-methylethylene sulphide.

3,4-Dimercaptobutan-2-ol, $HSCH_2CHSHCHOHCH_3$, is a viscous oil,
b.p. $65-66°/0.1mm$., prepared by the reaction of 3,4-dibromo-
butan-2-ol with methanolic sodium hydroxide saturated with
hydrogen sulphide at $0°C$ (E.P.Adams *et al.*, *loc. cit.*). The
reaction of 3,4-dibromobutan-2-ol with the sodium derivative

of benzyl thiol gives a mixture of 1,3-and 3,4-
dibenzylthiobutan-2-ol which on reduction gives a mixture
of 1,3- and 3,4-dimercaptobutan-2-ol.

1,4-*Dimercaptobutan-2-ol*, $HSCH_2CH_2CHOHCH_2SH$, b.p. 68-69°/0.02
mm., formed by the reduction of 1,4-dibenzylthiobutan-2-ol
obtained from 1,4-dibromobutan-2-ol and benzyl thiol. In
concentrated hydrochloric acid 1,4-dimercaptobutan-2-ol gives
3-mercaptothiophan (E.P.Adams *et al.*, *loc. cit.*).

2,3-Dimercaptobutan-1-ol, $CH_3CHSHCHSHCH_2OH$, b.p. 69-76°/0.01mm. is prepared from ethyl 2,3-di(acetylthio)butyrate by reduction with lithium aluminium hydride (E.P.Adams *et al.*, *loc. cit.*).

$$
\begin{array}{c}
CH_3 \\
| \\
CHSCOMe \\
| \\
CHSCOMe \\
| \\
COOEt
\end{array}
\quad \xrightarrow{\text{LiAlH}_4} \quad
\begin{array}{c}
CH_3 \\
| \\
CHSH \\
| \\
CHSH \\
| \\
CH_2OH
\end{array}
$$

2,4-Dimercaptobutan-1-ol, $HSCH_2CHSHCH_2CH_2OH$, b.p. 75-79°/0.05 mm., can be similarly made from methyl 2,4-di(acetylthio) butyrate (E.P.Adams *et al.*, *loc. cit.*).

$$
\begin{array}{c}
CH_2SCOMe \\
| \\
CH_2 \\
| \\
CHSCOMe \\
| \\
COOMe
\end{array}
\quad \xrightarrow{\text{LiAlH}_4} \quad
\begin{array}{c}
CH_2SH \\
| \\
CH_2 \\
| \\
CHSH \\
| \\
CH_2OH
\end{array}
$$

In concentrated hydrochloric acid it is converted into 3-mercaptothiophan which reacts with acetic anhydride to provide 3-acetylthiotetrahydrothiophen (E.P.Adams *et al.*, *loc. cit.*).

$$
\begin{array}{c}
CH_2SH \\
| \\
CH_2 \\
| \\
CHSH \\
| \\
CH_2OH
\end{array}
\quad \xrightarrow{\text{HCl}} \quad
\begin{array}{c}
CH_2\!-\!CHOH \\
| \qquad\quad | \\
CH_2 \quad CH_2 \\
\diagdown_S\diagup
\end{array}
\quad \xrightarrow{\text{Ac}_2O} \quad
\begin{array}{c}
CH_2\!-\!CHSAc \\
| \qquad\quad | \\
CH_2 \quad CH_2 \\
\diagdown_S\diagup
\end{array}
$$

2,5-Dimercaptopentan-1-ol, $HSCH_2CH_2CH_2CHSHCH_2OH$, b.p. 93-95°/
12mm., is formed by the reduction of ethyl 2,5-di(acetylthio)-
pentanoate with lithium aluminium hydride (E.P.Adams *et al.,*
loc. cit.).

$$
\begin{array}{ccc}
CH_2SCOMe & & CH_2SH \\
| & & | \\
CH_2 & & CH_2 \\
| & \xrightarrow{\text{LiAlH}_4} & | \\
CH_2 & & CH_2 \\
| & & | \\
CHSCOMe & & CHSH \\
| & & | \\
COOEt & & CH_2OH
\end{array}
$$

With concentrated hydrochloric acid it reacts to give
2-mercaptomethylthiophan.
1,6-Dimercaptohexan-2-ol, $HSCH_2CH_2CH_2CH_2CHOHCH_2SH$, b.p.
96-100°/0.005mm., is made by the reduction with sodium in
liquid ammonia of 1,6-dibenzylthiohexan-2-ol, prepared from
1,6-dibromohexan-2-ol and the sodium derivative of toluene
-ω-thiol (E.P.Adams *et al., loc. cit.*).

$$
\begin{array}{ccccc}
CH_2Br & & CH_2SCH_2Ph & & CH_2SH \\
| & & | & & | \\
CH_2 & & CH_2 & & CH_2 \\
| & \xrightarrow{\text{PhCH}_2\text{SNa}} & | & \xrightarrow{\text{Na/NH}_3} & | \\
CH_2 & & CH_2 & & CH_2 \\
| & & | & & | \\
CH_2 & & CH_2 & & CH_2 \\
| & & | & & | \\
CHOH & & CHOH & & CHOH \\
| & & | & & | \\
CH_2Br & & CH_2SCH_2Ph & & CH_2SH
\end{array}
$$

No products can be obtained by its reaction with concentrated
hydrochloric acid.
    2,3-Dimercaptopropan-1-ol reacts with 3-mercaptopropylene
sulphide in sodium hydroxide solution to provide 2,3-
dimercaptopropyl-3-hydroxy-2-mercaptopropyl sulphide
(E.P.Adams *et al., loc. cit.*).

$$\begin{array}{ccc}
\text{CH}_2\text{SH} & & \text{CH}_2\text{SH} \\
\text{CHSH} & + & \text{CH} \!\diagdown\!\! {}_S \\
\text{CH}_2\text{OH} & & \text{CH}_2\diagup
\end{array}
\longrightarrow
\begin{array}{cc}
\text{CH}_2\!-\!\text{S}\!-\!\text{CH}_2 & \\
\text{CHSH} & \text{CHSH} \\
\text{CH}_2\text{OH} & \text{CH}_2\text{SH}
\end{array}$$

Glyceryl nitrates.

Glycerol-1-nitrate is obtained by the oxidation of allyl nitrate with potassium permanganate (L.T.Eremenko and A.M. Korolev, Izv. Akad. Nauk. SSSR Ser. Khim., 1970, 1, 147). It can also be made by the nitration of 2,2-dialkyl-4-(hydroxymethyl)-1,3-dioxolane (R.Nec, Czech. Pat., 129,192; R.Nec and F.Krampera, Chem. Prum., 1968,18, 614).

$$\underset{\underset{\text{Me}\quad\text{Me}}{}}{\text{dioxolane-CH}_2\text{OH}} \xrightarrow[\text{0°C}]{\text{Ac}_2\text{O/HNO}_3} \underset{\underset{\text{Me}\quad\text{Me}}{}}{\text{dioxolane-CH}_2\text{ONO}_2} \cdot \xrightarrow{\text{HCl/EtOH}} \begin{array}{c}\text{CH}_2\text{OH}\\\text{CHOH}\\\text{CH}_2\text{ONO}_2\end{array}$$

cis-But-2-ene-1,4-diol dinitrate undergoes oxidation with potassium permanganate in acetone to meso-erythritol-1,4-dinitrate, whilst the trans isomer is converted into (±)-erythritol-1,4-dinitrate.

Glycerol-1- and 2-nitrate and glycerol-1,2- and 1,3-dinitrate can be differentiated by TLC of the hydrolysates of carbon-14 compounds on Silica Gel G plates using the following solvents:

n-Butanol/ammonia/water	4:1:3
Benzene/ethyl acetate/acetic acid	16:4:1
Ethyl acetate/n-heptane	9:1

(M.C.Crew and F.J.Dicarlo, J. Chromatogr., 1968, 35,506). Using the ethyl acetate/n-heptane mixture the following R_f values are obtained:

Glycerol-1,2-dinitrate   0.84
Glycerol-1,3-dinitrate   0.94
Glycerol-1-nitrate   0.37
Glycerol-2-nitrate   0.45

By use of this analytical method it can be shown that glycerol trinitrate on hydrolysis in 4M hydrochloric acid loses each nitro group at comparable rates.

A number of homologous glyceryl nitrates can be synthesised from the appropriate alkane triol by the following scheme (J.Legocki and J.Hackel, Przeim. Chem., 1967, $\underline{46}$, 214).

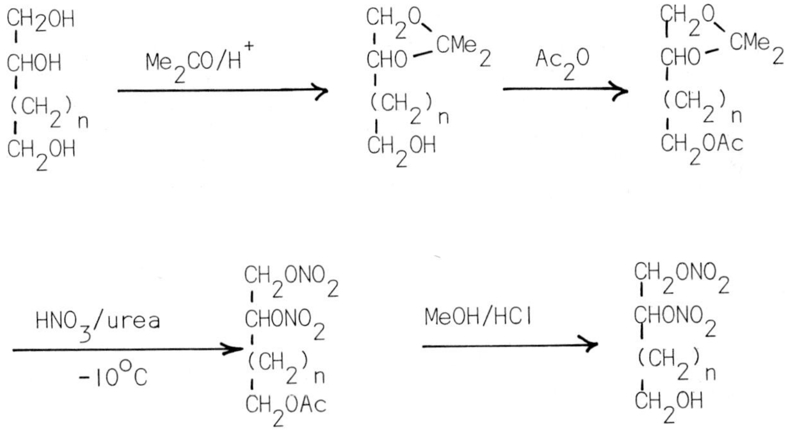

$$
\begin{array}{l}
CH_2OH \\
CHOH \\
(CH_2)_n \\
CH_2OH
\end{array}
\quad \xrightarrow{Me_2CO/H^+} \quad
\begin{array}{l}
CH_2O \\
CHO \diagdown CMe_2 \\
(CH_2)_n \\
CH_2OH
\end{array}
\quad \xrightarrow{Ac_2O} \quad
\begin{array}{l}
CH_2O \\
CHO \diagdown CMe_2 \\
(CH_2)_n \\
CH_2OAc
\end{array}
$$

$$
\xrightarrow[-10^\circ C]{HNO_3/urea}
\begin{array}{l}
CH_2ONO_2 \\
CHONO_2 \\
(CH_2)_n \\
CH_2OAc
\end{array}
\quad \xrightarrow{MeOH/HCl} \quad
\begin{array}{l}
CH_2ONO_2 \\
CHONO_2 \\
(CH_2)_n \\
CH_2OH
\end{array}
$$

where   n = 0, 1, 3.

Allyl compounds with the general formula, $RCH_2CH=CH_2$, where R is AcO, BzO, EtO, Ph, Cl or CN, on treatment with dinitrogen tetroxide in carbon tetrachloride in the presence of oxygen at $0^\circ C$ give the peroxynitrates

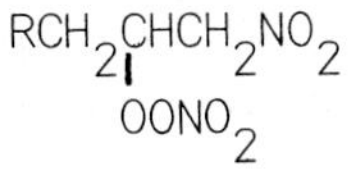

$$
\underset{OONO_2}{RCH_2CHCH_2NO_2}
$$

which are easily converted into the I-nitro-2-nitratopropanes

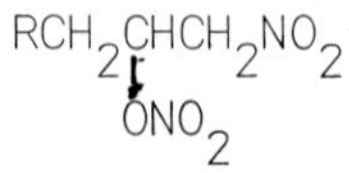

$$RCH_2CHCH_2NO_2$$
$$|$$
$$ONO_2$$

(W.M.Cumming, U.S.Pat., 3,910,987 and Ger.Pat., 2,157,648).
I,3-Dihydroxy-2-nitropropanes react quantitatively with
substituted benzaldehydes to provide the corresponding
nitrodioxanes (B.A.Arbuzov *et al.*, Izv. Akad. Nauk. SSSR
Ser. Khim., 1975, 12, 2746)

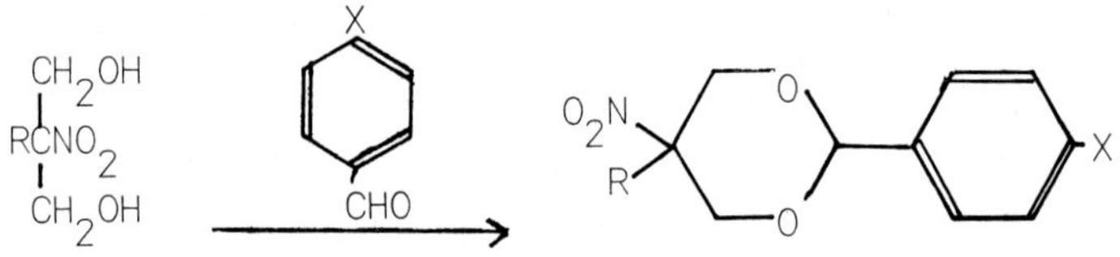

(R = H or Br;   X = H or Cl).
A number of I,5-dioxaspiro [5,5] -undecanes:

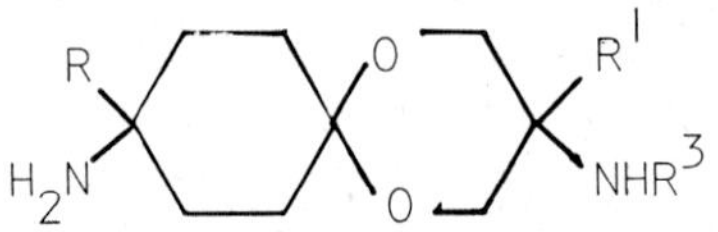

where R = H, Me;   R' = H, Me, Et; R^3 = H, Ac, 4-MeOC$_6$H$_4$CO
(G.Y.Lesher *et al.*, U.S.Pat., 3,901,920) can be made
from:

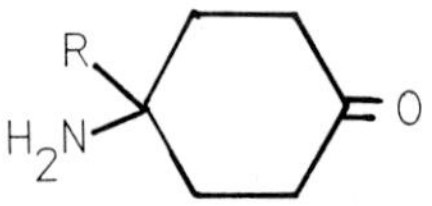

by ketalisation with I,3-dihydroxy-2-nitropropane, followed
by reduction of the nitro group to an amino group which can
then be acylated.

Cyclic Alkylidene Ethers, Acetals.

It has been long known that a series of compounds can be
obtained by condensing glycerol with aldehydes and ketones

(M.Schulz and B.Tollens, Ann., 1893, 289, 29). The simple compounds derived by condensing glycerol with either benzaldehyde or acetone have been particularly useful in the synthesis of glycerides. Interest has been increased in these compounds because of the reactions between carbohydrates and carbohydrate derivatives and aldehydes and ketones and their uses as drugs, pharmaceuticals, pesticides and other commercial products.

When glycerol is condensed with an aldehyde the compounds obtained have the following structures:

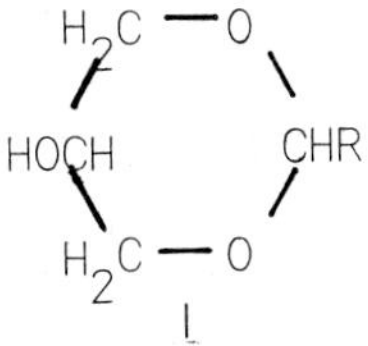

These can be referred to as 1,3-alkylidene or 1,3-arylidene glycerol for structure (1) but should be more correctly referred to as either 2-alkyl(or aryl)-5-m-dioxanol or 2-alkyl(or aryl)-5-hydroxy-m-dioxan, although the term m-dioxan is sometimes replaced by 1,3-dioxan or 1,3-dioxane. The structure (2) is often named as 2-alkyl(or aryl)-1,3-dioxolane-4-methanol although it is sometimes referred to as 2-alkyl(or aryl)-1,3-dioxolane-4-carbinol or 4-hydroxymethyl-1,3-dioxolane. With ketones the 2,2-disubstituted products a are obtained i.e. 2,2-dialkyl(or diaryl)-1,3-dioxolane-4-methanol and 2,2-dialkyl(or diaryl)-1,3-dioxan-5-ol.

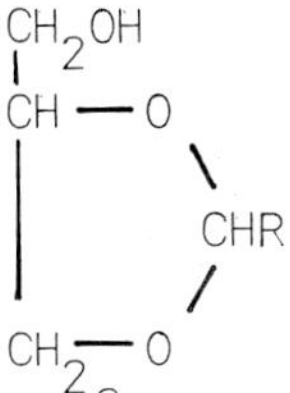

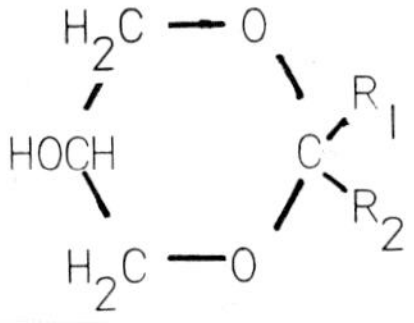

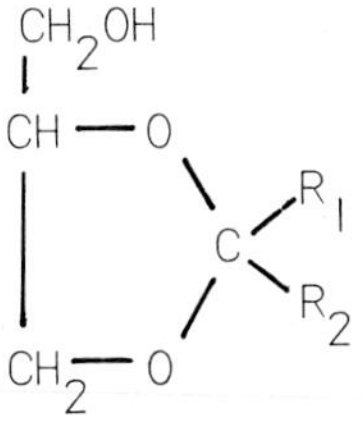

Condensation of glycerol with aldehydes.

1. The reaction between formaldehyde and glycerol provides

the corresponding 5-hydroxy-1,3-dioxan and 1,3-dioxolane-4-methanol (M.Trister and H.Hibbert, Canad. J. Res., 1937,14B, 415).

Aldehydes provide both 2-substituted-1,3-dioxans and 2-substituted-1,3-dioxolane-4-methanols. The formation of the dioxolane derivative is favoured by a low temperature of reaction, whilst the formation of the 1,3-dioxan derivative is favoured by the presence of electron-withdrawing groups in the aldehyde, *e.g.*, dibromomethyl by using dibromoacetaldehyde. The use of acid catalysts for the condensation prevents the formation of the dioxan derivative (H.Hibbert and J.Moorazain, Canad. J. Res., 1930, 2, 35,214).

2.   By use of a trans-acetalisation reaction only the 2-substituted-1,3-dioxolane-4-methanol derivatives are formed by the condensation of aliphatic acetals and glycerol in the presence of sulphosalicylic acid (W.R.Miller, E.H. Pryde and J.C.Cowan, J. polymer Sci., 1965, B3, 131). This reaction proceeds *via* a mixed acetal linked to C-1 of glycerol (C.Piantadosi *et al.*, J. org. Chem., 1963, 28, 242).

Pentane-1,2,5-triol reacts with paraformaldehyde in benzene in the presence of toluene-p-sulphonic acid and provides 4-(3-hydroxypropyl)-1,3-dioxolane which can be acetylated to 4-(3-acetoxypropyl)-1,3-dioxolane (A.B.Perstorp,

Brit. Pat., 1,272,509). Similarly, ethyl glycerate gives
rise to the formation of 4-(ethoxycarbonyl)-1,3-dioxolane.

$$
\begin{array}{c}
CH_2OH \\
| \\
CHOH \\
| \\
(CH_2)_2 \\
| \\
CH_2OH
\end{array}
\xrightarrow{HCHO/H^+}
$$

$$
\begin{array}{c}
(CH_2)_2CH_2OH \\
| \\
CH-O \\
\diagdown \\
CH_2 \\
\diagup \\
CH_2-O
\end{array}
\longrightarrow
\begin{array}{c}
(CH_2)_2CH_2OCOCH_3 \\
| \\
CH-O \\
\diagdown \\
CH_2 \\
\diagup \\
CH_2-O
\end{array}
$$

$$
\begin{array}{c}
CH_2OH \\
| \\
CHOH \\
| \\
COOEt
\end{array}
\xrightarrow{HCHO/H^+}
\begin{array}{c}
COOEt \\
| \\
CH-O \\
\diagdown \\
CH_2 \\
\diagup \\
CH_2-O
\end{array}
$$

3. When glycerol is heated with esters of *ortho*-acids a
mixture of the 1,3-dioxan and dioxolane derivatives can be
obtained. Thus, ethyl orthoformate reacts with glycerol to
give a mixture of the *cis*- and the *trans*-2-ethoxy-1,3-
dioxolane-4-methanol and 2-ethoxy-1,3-dioxan-5-ol
(G.Crank and F.W.Eastwood, Austral. J. Chem., 1964, 17, 1385).

$$
\begin{array}{c}
CH_2OH \\
| \\
CHOH \\
| \\
CH_2OH
\end{array}
\xrightarrow{HC(OEt)_3}
\begin{array}{c}
H_2C-O \\
\diagup \quad \diagdown \\
HOCH \qquad CHOEt \\
\diagdown \quad \diagup \\
H_2C-O
\end{array}
\qquad
\begin{array}{c}
CH_2OH \\
| \\
CH-O \\
\diagdown \\
CHOEt \\
\diagup \\
CH_2-O
\end{array}
$$

4. Dioxolanes can be prepared by the isomerisation of
unsaturated glycerol ethers using acetic acid (C.Piantadosi
et al., J. pharm. Sci., 1964, 53, 1024).

$$CH_2OH$$
$$CHOH$$
$$CH_2OCH=CHR$$
$$\longrightarrow$$

$$
\begin{array}{l}
CH_2OH \\
CH\!-\!O \\
\qquad\qquad CHCH_2R \\
CH_2\!-\!O
\end{array}
$$

(R = hexyl, octyl, or decyl).

5.   Both 2-methyl-1,3-dioxolane-4-methanol and 2-methyl-1,3-dioxan-5-ol can be produced by the reaction between glycerol and ethyne in the presence of mercury(II) sulphate and sulphuric acid (S.Fuzesi and J.V.Karabinos, Belg. Pat., 635,467). If this reaction is carried out in a solvent without the presence of a catalyst, only 2-methyl-1,3-dioxolane-4-methanol and 2-methyl-4-vinyloxy-1,3-dioxolane are obtained (J.J.Nedwick, Ind. eng. Chem., Process Design Develop., 1962, 1, 137).

Generally in these methods either the *cis*-dioxan or an equilibrium mixture of the *cis*- and *trans*-dioxan is obtained. The *trans*-2-alkyl-1,3-dioxan-5-ols can be prepared by the reaction of the *cis*-2-alkyl-1,3-dioxan-5-methanesulphonates with sodium benzoate in dimethylformamide solution when benzoate substitution occurs with inversion. Hydrolysis of the resulting *trans*-benzoates leads to the formation of the *trans*-dioxan derivative (N.Baggett *et al.*, J. chem. Soc., 1963, 4157)

If the preparation of one of the two possible structural isomers is particularly required, then a blocked glycerol is usually used. Thus, 1-chloro-2,3-dihydroxypropane condenses with an aldehyde in the presence of concentrated hydrochloric acid to give, after hydrolysis, only 2-alkyl-1,3-dioxolane-4 -methanol (G.S.Alieva, L.G.Truzhernikova and I.N.Belova, Zh. Obsch. Khim., 1962, 32, 3634). By blocking either the 1- or the 2-position of glycerol either by esterification (M.J.Egerton and T.Malkin, J. chem. Soc., 1953, 2800; R.J.Fischer and C.W.Smith, J. org. Chem., 1960, 25, 319), or by etherification (H.Hibbert and N.M.Carter, J. Amer. chem. Soc., 1928, 50, 3120), then 1,3-dioxolanes or 1,3-dioxans are effectively obtained.

When the preparation gives rise to both the 2-substituted
-1,3-dioxolane-4-methanol and the corresponding 1,3-dioxan-
5-ol their separation can be achieved by one of the following
methods.

    1. Fractional crystallisation (H.Hibbert and N.M.Carter,
J. Amer. chem. Soc., 1928, 50, 3120).
    2. By passing dry hydrogen chloride into the reaction
mixture the 1,3-dioxan-5-ol generally separates due to the
formation of its less soluble addition product (P.E.Verkade
and J.D. van Roon, Rec. trav. Chim., 1942, 61, 831).
    3. By tlc on silica gel (S.A.Kore, E.I.Shepelenkova and
E.M.Chernova, Maslob.-Zhir. Prom., 1962, 28, 32).

    The temperature of the reaction and the nature of the
substituents can influence the size of the ring formed when
aldehydes are condensed with glycerol. For each of these
structural isomers stereoisomers are possible. The products
obtainable by condensing glycerol with aldehydes are
illustrated below.

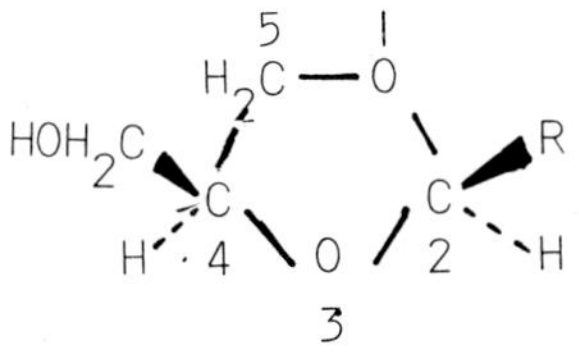

cis-2-alkyl-1,3-dioxolane-
4-methanol

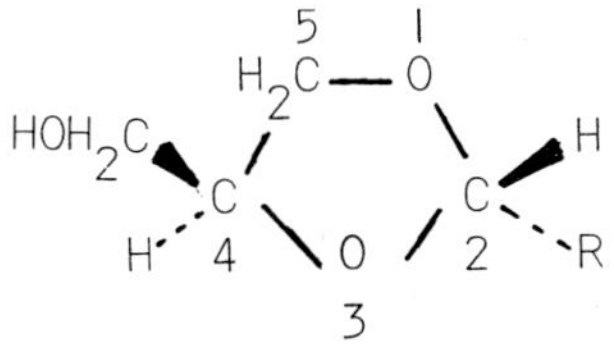

trans-2-alkyl-1,3-dioxolane-
4-methanol

In the dioxolane-4-methanols both the C-2 and C-4 centres are
chiral and thus there are four stereoisomers possible i.e.
cis-(+) and *cis*-(-) and *trans*-(+) and *trans*-(-) isomers.

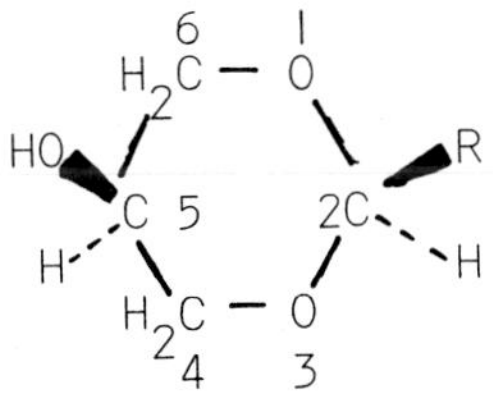

cis-2-alkyl-1,3-dioxan-5-ol

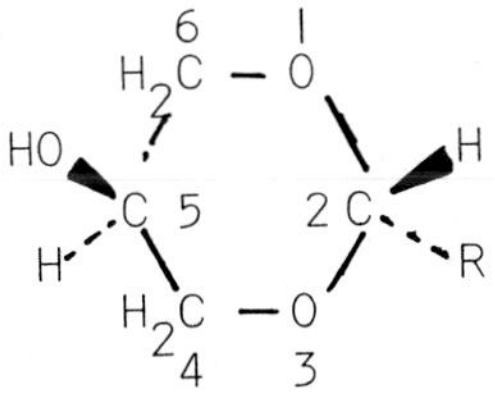

trans-2-alkyl-1,3-dioxan-5-ol

The *cis* isomer of this compound consists of an equilibrium mixture of the two conformations.

The *trans*-2-alkyl-1,3-dioxan-5-ol can best be represented as being in conformational equilibrium where most of the hydroxyl groups are non-bonded, but for the unsubstituted 1,3-dioxan-5-ol the reverse applies.

With the 1,3-dioxans since there is a plane of symmetry in both the *cis* and the *trans* isomer these are not optically active.

In the 6-membered ring dioxans the rings are non-planar. Thus, if no other factors are involved, the 1,4 - *trans* isomer would be expected to be more stable than the 1,4-*cis*-isomer since both substituents will be equatorial.

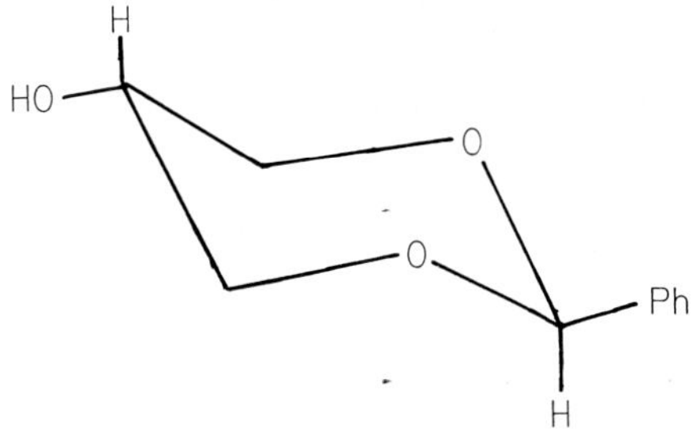

The reaction of benzaldehyde with glycerol at 140°C gives the more stable isomer, m.p. 82.5-83.5°. This was thought to be the *trans*-isomer (W.Gerhardt, Chem. Zentr., 1912,83, 1953). This compound, on acetylation and then treatment with sodium methoxide to bring about inversion, provides the less stable isomer, thought to be the *cis*-isomer (P.E.Verkade and J.D. van Roon, Rec. trav. Chim., 1942, 61, 801). However, the infra-red absorption spectrum of the stable isomer shows that the hydroxyl group is intramolecularly bonded and this can only occur if the hydroxyl group is axial. This cannot occur if the hydroxyl group is equatorial. Thus, if the phenyl group is equatorial , the more stable isomer, m.p. 82.5-83.5°, is the *cis*-isomer with the conformation shown below (N.Baggett *et al.*, J. chem. Soc., 1960, 2574).

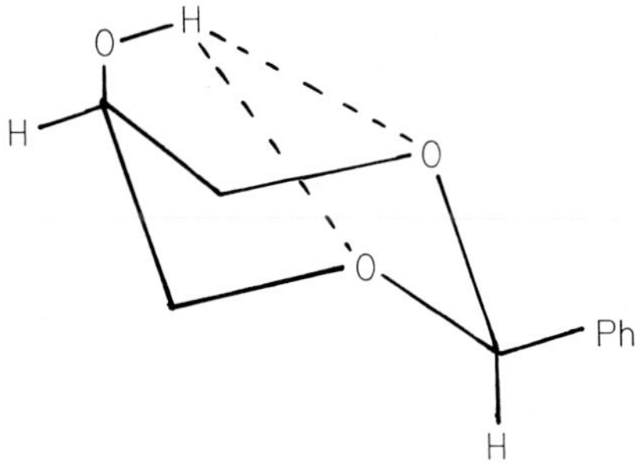

If the dioxan ring has the boat conformation an intramolecularly hydrogen bonded equatorial hydroxyl group is
possible. Such boat conformations always appear to be less
stable than the alternative chair conformations and do not
occur.

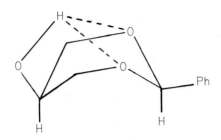

The higher m.p. 82.5-83.5° 2-phenyl-1,3-dioxan-5-ol
(the *trans* isomer) together with the other *trans*-2-alkyl-1,3-
dioxan-5-ols can be prepared from the *cis*-methanesulphonates by
inversion with sodium benzoate in dimethylformamide and
subsequent hydrolysis. These all exhibited infra-red absorption
at 3633-3634nm. and 3601-3604nm. indicating the presence of
free and intramolecularly bonded hydroxyl groups. The *cis*
isomers show no absorptions due to a free hydroxyl group
(N.Baggett and A.B.Foster, J. chem. Soc., 1963, 4157). The
alkyl group has little effect on the proportions of the two
conformational isomers. It has been suggested that either the
smaller alkyl groups do not exert the maximum steric effect
or that a third conformation is possible where the 2-
substituent exerts little steric effect (*ibid.*, *loc. cit.*).
When the substituents are in the 5-position the molecule
becomes less flexible, whilst substituents in the 2-position
make the molecule more flexible (H.Friebolin *et al.*,
Tetrahedron letters, 1962, 683) supporting the possibility
that certain molecules can take up the above conformations.

If the 5-hydroxyl group is replaced, hydrogen bonding is not
then possible. The NMR spectrum of *trans*-5-acetoxy-2-phenyl-
-1,3-dioxan indicates that the preferred conformation is the
chair form with both bulky groups equatorial (N.Baggett *et al.*,
Chem. Ind. (London), 1961, 106).

Pure *trans*-2-phenyl-1,3-dioxan-5-ol can be prepared by the lithium aluminium hydride reduction of 5-oxo-2-phenyl-1,3-dioxan (B.Dobinson and A.B.Foster, J. chem. Soc., 1961, 2338). *cis*- and *trans*-1,3-Dioxolanes have been separated as their acetates by a tedious multiple gas chromatographic fractionation (W.J.Baumann, J. org. Chem., 1971, 36, 2743).

The Raman and infra-red spectra of substituted 1,3-dioxolanes show that the 1,3-dioxolane ring is almost planar. The puckering of the ring is so small that this is of little steric consequence (S.A.Baker *et al.*, J. chem. Soc., 1959, 807). Thus, substituents occupy similar conformational positions irrespective of their position in the ring. Thus other steric factors influence the relative stability of the *cis* and the *trans* 2-alkyl-1,3-dioxolane-4-methanols.

When chloral is condensed with glycerol both the *cis* and the *trans* 2-trichloromethyl-1,3-dioxolane-4-methanol can be isolated as their toluene-p-sulphonyl derivatives (D.J.Triggle and B.Belleau, Canad. J. Chem., 1962, 40, 1201).

Condensation of glycerol with ketones.

When a ketone condenses with glycerol two possible products may result; either a 2,2-disubstituted-1,3-dioxolane-4-methanol (3).

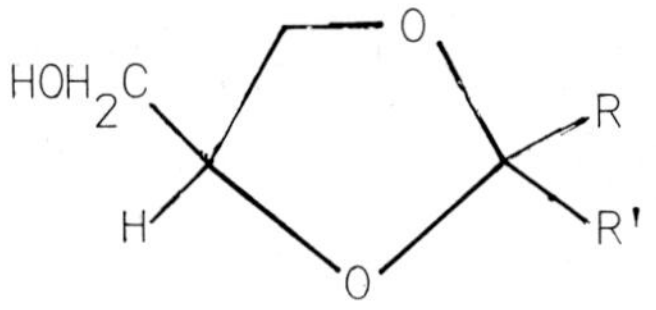

(3)

or a 2,2-disubstituted-1,3-dioxan-5-ol (4,5,6)

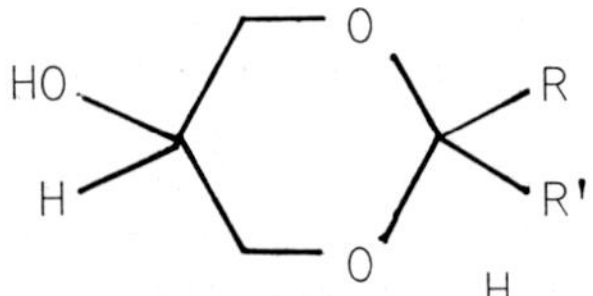

(4)

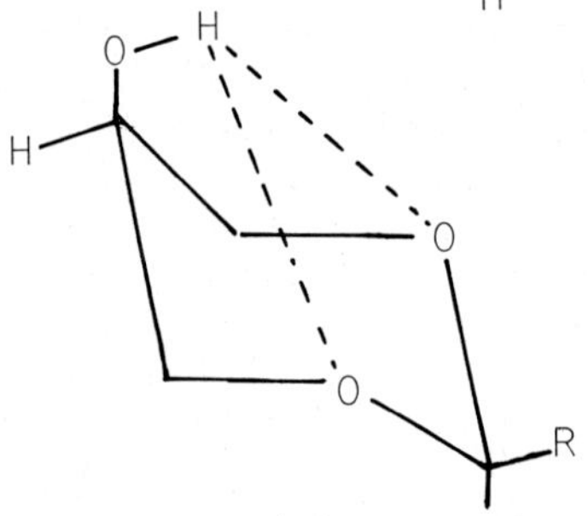

(5)   R'

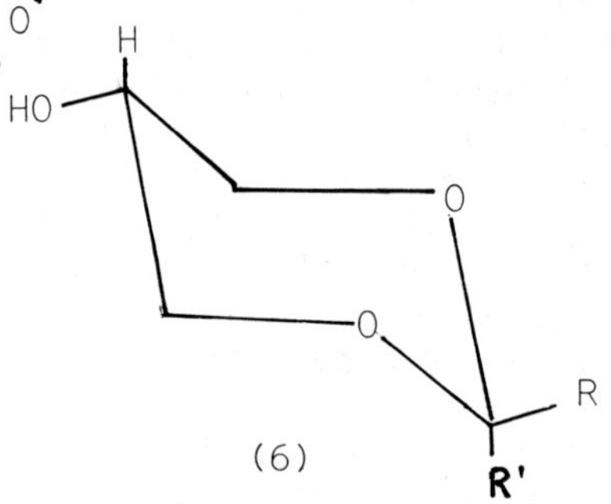

(6)

R'

In structure (5) above the hydroxyl group is axial and hydrogen bonded, whilst in the other form (6) it is stabilised by being in an equatorial position. However, in both (5) and (6) one of the substituents in the 2-position is always axial.

Thus, the dioxan structure is sterically unfavoured relative
to the dioxolane structure in which all groups occupy similar
conformational positions. Therefore, when ketones condense
with glycerol only the 2,2-disubstituted-1,3-dioxolane-4-
methanols are obtained.

In general, 2,2-disubstituted-1,3-dioxolane-4-methanols
can be synthesised as follows.

1. By the condensation of glycerol with dialkyl ketones
(J.C.Irvine, J.L.A.McDonald and C.W.Soutar, J. chem. Soc.,
1915, 107,337 ; D.R.McCoy, U.S. Pat., 3,909,460; J.Legocki
and J.Hackel, Przeim. Chem., 1967, 46, 214).

2. By the reaction of glycerol with ethyl ketal (C.
Piantadosi, C.E.Anderson, E.A.Brecht and C.L.Yarbro. J. Amer.
chem. Soc., 1958, 80, 6613; P.Calinaud and P.Gelas, Bull.
Soc. Chim. Fr., 1975, (5-6,pt2), 1228).

3. By condensation of glycidol with dialkyl ketones
(G.Crank and F.W.Eastwood, Austral. J. Chem., 1964, 17, 1385).

These methods are of general application and can be
achieved by the following techniques:

1. Azeotropic distillation (A.Dupire, Compt. rend., 1942
214, 359).

2. Use of catalysts, e.g. hydrogen chloride (G.S.Alieva,
L.G.Truzhernikova and I.N.Belova, Zh. Obsch. Khim., 1962, 32,
3634; T.Malkin and M.R. el Shurbagy, J. chem. Soc., 1936, 1628);
toluene-p-sulphonic acid (Consortium fur Elektrochem. Ind.
G.m.b.H., Brit. Pat., 948,084; A.Vystrcil and J.Vaiek, J.
Chem. Listy., 1956, 44, 204); phosphorus pentoxide (F.Fischer
and M.Lotzsch, J. prakt. Chem., 1962, 18, 86); iodine
(J.L.Harvey, U.S.Pat., 2,690,444); calcium carbide and
sodium alkylsulphonates (M.M.Maglio and C.A.Burger, J. Amer.
chem. Soc., 1946, 68, 529); sulphonated polystyrene resin
KU I (B.G.Yasnitskii, S.A.Sarkisyants and I.G.Ivanyuk, Zh.
Obsch. Khim., 1964, 34, 1940); boron trifluoride/ether
(F.G.Ponomarev, L.G.Esipova, O.G.Lamteva, M.F.Mizilina and
B. Sh.Farberova, Tr. Voronezsk. Gos. Univ., 1958,19, 9).

Since these 2,2-disubstituted-1,3-dioxolane-4-methanols
can also exist as optical isomers, they can be prepared
according to the following scheme (E.C.Baer, Biochem. Prep.,
1952, 2, 31; E.C.Baer and H.O.L.Fischer, J. Amer.chem. Soc.,
1945, 67, 338, 994; 1936, 61, 761).

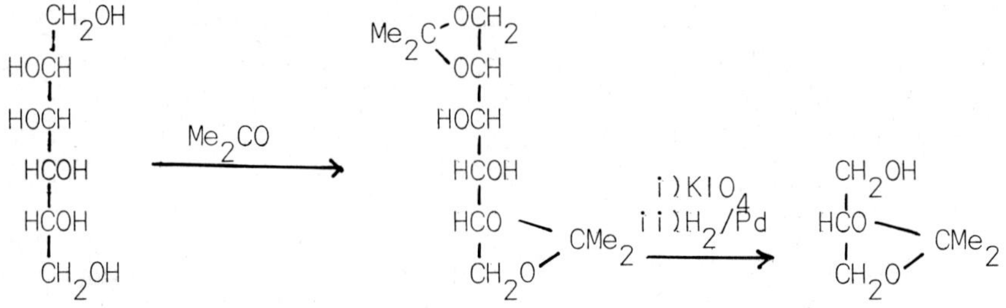

D-mannitol

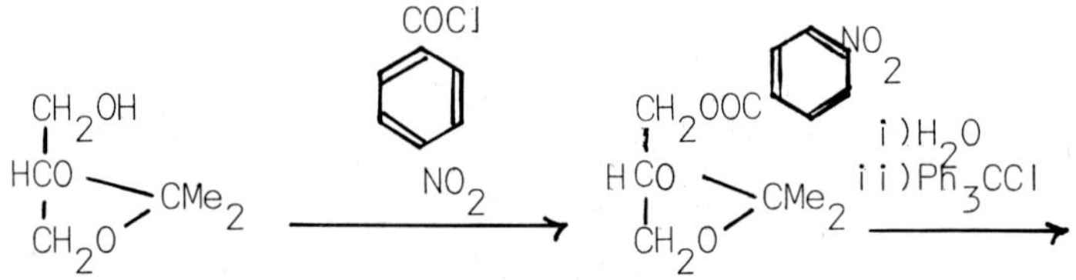

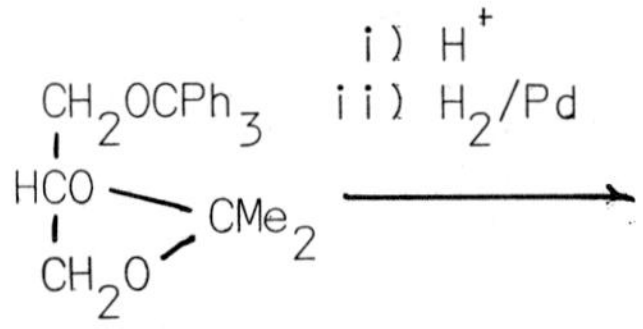

L-(+)-2,2-dimethyl-1,3-dioxolane-4-methanol

($\pm$) Erythro-1,2,3-butanetriol can be condensed with acetone to give the products shown below (P.Calinaud and J.Gelas, Compt. rend., 1973, 276, 1139).

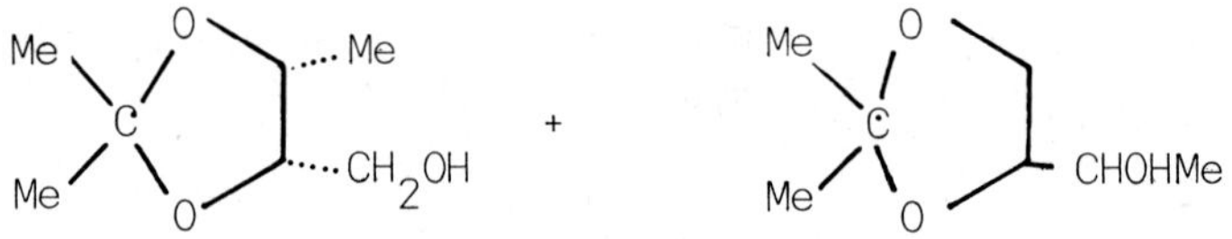

When only the 2,2-disubstituted-1,3-dioxan-5-ols are
required they can be made by the reaction of a ketone with
glycerol blocked in the 2-position. Thus, 2-glyceryl benzoate
reacts with acetone, methyl ethyl ketone, acetophenone or
cyclohexanone to give the corresponding 2,2-disubstituted
-5-benzoyloxy-1,3-dioxan (M.Bergmann and N.M.Carter,
Z. physiol. Chem., 1930, 191, 211).

A mixture of the dioxan and dioxolane is obtained by the
reaction of 1,1-diethoxyethane with glycerol in the presence
of sulphosalicylic acid. This mixture, on pyrolysis with
sodium hydrogen phosphate, gives a mixture of 2-methyl-5-
vinyloxy-1,3-dioxan and 2-methyl-4-vinyloxymethyl-1,3-
dioxolane (G.M.Nakaguchi, U.S.Pat., 3,714,202).

Reactions of 1,3-dioxans and 1,3-dioxolanes.

1. Hydrolysis with dilute acid liberates glycerol and the
corresponding aldehyde or ketone.

$$CH_2OH$$
$$CHO \diagdown_{CMe_2}$$
$$CH_2O \diagup$$

$$\xrightarrow{\text{dil. HCl}}$$

$$CH_2OH$$
$$CHOH \quad + \quad MeCOEt$$
$$CH_2OH$$

The estimation of the amount of aldehyde or ketone liberated can be used as a means of estimating the parent compound (F.Melson and H.Hofmann, Pharm. Zentralhalle, 1963, 102,59). In their chromatographic separations these compounds can be hydrolysed on the plate and the liberated aldehyde or ketone detected with a suitable reagent (J.F.G.Barnett and P.W.Kent, Nature, 1961, 192, 556).

2.   In catalytic hydrogenation the 2-substituent will be removed as a hydrocarbon from these compounds (M.Bergmann and N.M.Carter, Z. physiol. Chem., 1930, 181, 211).

$$\xrightarrow{H_2/Pd}$$

$$CH_2OH$$
$$CHOH \quad + \quad PhMe$$
$$CH_2OH$$

In general, 1,3-dioxans are resistant to attack by metal hydrides, but if an electron accepting is present on C-2, they readily undergo reduction with lithium aluminium hydride in the presence of aluminium chloride or boron trifuoride (B.E.Leggetter and R.K.Brown, Canad. J. Chem., 1964, 42, 990). With 1,3-dioxolanes electron-donor groups on C-2 accelerate the rate of reduction, but electron-acceptor groups slow the rate of reduction. Similar effects are seen with substituents at positions C-4 or C-5. In addition, donor groups at position C-4 favour cleavage of the C-2-O-bond remote from C-4; while acceptor groups tend to cause cleavage at the other C-2-O-bond, i.e. C-2-O-3 so that different proportions of primary and secondary alcohols can be obtained depending on the 4-position substituent.

$$\text{(2,2-dimethyl-1,3-dioxolane structure)} \xrightarrow{\text{LiAlH}_4/\text{AlCl}_3}$$

$$\begin{array}{c} \text{CH}_2\text{OH} \\ | \\ \text{HCOH} \\ | \\ \text{CH}_2\text{OCHMe}_2 \\ 81\% \end{array} \quad + \quad \begin{array}{c} \text{CH}_2\text{OH} \\ | \\ \text{CHOCHMe}_2 \\ | \\ \text{CH}_2\text{OH} \\ 19\% \end{array}$$

3. The hydroxyl group in both types of compounds can be acylated by reaction with acyl halides in the presence of pyridine (F.H.Mattson and R.A.Volpenhein, J. lipid Res., 1961, 2, 58; I.G. de Kuck *et al.*, Ind. Quim., 1971, 28, 149), or by direct reaction with carboxylic acids in trifluoroacetic acid (P.F.E.Cook and A.J.Showler, J. chem. Soc., 1965, 4594).

$$\text{(structure)} \xrightarrow{\text{RCOOH}} \begin{array}{c} \text{CH}_2\text{OCOR} \\ | \\ \text{CHO} \\ | \quad \text{CMe}_2 \\ \text{CH}_2\text{O} \end{array} \xrightarrow{\text{HCl}} \begin{array}{c} \text{CH}_2\text{OCOR} \\ | \\ \text{CHOH} \\ | \\ \text{CH}_2\text{OH} \end{array}$$

$$\text{HO}\text{(structure)}\text{Me}_2 \xrightarrow{\text{RCOOH}} \text{RCOO}\text{(structure)}\text{Me}_2 \rightarrow \begin{array}{c} \text{CH}_2\text{OH} \\ | \\ \text{CHOCOR} \\ | \\ \text{CH}_2\text{OH} \end{array}$$

The ORD and CD curves of 2,2-dimethyl-1,3-dioxolane-4-methyl acetate together with those of 3-acyl-*sn*-glycerol and triacyl-*sn*-glycerol have been measured to determine the effect of branching in the acyl groups which were MeCO, EtCO, PrCO, Me$_2$CHCO and Me$_3$CCO (S.Gronowitz *et al.*, Chem. phys. lipids,

1976, 17, 244). It is found that the branching in the acyl group in 1,2-dimyristoyl-3-acyl-*sn*-glycerols reverses the sign of rotation and the CD effect compared with the straight chain analogues.

4.   The reaction of the hydroxyl group with alkyl halides leads to the formation of ethers which can often be used as blocking groups.

Ethers can also be formed by reaction with vinyl ethers, but the initial product can undergo disproportionation to a symmetrical bis compound (M.F.Shostakovskii *et al.*, Izv. Sibirsk. Otd. Akad. Nauk. SSSR., 1965, 139).

Long-chain saturated and mono- and di-unsaturated l- and 2-glyceryl monoethers can be made by treating the potassium salts of 2,2-dimethyl-1,3-dioxan-5-ol and 2,2-dimethyl-1,3-dioxolane-4-methanol with alkyl halides for saturated compounds and with alkenyl toluene-p-sulphonates for unsaturated mono-ethers (R.Wood and F.Synder, Lipids, 1967, 2, 161), followed

by hydrolysis of the blocking groups with boric acid. This
method allows the synthesis of 2-glyceryl monoethers. Also,
when trichlorobenzene with sodium hydroxide reacts at 180°C
with 2,2-dimethyl-4-methoxy-1,3-dioxolane, 2,2-dimethyl-4-
methoxy-4-(2,5-dichlorophenyl)-1,3-dioxolane is formed
(W.Jacobs et al., Ger(East). Pat., 61,001). This can then be
hydrolysed with 50% sulphuric acid to provide 2,5-dichloro-
phenyl glycerol ether.

The synthesis of monopropenyl ethers of glycerol, which
are useful in controlling parasitic trichostrongylids and
ascarid worms, involves the initial reaction of 2,2-dimethyl-
1,3-dioxolane-4-methanol with sodium hydride, followed by
reaction with bromopropyne to give 2,2-dimethyl-4-
2-propenyl oxymethyl)-1,3-dioxolane (J.D.Bunger and W.L.Howard,
U.S. Pat., 3,290,388). Hydrolysis of this compound with
methanolic hydrogen chloride gives 3-(2-propenyloxy)-1,2-
propane-diol.

5.   The internal cyclisation of 2,2-disubstituted-1,3-
dioxolane-4-methanol with hydrogen chloride gives a
trioxabicyclo [3.1.1] bicycloheptane   (J.Gelas, Bull. Soc.
Chim. Fr., 1970, 3721).

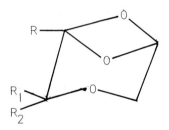

Similarly, the internal cyclisation of 2-substituted-2-
vinyl-1,3-dioxolane-4-methanol leads to the formation of
a bicyclo [4.2.1] nonane.

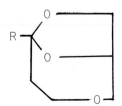

If, instead, 2-alkyl-1,3-dioxolane-4-chloromethane is distilled over potassium carbonate it is dehydrochlorinated to 2-alkyl-1,3-dioxolane-4-methylene (*idem ibid.*, 1970, 11, 4046). This monomer can be polymerised with $BF_3.OEt_2$, $FeCl_3$, or $AlCl_3$ at $-82°C$ to give the polymer, poly(4-methylene-1,3-dioxolane (S.I.Bagaev and K. Yu Yesnin, Khim. Teknol. Polim., 1972, 10).

The prostaglandin-like smooth muscle contracting effect of lipophilic glycerol acetal phosphates, the physiologically active principle of "Darmstoff", has generated considerable interest in efficient syntheses of isomeric long-chain cyclic glycerol acetals (W.Vogt, Arch. exp. Pathol. Pharmakol., 1949, 206, 1; J. physiol. (London), 1957, 137, 154; Nature, 1957, 170, 300; Pharmacol. Rev., 1958, 3, 407; Arzneim. Forsch., 1958, 8, 253; Biochem. Pharmacol., 1963, 4, 15; G.G.Gray, J. pharmacol. Exp. Ther., 1964, 146, 215; J.B.Lee *et al.*, Biochem. J., 1967, 1251; C.Dyer and E.J.Walaszek, J. pharmacol. Exp. Ther., 1966, 160, 360; D.D.Sumner and E.J.Walaszek, Lipids, 1970, 5, 803). The specific synthesis of *cis*- and of *trans*-2-pentadecyl- and 2-(*cis*-8'-heptadecenyl) -4-hydroxymethyl-1,3-dioxolanes can be achieved. The acid-catalysed condensation of methyl glycerate with hexadecanal or *cis*-9-octadecenal to give a mixture of the geometrical isomers of 2-pentadecyl- and 2-(*cis*-8'-heptadecenyl)-4-methoxycarbonyl-1,3-dioxolanes which, on reduction with lithium aluminium hydride yield the corresponding 4-hydroxy -methyl-1,3-dioxolanes. 1,3-Dioxan formation is avoided through the use of methyl glycerate as the three carbon backbone. More important the stereoisomeric glycerate acetals can be reduced quantitatively by lithium aluminium hydride to the corresponding glycerol acetals.

Although the most important uses of 1,3-dioxan and 1,3-dioxolane derivatives are in the syntheses of glycerides, these compounds have been found useful in other areas. A mixture of 1,3-dioxan-5-ol and 1,3-dioxolane-4-methanol can be used as a solvent for pharmaceuticals (J.Matzke, Austrian Pat., 214,073) since these compounds are non-toxic to humans (D.M.Sanderson, J. pharm. Pharmacol., 1959, 11, 150). 2,2-Dimethyl-1,3-dioxolane-4-methanol can act as a bactericide, as a solvent for cosmetics and for disinfectants and aerosols (F.Melson and K.Luedde, Pharmazie, 1962, 17, 614; E.Mikschik, Mitt. Chem. Forschungs-inst. Wirtsch. Oesterr., 1955, 9, 153; 1954, 8, 149). The 2-substituted derivatives of both type have insecticidal properties and mosquito repellant properties (C.N.Smith and D.Burnett, J. econ. Entomol., 1949, 42, 439; B.V.Travis, F.A.Morton, H.A.Jones and J.H.Robinson, *ibid.*, 1949, 42, 686).

An interesting and important reaction of the cyclic acetals of glycerol is that of acetal migration. The isomeric methylidene, benzylidene and p-nitrobenzylidene cyclic acetals are readily interconverted in the presence of hydrogen chloride (H.S.Hill, M.S.Whelan and H.Hibbert, J. Amer. chem. Soc., 1928, 50, 2235; H.Hibbert and N.M.Carter, *ibid.*, 1928, 50, 3120). By following the change in the signal pattern in the benzylic proton region of the ^{1}H NMR spectrum the acetal migration can be studied (N.Baggett *et al.*, J. chem. Soc., 1966, 212; F.S. Al-Jeboury *et al.*, Chem. Comm., 1965, 222).

Glycerol esters of aromatic carboxylic acids.

Glycerol triesters can be prepared from allyl benzoate by its reaction with the complex formed by the reaction of iodine with the silver salt of substituted benzoic acids. This method does not provide the expected product, 1,2-diaroyl-3-benzoyl esters of glycerol. Transposition occurs in this reaction to provide the symmetrical 1,3-diaroyl-2-benzoates of glycerol.

1,2-Di-O-benzoyl-3-O-benzyl-(±)-glycerol, $PhCH_2OCH_2CHOCOPh-CH_2OCOPh$, b.p. 214-218°/0.1mm., is prepared from 1-O-benzyl-(±)-glycerol by its reaction with benzoyl chloride in the presence of dry pyridine (E.J.Hedgley and N.H.Leon, J. chem. Soc., 1970, 467).

1,2-Di-O-benzoyl-(±)-glycerol, $HOCH_2CH(OCOPh)CH_2OCOPh$, m.p. 57.5-58°, can be made by hydrogenation of 1,2-di-O-benzoyl-

-3-*O*-benzyl-(±)-glycerol (Hedgley and Leon,*loc. cit.*;
B.F.Daubert and C.G.King, J. Amer. chem. Soc., 1939, 61, 5328).
2,3-Di-O-*benzoyl-1*-O-*thiobenzoyl-(±)-glycerol*, PhCOOCH$_2$CH-
(OCOPh)CH$_2$OCSPh, yellow needles, m.p. 74.5-75°, is obtained
by reaction of 2,3-di-*O*-benzoyl-(±)-glycerol with thiobenzoyl
chloride. 2,3-Di-*O*-benzoyl-1-*O*-thiobenzoyl-(±)-glycerol when
treated with Raney nickel gives a high yield of 2,3-di-*O*-
benzoyl-(±)-glycerol. This reaction illustrates the value of
selective and quantitative dethiobenzoylation.

$$\begin{array}{c} CH_2OCOPh \\ | \\ CHOCOPh \\ | \\ CH_2OCSPh \end{array} \xrightarrow{\text{Ni}/H_2} \begin{array}{c} CH_2OCOPh \\ | \\ CHOCOPh \\ | \\ CH_2OH \end{array}$$

When 2,3-di-*O*-benzoyl -1-*O*-thiobenzoyl-(±)-glycerol is
heated at 210°C it will isomerise completely to *O,O,S-*
tribenzoyl-(±)-1-thioglycerol, colourless crystals, m.p.
88-89°. This isomerisation also occurs during uv-irradiation.
The mechanistic alternatives are either S$_N$i (not precluding
a type of internal reaction:

$$\begin{array}{c} CH_2OCOPh \\ | \\ CHOCOPh \\ | \\ CH_2 \end{array} \xrightarrow{210°C} \begin{array}{c} CH_2OCOPh \\ | \\ CHOCOPh \\ | \\ CH_2SCOPh \end{array}$$

or by rearrangement with migration:

The latter course, though considered unlikely, is not without precedent. An unequivocal example of a 1,3-migration is provided by the isomerisation of some allylic thiobenzoates where, however, a monocyclic transition state is implicit (S.G.Smith, J. Amer. chem. Soc., 1961, 83, 4328). In the above concerted rearrangement a bicyclic transition state is required.

2,3-Di-O-benzoyl-1-O-thiobenzoyl-($\pm$)-glycerol reacts rapidly with silver nitrate in acetone solution to provide glycerol tribenzoate.

$$\begin{array}{c} CH_2OCSPh \\ | \\ CHOCOPh \\ | \\ CH_2OCOPh \end{array} \xrightarrow{AgNO_3} \begin{array}{c} CH_2OCOPh \\ | \\ CHOCOPh \\ | \\ CH_2OCOPh \end{array}$$

1,3-Di-O-benzoyl-2-O-thiobenzoyl glycerol, PhCOOCH$_2$CH(OCSPh)-CH$_2$OCOPh, yellow needles, m.p. 75-76°, is prepared from 1,3-di-O-benzoyl glycerol by its reaction with thiobenzoyl chloride (O.T.Schmidt and W.Black, Ber., 1956, 89, 283; H.L.White, J. Amer. chem. Soc., 1952, 74, 3451). This thiobenzoate on heating at 210° gives ($\pm$)-1-thioglycerol tribenzoate, m.p. 88-89°.

$$\begin{array}{c} CH_2OCOPh \\ | \\ CHOCSPh \\ | \\ CH_2OCOPh \end{array} \xrightarrow{210°C} \begin{array}{c} CH_2OCOPh \\ | \\ CHOCOPh \\ | \\ CH_2SCOPh \end{array}$$

This reaction clearly involves ester group migration. No other derivatives of 2-thioglycerol are formed.

2-O-Benzoyl-1,3-di-O-thiobenzoyl glycerol, $PhCSOCH_2CH(OCOPh)$-CH_2OCSPh, yellow, m.p. 57.5-58.5°, is obtained by thiobenzoylation of 2-O-benzoyl glycerol. This bis-benzoate with Raney nickel in ethanol gives 2-O-benzoyl glycerol. This reaction may be monitored by observing the change in the NMR spectrum. The initial pattern in the aliphatic proton signals at δ 5.13 and 6.17 steadily diminishes in intensity and a pattern at δ 3.59 and 5.52 replaces it which is characteristic of 1,3-dithioglycerol tribenzoate. However, if the bis-thiobenzoate is heated at 300°C for 5 minutes, chromatographic purification of the product yields unchanged starting material and 2-O-benzoyl-S-benzoyl-3-O-thiobenzoyl-(±)-1-thioglycerol, which on reaction with silver nitrate in acetone solution gives (±)-1-thioglycerol tribenzoate, and on heating at 180°C for 14 hours gives 1,3-dithioglycerol tribenzoate.

These reactions confirm that the thermal rearrangement of 2-O-benzoyl-1,3-di-O-thiobenzoylglycerol into 1,3-dithio-glycerol tribenzoate proceeds not by ester migration but by the sequential process, since the intermediate can only arise by an $S_N i$ route.

It thus seems likely that 2,3-di-*O*-benzoyl-1-*O*-thiobenzoyl-(±)-glycerol rearranges by an $S_N i$ route.

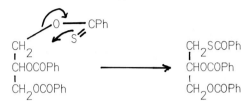

$$\begin{array}{ccc}
\text{CH}_2 & & \text{CH}_2\text{SCOPh} \\
| & & | \\
\text{CHOCOPh} & \longrightarrow & \text{CHOCOPh} \\
| & & | \\
\text{CH}_2\text{OCOPh} & & \text{CH}_2\text{OCOPh}
\end{array}$$

Tri-O-thiobenzoyl glycerol, PhCSOCH$_2$CH(OCSPh)CH$_2$OCSPh, yellow crystals, m.p. 88-89°, is obtained by the thiobenzoylation of glycerol.

(±)-1-Thioglycerol tribenzoate, PhCOSCH$_2$CH(OCOPh)CH$_2$OCOPh, colourless needles, m.p. 72.5-73.5°, can be made by the reaction of 1,3-dithioglycerol with benzoyl chloride in the presence of pyridine (E.P.Adams *et al.*, J. chem. Soc., 1960, 2649).

S-Thiobenzoyl-(±)-thioglycerol, PhCSSCH$_2$CHOHCH$_2$OH, a scarlet oil, is made by the reaction of (±)-1-thioglycerol with excess methyl dithiobenzoate, followed by tlc purification on silica gel. On reaction with benzoyl chloride in pyridine it yields 2,3-di-*O*-benzoyl-*S*-thiobenzoyl-(±)-1-thioglycerol as pink coloured needles with m.p. 67.5°. The latter on reaction with aqueous silver nitrate is converted into (±)-1-thioglycerol tribenzoate. This method, thus, provides a practical method of converting the dithiobenzoate group into the *S*-benzoyl group (CH-S-CS-Ph $\longrightarrow$ CH-S-CO-Ph). In addition, the selective thiobenzoylation of the mercapto group in (±)-thioglycerol in the presence of hydroxyl groups illustrates the greater nucleophilicity of sulphur compared to oxygen (A.Streitweiser,jun., Chem. Rev., 1956, **56**, 571; S.G.Smith, Tetrahedron Letters, 1962, 979).

Glycerol esters of fatty acids. Glycerides.

Triacyl glycerides make up a major portion of the neutral lipids found in most living organisms. They may be depicted by the following formula.

$$\begin{array}{ll} \alpha & CH_2OCOR_1 \\ & | \\ \beta & HCOCOR_2 \\ & | \\ \alpha' & CH_2OCOR_3 \end{array}$$

If the alkyl groups, R_1 and R_2, are different, then the central carbon of the glycerol fragment is asymmetric. The glyceride is therefore chiral and can exist in enantiomeric forms. It has been shown (W.Schlenk,jnr., J. Amer. Oil Chemists' Soc., 1965, 42, 945) that the enantiomeric monoglycerides will exhibit optical rotatory dispersion whilst the diglycerides and the triglycerides will not unless the alkyl chains differ markedly in length.

Although glycerol has always been considered to be a symmetrical molecule because it is optically inactive, its primary hydroxyl groups can be differentiated in biological systems (H.Hirschmann, J. biol. Chem., 1960, 235, 2762). Thus, when glycerol-1-^{14}C is fed to rats the glycogen formed from this substrate has glucose units with the ^{14}C label mainly at the C-3 and the C-4 positions (P.Schambye et al., J. biol. Chem., 1954, 206, 883) and it has been found that, after treating glycerol-1-^{14}C with glycerokinase, almost optically pure glycerophosphoric acid is obtained (C.Bublitz and E.P.Kennedy, J. biol. Chem., 1954, 211, 963). These observations provide evidence for the biological asymmetry of glycerol (P.Schwartz and H.E.Carter, Proc. nat. Acad. Sci. Wash., 1954, 40, 499).

Nomenclature for chiral glycerides.

Since D-glyceraldehyde has been used as the standard of reference for the Fischer convention (L.F.Fieser and M.Fieser, 'Advanced Organic Chemistry', 1961, Reinhold, New York, and E.L.Eliel, 'The Stereochemistry of Carbon Compounds', 1962

McGraw-Hill, New York) it seems logical that this convention
should be applied to glycerides. However, it does not specify
the positional numbering of glycerol. A rule has been devised
(E.Baer and H.O.L.Fischer, J. biol. Chem., 1939, 128, 475)
where an **α**-monoglyceride is put into the same category as
that of glyceraldehyde into which it can be transformed by
oxidation without any alteration or removal of substituents.
Thus, if the monoglyceride can be oxidised into 3-acyl-D-
glyceraldehyde, then it has the same configuration as the
D-glyceraldehyde.

Under these terms the compound (7) is called L-**α**-glycero-
phosphoric acid or L-glycerol-3-phosphate. This convention
depends upon the interconversion of compounds and can thus
lead to ambiguities.

(7)

The Cahn-Ingold-Prelog convention adopts an approach based
on sequence rules (R.S.Cahn, C.K.Ingold and V.Prelog,
Experientia, 1956, 12, 81; *ibid.*, Angew. Chem. internat. Edn.,
1966, 5, 385; R.S.Cahn, J. chem. Ed., 1964, 41, 116) which
uses the prefixes, R and S, to designate optical isomers. This
system can be extended to distinguish identical groups

attached to a *meso*-carbon atom which are distinguishable by
enzymatic reactions. Such a carbon atom can be called
prochiral (K.R.Hanson, J. Amer. chem.Soc., 1966, 88, 2731).
The term, enantiotropic, can be used to describe two
chemically equivalent substituents attached to a prochiral
centre such as carbon-2 of glycerol (K.Mislow and M.Raban,
Topics in Stereochemistry, 1967, 1, 1).

The present day system for glyceride nomenclature,
adopted by IUPAC-IUB, is based on the convention proposed
by H.Hirschmann (J. biol. Chem., 1960, 235, 2762). By this
convention, in a compound, Caabc, the four substituents,aabc,
are selected so that b has a higher priority than c under the
Cahn-Ingold-Prelog system. A model of the molecule is viewed
from the side opposite the group c as in the following
figures

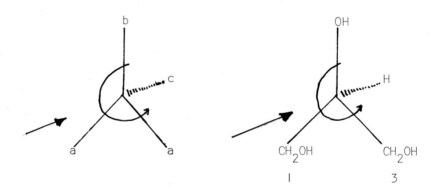

Moving counter-clockwise from the group b, the first
enantiotropic group a is given the number 1; the remaining
group a receives the number 3 as in the diagrams above.
Alternatively, the numbering may be assigned by reference
to a Fischer projection which shows the higher priority
substituent b to the left of the subsituent c to the right.
The group a above the central carbon atom then receives the
lower number. This is illustrated below for the general
case and for glycerol.

Accordingly, glycerol has prochirality; the hydroxymethyl group at position 1 has a pro-S-configuration and the one on position 3 has a pro-R-configuration. The IUPAC-IUB convention uses this stereospecific numbering but also uses the prefix sn before the word glycerol to indicate that the derivative is stereospecifically numbered with the C-2 hydroxyl group or derived substituent always placed on the left.

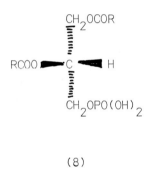

(8)

Compound (8) is named 1,2-diacyl-sn-glycerol-3-phosphate.
In the early work, synthetic monoglycerides showed measurable rotations in pyridine solution, whilst triglycerides containing only long-chained acyl groups appeared to be optically inactive (E.Baer and H.O.L.Fischer, J. biol. Chem., 1939, 128, 475). However, with improved spectropolarimetric methods (+)-1-lauroyl-2,3-dipalmitoyl-sn-glycerol has a small measurable rotation (W.Schlenk, J. Amer. oil Chem. Soc., 1965, 42, 945).

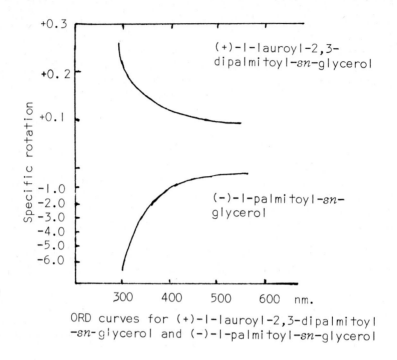

ORD curves for (+)-I-lauroyl-2,3-dipalmitoyl
-*sn*-glycerol and (-)-I-palmitoyl-*sn*-glycerol

With 3-monoacyl-*sn*-glycerols it has been found that the optical
rotations are at a maximum for acetyl derivatives and decline
as the size of the acyl group increases (E.Baer and H.O.L.
Fischer, J. Amer. chem. Soc., 1945, 67, 2031). The magnitude
of rotation usually increases as the wavelength becomes
shorter (C.R.Smith, Topics in lipid Chemistry, 1970, 1, 277).

Synthesis of Glycerides.

The synthesis of glycerides has been reviewed by F.H. Mattson and R.A.Volpenhein (J. lipid Res., 1962, 3, 281); L.Hartmann (Chem. Rev., 1958, 58, 845); J.G.Quinn, J.Sampugna and R.G.Jensen (J. Amer. oil Chem. Soc., 1967, 44,439).

Monoglycerides.

There are a number of procedures available for the synthesis of monoglycerides. The choice of procedure is determined by the isomer required,i.e. the 1- or 2-acyl glycerol. The most widely used method starts with isopropylideneglycerol (2,2-dimethyl-1,3-dioxolane-4-methanol) which can be made by condensing glycerol with acetone in the presence of an acid catalyst. The free primary hydroxyl group is then esterified using a free carboxylic acid, its acid chloride or its ester (E.Fischer, M.Bergmann and E.Barwind, Ber., 1920, 53, 1589).

$$\begin{array}{c} CH_2OH \\ | \\ CHOH \\ | \\ CH_2OH \end{array} \xrightarrow{Me_2CO/H^+} \begin{array}{c} CH_2OH \\ | \\ CHO \\ | \\ CH_2O \end{array}\!\!\!>\!CMe_2 \xrightarrow{RCOCl}$$

$$\begin{array}{c} CH_2OCOR \\ | \\ CHO \\ | \\ CH_2O \end{array}\!\!\!>\!CMe_2 \xrightarrow{dil.\ HCl} \begin{array}{c} CH_2OCOR \\ | \\ CHOH \\ | \\ CH_2OH \end{array}$$

A number of modifications have been devised for the esterification step. The condensation and esterification reactions may be, for instance, performed in a single step by using toluene-p-sulphonic acid as a catalyst and chloroform or benzene as a carrier to remove the water (L.Hartman, Chem.

Ind., 1960, 711; J.R.Afinsen and E.G.Perkins, J. Amer. oil
Chem. Soc., 1964, 41, 779; J.G.Quinn *et al.*, *ibid.*, 1968,
45,581).

$$\begin{array}{c} CH_2OH \\ | \\ CHOH \\ | \\ CH_2OH \end{array} \quad + \quad Me_2CO \quad \xrightarrow{TsOH} \quad \begin{array}{c} CH_2OH \\ | \\ CHO \\ | \\ CH_2O \end{array}\!\!\!\!\searrow\!\!\!\!CMe_2$$

$$\xrightarrow[\text{TsOH}]{\text{RCOOH}} \quad \begin{array}{c} CH_2OCOR \\ | \\ CHO \\ | \\ CH_2O \end{array}\!\!\!\!\searrow\!\!\!\!CMe_2 \quad \xrightarrow{H_3BO_3/H_2O} \quad \begin{array}{c} CH_2OCOR \\ | \\ CHOH \\ | \\ CH_2OH \end{array}$$

After the acylation the removal of the blocking acetal
group can be achieved by treatment with either 10% aqueous
acetic acid or cold concentrated hydrochloric acid (E.Baer
and H.O.L.Fischer, J. Amer. chem. Soc., 1945, 67, 2031) or
boric acid (J.B.Martin, *ibid.*, 1953, 75, 5483)
The 2-acyl glycerols can be synthesised from 1,3-
benzylidene glycerol (2-benzyl-1,3-dioxolane-4-methanol) which
is obtained from the reaction of glycerol with benzaldehyde
(F.H.Mattson and R.A.Volpenhein, J. lipid Res., 1962, 3, 281).
Glycerol-2-oleate, 2-elaidate and 2-linoleate can be made by
this method.

$$\begin{array}{c} CH_2OH \\ | \\ CHOH \\ | \\ CH_2OH \end{array} \quad + \quad PhCHO \quad \xrightarrow{TsOH} \quad \begin{array}{c} CH_2O \\ | \\ HOCH \\ | \\ CH_2O \end{array}\!\!\!CHPh$$

$$\xrightarrow[C_5H_5N]{RCOCl} \quad \begin{array}{c} CH_2O \\ | \\ RCOOCH \\ | \\ CH_2O \end{array}\!\!\!CHPh \quad \xrightarrow{H_3BO_3/H_2O} \quad \begin{array}{c} CH_2OH \\ | \\ RCOOCH \\ | \\ CH_2OH \end{array}$$

The 2-acyl group is capable of migrating fairly readily. Thus, the yields of 2-acyl glycerols are relatively low. In most cases most di- and triglycerides can be prepared from the 1-acyl glycerol and the synthesis of the 2-acyl glycerol may not always be necessary (F.H.Mattson and R.A.Volpenhein, j. lipid Res., 1962, 3, 281).

Diglycerides.

There are a number of procedures available for the synthesis of diglycerides. The choice of method is determined by whether the 1,2 or the 1,3-diglyceride is required. The synthesis of the 1,2-diglycerides is the more difficult to achieve.

a).   The direct acylation of 1-monoglycerides with an acid chloride is one of the simplest methods (F.H.Mattson and R.A.Volpenhein, J. lipid Res., 1962, 3, 281). This method provides the 1,3-diglyceride.

$$
\begin{array}{ccc}
CH_2OH & & CH_2OCOR' \\
| & \xrightarrow{R'COCl/C_5H_5N} & | \\
CHOH & & CHOH \\
| & & | \\
CH_2OCOR & & CH_2OCOR
\end{array}
$$

b). The acylation of the primary hydroxyl groups of 1,3-dihydroxyacetone with acyl chloride, followed by reduction with sodium borohydride of the carbonyl group. This procedure has an added advantage, since it can be applied to the synthesis of unsaturated 1,3-diglycerides (P.H.Bentley and W.McCrae, J. org. Chem., 1970, 35, 2082).

$$
\begin{array}{ccccc}
CH_2OH & & CH_2OCOR & & CH_2OCOR \\
| & \xrightarrow{RCOCl/C_5H_5N} & | & \xrightarrow{NaBH_4} & | \\
C=O & & C=O & & CHOH \\
| & & | & & | \\
CH_2OH & & CH_2OCOR & & CH_2OCOR
\end{array}
$$

c). 1,2-Isopropylidene glycerol (2,2-dimethyl-1,3-dioxolane-4-methanol) is benzylated with benzyl chloride to block the 3-position. The ketal group is then cleaved with acid to free the 1- and the 2-positions which can then be acylated with an acid chloride. The removal finally of the blocking benzyl group is best done by hydrogenolysis (R.J.Howe and T.Malkin, J. chem. Soc., 1951, 2663).

d). 2-Monoglycerides can be partially acylated (F.H.Mattson and R.A.Volpenhein, J. lipid Res., 1962, 3, 281).
e). Triglycerides can be converted into 1,2-diglycerides by pancreatic lipolysis (A.E.Thomas, J.E.Scharoun and H. Ralston, J. Amer. oil Chem. Soc., 1965, 42, 789).
f). The Cl tetrahydropyranyl ether of glycerol may be acylated with an acid chloride and the 1,2-diglyceride obtained by removal of the tetrahydropyranyl blocking group by limited exposure to hydrochloric acid (L.Krabisch and B.B Borgstrom, J. lipid Res., 1965, 6, 156; D.L.Turner *et al.*, Lipids, 1968, 3, 228).

The tetrahydropyranyl glycerol can be synthesised from
allyl alcohol and dihydropyran (P.J.Barry and B.M.Craig,
Canad. J. Chem., 1955, 33, 716).

g). 3-Benzyl glycerol-1,2-carbonate can be used as the
starting material and this is prepared .by refluxing together
1-benzylglycerol, ethyl carbonate and sodium hydrogen
carbonate. This, on hydrogenation, gives glycerol-1,2-
carbonate which can be converted into the tetrahydropyranyl
ether by using dihydropyran. The carbonate blocking group
can then be removed by heating with potassium hydroxide
(J.Cunningham and R.Gigg, J. chem. Soc., 1965, 1553;
J.Gigg and R.Gigg, *ibid.*, 1967, 431 and 1865). This can now
be used to synthesise 1,2-diglycerides.

$$
\begin{array}{c}
\text{CH}_2\text{OH} \\
\text{CHOH} \\
\text{CH}_2\text{OCH}_2\text{Ph}
\end{array}
\xrightarrow[\text{NaHCO}_3]{(\text{EtO})_2\text{CO}}
\begin{array}{c}
\text{CH}_2\text{O} \\
\text{CHO} \hspace{-0.3em}>\hspace{-0.3em}\text{C=O} \\
\text{CH}_2\text{OCH}_2\text{Ph}
\end{array}
\xrightarrow{\text{H}_2/\text{Pd}}
\begin{array}{c}
\text{CH}_2\text{O} \\
\text{CHO} \hspace{-0.3em}>\hspace{-0.3em}\text{C=O} \\
\text{CH}_2\text{OH}
\end{array}
$$

$$
\xrightarrow{\text{(dihydropyran)}}
\begin{array}{c}
\text{CH}_2\text{O} \\
\text{CHO} \hspace{-0.3em}>\hspace{-0.3em}\text{C=O} \\
\text{CH}_2\text{O}\text{(THP)}
\end{array}
\xrightarrow{\text{KOH}}
\begin{array}{c}
\text{CH}_2\text{OH} \\
\text{CHOH} \\
\text{CH}_2\text{O}\text{(THP)}
\end{array}
$$

$$
\xrightarrow{\text{RCOCl}/\text{C}_5\text{H}_5\text{N}}
\begin{array}{c}
\text{CH}_2\text{OCOR} \\
\text{CHOCOR} \\
\text{CH}_2\text{O}\text{(THP)}
\end{array}
\xrightarrow{\text{pancreatic lipase}}
\begin{array}{c}
\text{CH}_2\text{OH} \\
\text{CHOCOR} \\
\text{CH}_2\text{O}\text{(THP)}
\end{array}
$$

$$
\xrightarrow{\text{R''COCl}/\text{C}_5\text{H}_5\text{N}}
\begin{array}{c}
\text{CH}_2\text{OCOR'} \\
\text{CHOCOR} \\
\text{CH}_2\text{O}\text{(THP)}
\end{array}
\xrightarrow{\text{H}_3\text{BO}_3}
\begin{array}{c}
\text{CH}_2\text{OCOR'} \\
\text{CHOCOR} \\
\text{CH}_2\text{OH}
\end{array}
$$

h). 2-Benzylglycerol can be used directly as the precursor of 1,3-diglycerides (F.R.Pfeiffer, C.K.Mico and J.A.Weibach, J. org. Chem., 1970, 35, 221). The catalytic hydrogenation used to remove the blocking benzyl group limits the application of this method to saturated 1,3-diacylglycerides

Triglycerides.
    These are readily available for most of the common fatty acids and can be fairly easily prepared.
a). By the acylation of glycerol with an acid chloride (F.H.Mattson and R.A.Volpenhein, J. lipid Res., 1962, 3, 281).

b). By the reaction of sodium glyceroxide with the methyl
esters of fatty acids (E.S.Lutton and A.J.Fehl, Lipids,
1970, 5, 90).
c). By the acylation of glycerol with acid anhydrides
using perchloric acid catalyst (F.H.Mattson, R.A.Volpenhein
and J.B.Martin, J. lipid Res., 1964, 5, 374).

Stereospecific synthesis of Glycerides.
   The stereospecific synthesis of glycerides has been
reviewed (E.Baer and H.O.L.Fischer, Chem. Rev., 1941,29, 287;
T.Malkin and T.H.Brown, Progress in the Chemistry of Fats
and other Lipids, 1957, 4, 64; F.H.Mattson and R.A.Volpenhein,
J. lipid Res., 1962, 3, 281; L.Hartmann, Chem. Rev., 1958,
58, 845; E.Baer, J. Amer. oil Chem. Soc., 1965, 42, 357).
   The methods for obtaining optically active glycerides
involving the resolution of reaction intermediates are
usually unsatisfactory because of racemisation at a later
stage in the synthesis.
   The most successful stereospecific synthesis of glycerides
involves the initial preparation of 1,2-isopropylidene-
sn-glycerol from D-mannitol as outlined in the following
scheme (E.Baer and H.O.L.Fischer, J. biol. Chem., 1939, 128,
463).

A similar procedure can be used for the other enantiomer and involves the preparation of 2,3-isopropylidene-*sn*-glycerol from L-mannitol (*idem.*, J. Amer. chem. Soc., 1939, 61, 761). Starting with these intermediates, various optically active monoglycerides can now be synthesised by procedures considered earlier for the preparation of racemic monoglycerides.

   Optically active diglycerides use 1,2-isopropylidene-*sn*-glycerol as the starting material. This can be benzylated and the isopropylidene group can be removed by acid hydrolysis to provide 3-*O*-benzyl-*sn*-glycerol. This can now be acylated with an acid chloride and the benzyl group removed by hydrogenolysis to yield 1,2-diacyl-*sn*-glycerol (J.Sowden and H.O.L.Fischer, J. Amer. chem. Soc., 1941, 63 , 3244).

$$Me_2C\overset{OCH_2}{\underset{CH_2OH}{\diagdown OCH}} \xrightarrow{PhCH_2Cl/Na} Me_2C\overset{OCH_2}{\underset{CH_2OCH_2Ph}{\diagdown OCH}} \xrightarrow{H^+} \overset{CH_2OH}{\underset{CH_2OCH_2Ph}{HO\ CH}}$$

$$\xrightarrow{RCOCl/C_5H_5N} \overset{CH_2OCOR}{\underset{CH_2OCH_2Ph}{RCOOCH}} \xrightarrow{H_2/Pd} \overset{CH_2OCOR}{\underset{CH_2OH}{RCOOCH}}$$

This method can be modified by transforming the 3-*O*-benzyl-*sn*-glycerol into the enantiomeric 1-*O*-benzyl-*sn*-glycerol *via* its 1,2-ditosylate which can be converted into the 1-*O*-2,3-diacetate with potassium acetate by nucleophilic displacement accompanied by inversion at C-1 and C-2. The acetate groups can then be removed by base hydrolysis to provide 1-*O*-benzyl-*sn*-glycerol (W.E.M.Lands and A.Zschocke, J. lipid Res., 1965, 6, 324).

$$\begin{array}{c} CH_2OH \\ | \\ HOCH \\ | \\ CH_2OCH_2Ph \end{array} \quad \xrightarrow{\ TsCl/C_5H_5N\ } \quad \begin{array}{c} CH_2OTs \\ | \\ TsOCH \\ | \\ CH_2OCH_2Ph \end{array}$$

$$\xrightarrow{\ MeCOO^- K^+\ } \quad \begin{array}{c} CH_2OCOMe \\ | \\ MeCOOCH \\ | \\ CH_2OCH_2Ph \end{array}$$

Methods for the synthesis of mixed diglycerides are outlined below.

a). From 3-O-benzyl-*sn*-glycerol (D.Buchnea and E.Baer , J. lipid Res., 1960, 1, 405; Y.G.Molotkovskii *et al.*, Izv. Akad. Nauk. SSSR Ser. Khim., 1967, 4, 927).

$$\begin{array}{c} CH_2OH \\ | \\ HOCH \\ | \\ CH_2OCH_2Ph \end{array} \quad \xrightarrow{\ Ph_3CCl\ } \quad \begin{array}{c} CH_2OCPh_3 \\ | \\ HOCH \\ | \\ CH_2OCH_2Ph \end{array} \quad \xrightarrow{\ RCOCl\ }$$

$$\begin{array}{c} CH_2OCPh_3 \\ | \\ RCOOCH \\ | \\ CH_2OCH_2Ph \end{array} \quad \xrightarrow[\ ii)\ RCO\text{-migration}\ ]{\ i)\ Detritylate\ } \quad \begin{array}{c} CH_2OCOR \\ | \\ HOCH \\ | \\ CH_2OCH_2Ph \end{array}$$

$$\xrightarrow{\ R'COCl/C_5H_5N\ } \quad \begin{array}{c} CH_2OCOR \\ | \\ R'COOCH \\ | \\ CH_2OCH_2Ph \end{array} \quad \xrightarrow{\ H_2/Pd\ } \quad \begin{array}{c} CH_2OCOR \\ | \\ R'COOCH \\ | \\ CH_2OH \end{array}$$

210

b). From *sn*-glycerol-1,2-carbonate (J.Gigg and R.Gigg, J. chem. Soc.(C), 1967, 431).

c).   From *sn*-glycerol-2,3-carbonate which can be made from
1,2-isopropylidene-*sn*-glycerol and 2,2,2-trichloroethyl
chloroformate. Removal of the isopropylidene group by acid
hydrolysis, followed by heating in pyridine leads to the
displacement of the trichloroethoxy group to form *sn*-
glycerol-2,3-carbonate (E.R.Pfeiffer *et al.*, Tetrahedron
Letters, 1968, 3549; J. org. Chem., 1970, 35, 221). The
acyclic carbonate can now be used as a suitable substrate
for direct acylation in another approach to the synthesis
of optically active *sn*-glycerol 1,2-diacylates. After
removal of the carbonate moiety with zinc in acetic acid,
the 3-position is now available for the introduction of
any additional acyl group desired.

$$Me_2C \overset{OCH_2}{\underset{CH_2OH}{\overset{|}{\underset{|}{OCH}}}} \quad \xrightarrow{ClCOOCH_2CCl_3} \quad \overset{CH_2OH}{\underset{CH_2OCOOCH_2CCl_3}{\overset{|}{\underset{|}{HOCH}}}}$$

$$\xrightarrow[\text{heat in } C_5H_5N]{} \quad O=C \overset{OCH}{\underset{OCH_2}{\overset{\diagup}{\diagdown}}} \overset{CH_2OH}{\underset{}{\overset{|}{}}}$$

d).   From 2-*O*-dimethyl-t-butylsilyl glycerol which can be made
by the reaction of *cis*-2-phenyl-1,3-dioxan-5-ol and dimethyl
-butylsilyl chloride, followed by hydrogenolysis G.H.Dodd,
B.T.Golding and P.V.Ioannone, J. chem. Soc. Perkin I, 1976,
2273).

The optically active 3-O-dimethyl-t-butylsilyl glycerol can be prepared by the silylation and hydrogenolysis of 1,2-dibenzyl-*sn*-glycerol.

These silyl derivatives can be acylated and phosphorylated without disturbing the blocking group. However, it has not been possible to apply these derivatives to the synthesis of 1,3-diacyl-*sn*-glycerols *via* the intermediate (8).

(8)

Suitable conditions have not been found for the removal of the dimethyl-t-butylsilyl (DMTBS) blocking group without causing acyl group migration. The recommended conditions for the acidic hydrolysis of these DMTBS ethers are not effective, and more strongly acidic conditions, although partially effective, also cause acyl group migration and deacylation (K.K.Ogilvie, Canad. J. Chem., 1973, 51, 3799). Fluoride ions, from tetra-n-butylammonium fluoride in tetrahydrofuran, will cleave an Si-O bond in, for example (9). This reaction is fast but provides an equilibrium mixture of 1,2- and 1,3-diacyl glycerols.

$$Me_3CMe_2SiOCH \begin{array}{c} CH_2OCOR \\ | \\ CH_2OCOR \end{array} \xrightarrow{Bu_4NF/THF} \begin{array}{c} CH_2OCOR \\ | \\ HOCH \\ | \\ CH_2OCOR \end{array} \; + \; \begin{array}{c} CH_2OCOR \\ | \\ RCOOCH \\ | \\ CH_2OH \end{array}$$

(9)

Hydroxy compounds can be selectively silylated with compounds of general formula, R_nSiX_{4-n}, where R = Me or $CH_2=CH$ and X = OOCH, OAc or Cl, n = 2 or 3 (E.P.Plueddemann, Ger. Pat., 2,117,028). Thus glycerol can be silylated selectively with Me_3SiOAc. The course of reaction of glycerol with hexamethyldisilazane or trimethylsilyl chloride can be studied by gas chromatography (O.Mlejnek, Coll. Czech. chem. Comm., 1969, 34, 1777). Silylation of the primary hydroxyl groups occurs very much faster than that of the secondary hydroxyl group.

Naturally occurring cardiolipins (e.g. 10) are optically active diphosphatidyl glycerols and may be synthesised from an appropriate diglyceride, a cyclic enediol pyrophosphate and glycerol where the 2-hydroxyl group is protected (F.Ramirez, Synthesis, 1976, 11, 769). An example of such a synthesis is given below.

214

CH$_2$OCO(CH$_2$)$_{16}$CH$_3$
CHOCO(CH$_2$)$_{16}$CH$_3$ $\quad$ cyclic bis(dimethylvinylene) phosphate $\longrightarrow$
CH$_2$OH

CH$_2$OCO(CH$_2$)$_{16}$CH$_3$ $\qquad$ CH$_2$OH
CHOCO(CH$_2$)$_{16}$CH$_3$ $\qquad$ CHOSiMe$_2$CMe$_3$ and Et$_3$N
CH$_2$OR' $\qquad$ CH$_2$OH $\longrightarrow$
(R' = cyclic dimethyl-
vinylene phosphate)

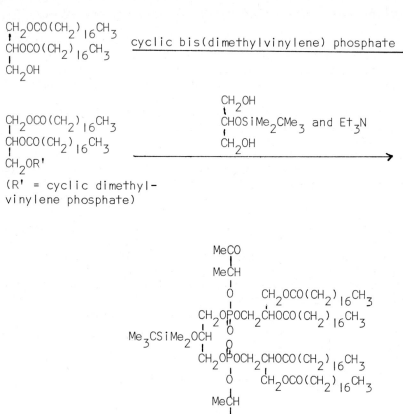

hydrolysis $\longrightarrow$

Determination of purity and structure of glycerides.

a). *Melting point.* All triglycerides containing acyl chains of various lengths and differing amounts of unsaturation exhibit polymorphism as shown by their multiple melting points (D.Chapman, 'Introduction to Lipids', McGraw-Hill, London, 1969). The possibility of several melting points means that the crystals must always be prepared in the same way if the melting point is to have any meaning.

b). *Optical rotatory dispersion.* With the advent of recording spectropolarimeters precise measurements can now be made. ORD curves show that the magnitude of rotation usually increases as the wavelength becomes shorter (C.R.Smith, Topics in Lipid Chemistry, 1970, $\underline{1}$, 277). Also the rotation of, for example, 1,2-didecanoyl-*sn*-glycerol varies both with the solvent used and the concentration of the glyceride in solution (E.Baer and V.Mahadevan, J. Amer. chem. Soc., 1959,$\underline{81}$, 2494). The results of ORD measurements have to be treated with caution but can be used for checking precursors.

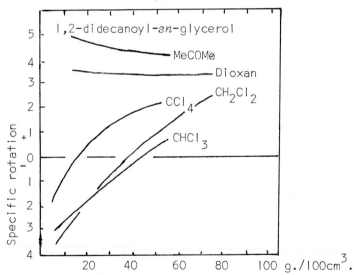

c). *Gas-liquid chromatography.* This technique is particularly useful for the identification and quantitative determination of fatty acids derived from glycerides, and also for the separation of triglycerides (R.G.Ackman,

Methods in Enzymology, 1969, 14, 329).

d). *Thin-layer chromatography*. The application of TLC is useful in the separation and identification of glyceride systems and has been reviewed (V.P.Skipski and M.Barclay, Methods of Enzymology, 1969, 14, 530).

e). *Enzymatic methods*. The positional distribution of the various fatty acids within glycerides has been considered in some detail (R.J. van der Wal, J. Amer. oil Chem. Soc., 1960, 37, 18; F.H.Mattson and E.S.Lutton, J. biol. Chem., 1958, 233, 868; P.Savary and P.Desnuelle, Biochem. biophys. Acta., 1961, 50, 319; F.D.Gunstone, Chem. Ind., 1962, 1214; T.P.Hilditch and P.N.Williams, 'The Chemical Constitution of Natural Fats', 4th. Edition, John Wiley, New York, 1964; C.D.Evans *et al.*, J. Amer. oil Chem. Soc., 1969, 46, 421).

The most useful method for the structural analysis of glycerides involves the use of enzymes. The enzymes used are very specific for both position and stereochemical configuration and do not have the limitations of most physical methods.

(i) *Pancreatic lipase* is very specific for primary fatty acids esters (F.H.Mattson and L.W.Beck, J. biol. Chem., 1956, 219, 735). It can be used to determine the positional structure of synthetic triglycerides (L.J.Morris, Biochem. Biophys. res. Comm., 1965, 18, 495; N.R.Bottino, C.A. Vandenberg and R.Reiser, Lipids, 1967, 2, 489; R.Kleiman *et al.*, *ibid.*, 1970, 5, 513; R.H.Barford, F.E.Luddy and P.Magidman, *ibid.*, 1966, 1, 287; F.E.Luddy *et al.*, J. Amer. oil Chem. Soc., 1964, 41, 693).

$$
\begin{array}{c}
CH_2OCOR_1 \\
| \\
R_2COOCH \\
| \\
CH_2OCOR_3
\end{array}
$$

$\downarrow$ pancreatic lipase

$$
\begin{array}{c}
CH_2OCOR_1 \\
| \\
R_2COOCH \\
| \\
CH_2OH
\end{array}
\quad + \quad
\begin{array}{c}
CH_2OH \\
| \\
R_2COOCH \\
| \\
CH_2OCOR_3
\end{array}
$$

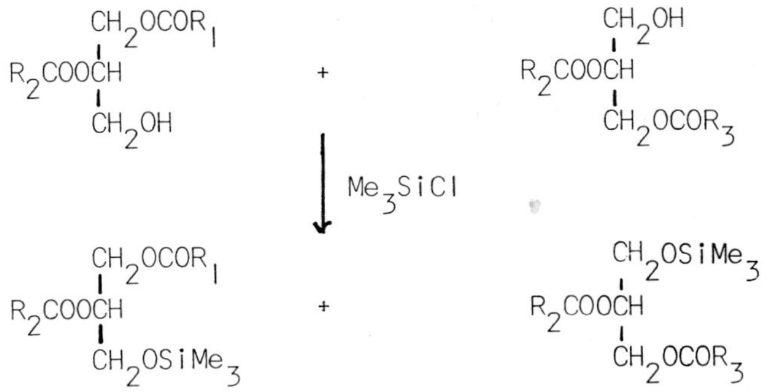

$$
\begin{array}{ccc}
\begin{array}{c}
CH_2OCOR_1 \\
| \\
R_2COOCH \\
| \\
CH_2OH
\end{array}
&
+
&
\begin{array}{c}
CH_2OH \\
| \\
R_2COOCH \\
| \\
CH_2OCOR_3
\end{array}
\end{array}
$$

$$\downarrow Me_3SiCl$$

$$
\begin{array}{ccc}
\begin{array}{c}
CH_2OCOR_1 \\
| \\
R_2COOCH \\
| \\
CH_2OSiMe_3
\end{array}
&
+
&
\begin{array}{c}
CH_2OSiMe_3 \\
| \\
R_2COOCH \\
| \\
CH_2OCOR_3
\end{array}
\end{array}
$$

In the above scheme the triglyceride sample is treated with pancreatic lipase to generate a mixture of diglycerides. This mixture of glycerides is then acylated with sorbyl chloride (hexa-2,4-dienoyl chloride) to form optically active triglycerides with enhanced activity. This procedure hasbeen modified by isolating trimethylsilyl derivatives and measuring their optical activity. A number of naturally occurring triglycerides can be seen to be composed predominantly of one enantiomer (L.J.Morris, Biochem. Biophys. Res. Commun., 1965, 20, 340).

Stereospecific analysis of triglycerides using Pancreatic using Pancreatic Lipase.

Source	Triglyceride structure acid groups joined to sn-glycerol.	Rotation of diacyl TMS derivatives, 1,2-Diacyl	2,3-Diacyl
Lard	stearic — [stearic / oleic]	$+1.6°$	$-1.3°$
Palm oil	oleic — [oleic / ,stearic]	$+2.5°$	$-2.3°$

(ii) *Phospholipase A enzyme* is specific for the *sn*-2-acid of *sn*-3-phospholipids or *sn*-1-acid of *sn*-2-phospholipids (G.H.de Haas and L.L.M. van Deenen, Biochem. biophys. Acta., 1964, 84, 469) and can be used to help identify the fatty acids in the *sn*-positions,1, 2 and 3, of triglycerides (H.Brockerhoff, J. lipid Res., 1967,8, 167; M.Yurkowski and H.Brockerhoff, Biochem. Biophys. Acta., 1966, 125, 55; G.H. de Haas and L.L.M. van Deenen, *ibid.*, 1964, 84, 469). The scheme is outlined in the following diagram.

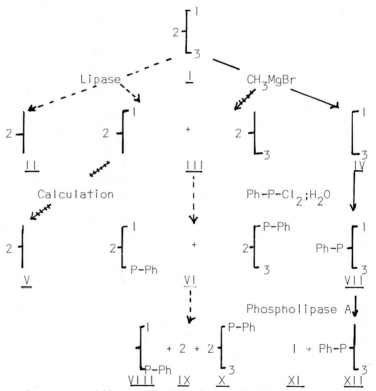

Stereospecific analysis of a triglyceride according to Brockerhoff's method 1 (- - - - -▶) and method 2 (———▶), and possible calculation of the fatty acid composition at position 2 (⊹⊹⊹⊹▶).

(iii). *Diglyceride kinase* from *Escherichia coli* is a
stereospecific phosphorylating enzyme which acts only at
the 3-position and yields 1,2-diacyl-*sn*-glycerol-3-phosphate
leaving the 2,3-diglyceride unaffected (W.E.M.Lands *et al.*,
Lipids, 1966,1, 444). Thus, a mixture of 1,2- and 2,3-
diglycerides, generated by pancreatic lipase, can be
isolated and treated with diglyceride kinase which will only
act on the 1,2-diglyceride to give 1,2-diacyl-*sn*-glycerol-
3-phosphate. The 2,3-diacyl-*sn*-glycerol remains unaffected
(R.A.Pieringer and R.S.Kunnes, J. biol. Chem., 1965, 240,
2833).
(iv).   *Geotrichum candidum lipase* is extremely specific
for acids containing *cis*-9-unsaturation regardless of their
position in the molecule. This lipase can be used to help
resolve racemic pairs of triglycerides (J.Sampugna and
R.G.Jensen, Lipids, 1968, 3, 519). The enantiomeric
triglycerides are treated with pancreatic lipase to give a
mixture of *sn*-1,2- and 2,3-diglycerides. These are converted
into phosphatidylphenols and are then treated with
phospholipase A which digests only the *sn*-3-phosphatidyl-
phenol, derived from the *sn*-1,2-diglyceride. The unhydrolysed
sn-1-phosphatidylphenol and the products of hydrolysis are
separated by TLC and the component acids analysed by GLC.
Thus, the acids located at *sn*-positions 1-, 2- and 3- can
be identified. This technique can be used to determine the
structure of triglycerides with oleic acid components.

Triglycerides breakdown into a number of products on
irradiation with γ-radiation. Many studies have been carried
out on natural fats in the presence of air. It is thus
difficult to distinguish between radiation-induced oxidative
changes and changes resulting directly from exposure to
high energy radiation. The major effect of radiation is the
formation of free radicals (F.K.Truby, J.P.O'meara and
T.M.Shaw, Rept. No. 16 to U.S. Quartermaster Food and
Container Institute, Chicago Contract No. 18-129-QM-378,
1957) Radiation of various lipid systems including methyl
stearate, methyl oleate and tristearin gives a series of
n-alkanes up to C_{13} and n-alkenes up to C_{10} and several
alkynes. These radiation products are primarily the result of
direct bond cleavage which is a random process and not
selective (C.Merritt *et al.*, Adv. Chem. Series, 1966,
56, 225). The radiolysis products of triglycerides are
mainly hydrocarbons and some oxygen-containing compounds
(M.F.Dubravcic and W.W.Nawar, J. Amer. oil Chem., 1968, 656).

In all four triglycerides the dominant alkane has one
carbon less than the parent fatty acid. In addition,
irradiationof these triglycerides produced four oxygen-
containing compounds: a saturated and an unsaturated
aldehyde containing the same number of carbon atoms as the
parent fatty acids and also the corresponding methyl
and ethyl ester.

In addition to glycerides containing acyloxy groups
there are a large number of glycerolipids where the acyloxy
groups are replaced by ether groups. The total lipids of
bovine heart muscle give a mixture of lipids which contains
optically active l-(*cis*-l-alkenyl)glycerol ethers
(W.J.Baumann, V.Mahadevan and H.K.Mangold, Z. physiol.
Chem., 1966, 347, 52). This group of lipids has been
reviewed (E.Klenk and H.Debuch, Progress in the Chemistry of
Fats and other Lipids, 1963, 6, 1; R.Wood, *ibid.*, 1969,
10, 287). There are two main forms of glyceryl ethers.
    a).   Aldehydogenic alk-*cis*-l-enyl glyceryl ethers
containing double bonds in the alcohol fragment, usually
C_{16} or C_{18}.

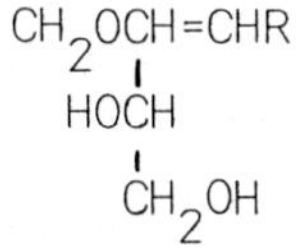

$$CH_2OCH=CHR$$
$$HOCH$$
$$CH_2OH$$

b).   Alkoxyglycerides which have no double bonds in the
chain close to the ether linkage.

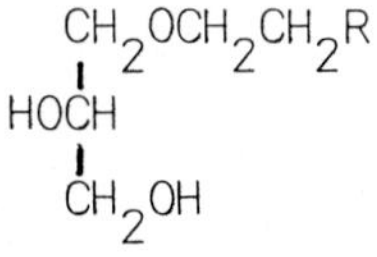

$$CH_2OCH_2CH_2R$$
$$HOCH$$
$$CH_2OH$$

Batyl alcohol (1-O-octadecyl-sn-glycerol) can be made from 2,3-isopropylidene-sn-glycerol by condensation with octadecyl iodide in the presence of sodium, followed by hydrolysis (E.Baer and H.O.L.Fischer, J. biol. Chem., 1941, 140, 397).

Both chimyl alcohol (1-O-hexadecyl-sn-glycerol) and selachyl alcohol (1-O-octadeca-cis-9-enyl-sn-glycerol) can be similarly prepared. It is also possible to convert 3-O-alkyl derivatives into the corresponding 1-O-alkyl derivatives by a Walden inversion at C-2 (W.E.M.Lands and A.Zschocke, J. lipid Res., 1965, 6, 324).

A series of glyceryl ethers can be synthesised from 1,2-isopropylidene-sn-glycerol by alkylation with eicosyl mesylate in the presence of potassium hydroxide to give 3-O-eicosyl-sn-glycerol. The isopropylidene group is then removed by hydrolysis to give a product which can then be further alkylated or acylated (W.J.Baumann *et al.*, Z. physiol. Chem., 1966, 347, 52).

$$\underset{\substack{\text{Me}_2\text{C}}}{\overset{\displaystyle \text{OCH}_2}{\underset{\displaystyle \text{OCH}}{\diagdown}}}\quad \xrightarrow{\text{C}_{20}\text{H}_{41}\text{SO}_3\text{Me/KOH}} \quad \underset{\substack{\text{Me}_2\text{C}}}{\overset{\displaystyle \text{OCH}_2}{\underset{\displaystyle \text{OCH}}{\diagdown}}} \quad \xrightarrow{\text{MeCOOH}}$$

with pendant CH_2OH on left structure and $\text{CH}_2\text{OC}_{20}\text{H}_{41}$ on right structure, m.p. 37–38°

$$\underset{\substack{\text{CH}_2\text{OC}_{20}\text{H}_{41}}}{\overset{\displaystyle \text{CH}_2\text{OH}}{\underset{\displaystyle \text{HOCH}}{\big|}}} \quad \xrightarrow{\text{Ph}_3\text{CCl}} \quad \underset{\substack{\text{CH}_2\text{OC}_{20}\text{H}_{41}}}{\overset{\displaystyle \text{CH}_2\text{OCPh}_3}{\underset{\displaystyle \text{HOCH}}{\big|}}} \quad \xrightarrow{\text{C}_{20}\text{H}_{41}\text{SO}_3\text{Me}}$$

m.p. 75–76°

$$\underset{\substack{\text{CH}_2\text{OC}_{20}\text{H}_{41}}}{\overset{\displaystyle \text{CH}_2\text{OCPh}_3}{\underset{\displaystyle \text{H}_{41}\text{C}_{20}\text{OCH}}{\big|}}} \quad \xrightarrow{\text{MeOH/HCl}} \quad \underset{\substack{\text{CH}_2\text{OC}_{20}\text{H}_{41}}}{\overset{\displaystyle \text{CH}_2\text{OH}}{\underset{\displaystyle \text{H}_{41}\text{C}_{20}\text{OCH}}{\big|}}}$$

m.p. 48–49° m.p. 63–63.5°

$$\xrightarrow{\text{C}_{19}\text{H}_{39}\text{COCl}} \quad \underset{\substack{\text{CH}_2\text{OC}_{20}\text{H}_{41}}}{\overset{\displaystyle \text{CH}_2\text{OCOC}_{19}\text{H}_{39}}{\underset{\displaystyle \text{H}_{41}\text{C}_{20}\text{OCH}}{\big|}}}$$

m.p. 59–60°

and

$$\underset{\substack{\text{CH}_2\text{OC}_{20}\text{H}_{41}}}{\overset{\displaystyle \text{CH}_2\text{OH}}{\underset{\displaystyle \text{HOCH}}{\big|}}} \quad \xrightarrow{\text{C}_{19}\text{H}_{39}\text{COCl}} \quad \underset{\substack{\text{CH}_2\text{OC}_{20}\text{H}_{41}}}{\overset{\displaystyle \text{CH}_2\text{OCOC}_{19}\text{H}_{39}}{\underset{\displaystyle \text{C}_{19}\text{H}_{39}\text{COOCH}}{\big|}}}$$

m.p. 65–66.5°

Long chain saturated and mono- and di-unsaturated 1- and 2-glyceryl mono-ethers are made by treating the potassium salts of 1,2-isopropylidene and 1,3-benzylidene-*sn*-glycerol with alkyl halides for the saturated compounds, and alkenyl toluene-p-sulphonates for the unsaturated compounds. This is followed by hydrolysis of the blocking groups with boric acid. This method allows the synthesis of 2-glyceryl ethers (R.Wood and F.Synder, Lipids, 1967, 2, 161).

The synthesis of monopropenyl ethers of glycerol involves the initial reaction of 1,2-isopropylidene-*sn*-glycerol with 3-bromoprop-1-ene in the presence of sodium hydride according to the following scheme (J.D.Bunger and W.L.Howard, U.S. Pat., 3,290,388).

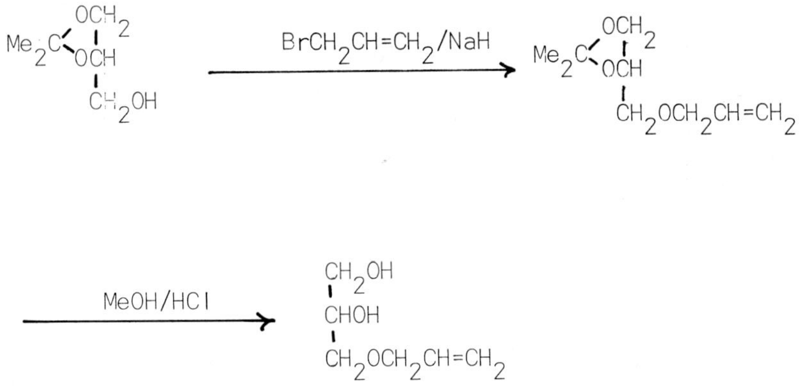

The absolute configuration of natural plasmalogens can be correlated with that of chimyl and batyl alcohols. Thus, 1-alkenyl-2-acyl-*sn*-glycerol from pig heart plasmalogen can be converted into an *O*alkyl glycerol by hydrolysis and hydrogenation. The diacetates of both these diols have been made and their ORD curves compared with those of chimyl and batyl diacetate and distearate. The similarity of the ORD curves show that these compounds have the same absolute configuration (J.Cymerman-Craig *et al.*, Tetrahedron, 1966, 22, 175). It can be concluded that natural alkoxyglycerides are 1-*O*-alkyl-2,3-diacyl-*sn*-glycerol and have the stereochemistry indicated by the structure (II).

$$CH_2OR$$
$$RCOOCH$$
$$CH_2COOR$$

(I I)

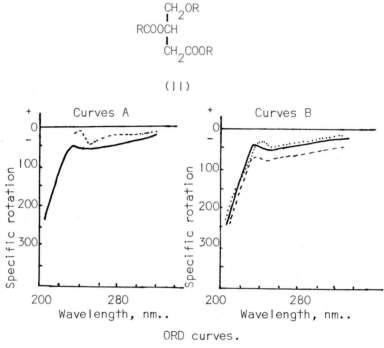

ORD curves.

Curves A   alkenylglycerol diacetate(- - - -) and
             reduced alkenylglycerol diacetate(———).

Curves B   chimyl diacetate(———), batyl diacetate(- -)
             batyl distearate(· · · · · · ·).

Dihydroxyacetone, 1,3-dihydroxypentan-2-one, $HOCH_2COCH_2OH$.

Dihydroxyacetone can be condensed with diaminonitrile to
yield 2-methyl-pyrazine-5,6-dicarbonitrile (Y.Ohtsuka,
Japan Pat., 75 77,379).

Di-*O*-benzoyl dihydroxyacetone and other diesters can be synthesised by reaction of dihydroxyacetone with acyl chlorides in pyridine solution (P.H.Bentley and W.McCrae, J. org. Chem., 1970, 35, 2082). The reaction of dihydroxyacetone with acyl chlorides alone provides the dihydroxyacetone mono-esters.

$$
\begin{array}{c}
CH_2OH \\
| \\
C=O \\
| \\
CH_2OH
\end{array}
$$

RCOCl
$$
\begin{array}{c}
CH_2OCOR \\
| \\
C=O \\
| \\
CH_2OH
\end{array}
$$

$RCOCl/C_5H_5N$
$$
\begin{array}{c}
CH_2OCOR \\
| \\
C=O \\
| \\
CH_2OCOR
\end{array}
$$

Dihydroxyacetone reacts with RC=NH(OEt) in liquid ammonia under pressure to produce substituted imidazole-4-methanols (D.P.Johannes and W.Schunack, Arch. Pharm., 1974, 307, 470, 46). Its reaction with ammonia and $NH_2CH=\overline{NH}.AcOH$ gives the unsubstituted imidazole-4-methanol.

$$
RC(=NH)OEt \quad + \quad \begin{array}{c} CH_2OH \\ | \\ C=O \\ | \\ CH_2OH \end{array}
$$

$$\downarrow$$

The formation of di-*O*-benzoyl dihydroxyacetone dialkyl mercaptals can be achieved by the reaction of di-*O*-benzoyl dihydroxyacetone with alkyl mercaptans (RSH) where R = Me, Et or $CHMe_2$, in the presence of zinc chloride (H.Zinner *et al.*, J. prakt. Chem., 1972, 31). These mercaptals can be oxidised with hydrogen peroxide.

$$\underset{\substack{|\\ \text{C=O}\\ |\\ \text{CH}_2\text{OCOPh}}}{\text{CH}_2\text{OCOPh}} \xrightarrow{\text{RSH/ZnCl}_2} \underset{\substack{\text{RS}\diagdown\ |\\ \text{C}\\ \text{RS}\diagup\ |\\ \text{CH}_2\text{OCOPh}}}{\text{CH}_2\text{OCOPh}}$$

$$\xrightarrow{\text{H}_2\text{O}_2} \underset{\substack{\text{ROS}\diagdown\ |\\ \text{C}\\ \text{ROS}\diagup\ |\\ \text{CH}_2\text{OCOPh}}}{\text{CH}_2\text{OCOPh}} \quad \text{or} \quad \underset{\substack{\text{RO}_2\text{S}\diagdown\ |\\ \text{C}\\ \text{RO}_2\text{S}\diagup\ |\\ \text{CH}_2\text{OCOPh}}}{\text{CH}_2\text{OCOPh}}$$

Diethyl oxomalonate, EtOCOCOCOOEt, can be made from diethyl malonate by bromination, followed by reaction with potassium acetate and then finally heating at 270°C (J.Faust and R. Mayer, Synthesis, 1976, <u>6</u>, 411).

$$\underset{\substack{\text{COOEt}\\ |\\ \text{CH}_2\\ |\\ \text{COOEt}}}{} \xrightarrow{\text{Br}_2} \underset{\substack{\text{COOEt}\\ |\\ \text{CBr}_2\\ |\\ \text{COOEt}}}{} \xrightarrow[\text{ii)heat}]{\text{i)MeCOOK}} \underset{\substack{\text{COOEt}\\ |\\ \text{C=O}\\ |\\ \text{COOEt}}}{}$$

It undergoes reaction with primary aromatic amines to yield the corresponding iminomalonates which, on reaction with an acid chloride inthe presence of triethylamine, give β-lactams (A.K.Bose, M.Tsai and J.C.Kapur, Tetrahedron letters, 1974, <u>40</u>, 3547).

COOEt
|
C=O   +
|
COOEt

PhOCH₂COCl
———————————→

$$ PhOCH_2COCl $$

Anumber of substituted aromatic amino-carboxylic acids
or amides react with diethyl oxomalonate to give addition
compounds which do not cyclise to quinazolines or
benzoxazines on refluxing with polyphosphoric acid
(K.Wasti, T.F.Lemke and M.W.Mosher, Synthesis, 1974, **8**, 570).

Similarly, condensation of diethyl oxomalonate can occur with
benzofuran-2(3H)-one, benzothiophen-2(3H)-one and 2-indoline
(H.Wolfers, U.Kraatz and F.Korte, Heterocycles, 1975, **3**, 571).

When diethyl malonate is left in a solution of an aromatic diazonium salt it provides the aryl hydrazones of diethyl oxomalonate (E.P.Nesynov, M.M.Besprozvannaya and P.S.Pel'kis, Zh. org. Khim., 1966, **2**, 1213).

The phenylhydrazone derivative may also be prepared by the reaction of diethyl oxomalonate with phenylhydrazine. This can then be converted to the dicarboxylic acid and then to the diacid chloride which can be cyclised with titanic chloride to 4-hydroxy-cinnoline-3-carboxylic acid (H.J.Barber, E.Lunt, K.Washbourn and W.R.Wragg, J. chem. Soc.(C), 1967, 1657).

When diethyl oxomalonate reacts with the morpholine enamine of pentan-3-one in benzene at $25^{\circ}C$ and then with sodium acetate in aqueous acetic acid it provides a substituted hydroxymalonate (12) which can be converted into α-carbethoxy-β-methyl-$\gamma_{\alpha\beta}$-ethylidene-$\Delta^{\alpha\beta}$-

butenolide (13) (A.G.Schultz and Y.K.Yee, J. org. Chem.,
1976, 41, 561).

If, however, the pyrrolidine enamine of pentan-3-one is used
the product (14) is obtained.

When diethyl oxomalonate is reacted with ketones in the presence of phosphorus pentoxide in methane sulphonic acid the butenolide can be obtained directly.

The Wittig reaction between o-$(Ph_3P=CH)_2C_6H_4$ and diethyl oxomalonate yields ethyl 3-ethoxy-2-naphthoate (W.H.Ploder and D.F.Tavares, Canad. J. Chem., 1970, 48, 2440).

The reaction of diethyl oxomalonate with butadiene derivatives, such as $H_2C=CMeCH=CHOAc$, dihydropyran derivatives can be prepared (J.F.W.Keana and P.E.Eckler, J. org. Chem., 1976, 41, 2850).

Hydroxymalonic acid, HOCH(COOH)$_2$, can be silylated with trimethyl chlorosilanes to give initially the bis(trimethylsilyl) ester which can be further silylated to the tris(trimethylsilyl) ester (O.A.Mamer and S.S.Tjoa, Clin. Chem., 1973, 19, 58).
Mesoxalic semialdehyde, CHOCOCOOH, can be made from dihydroxyfumaric acid by its oxidation with iodine (M.B.Fleury, Bull. Soc. Chim. Fr., 1966, 9, 2899)

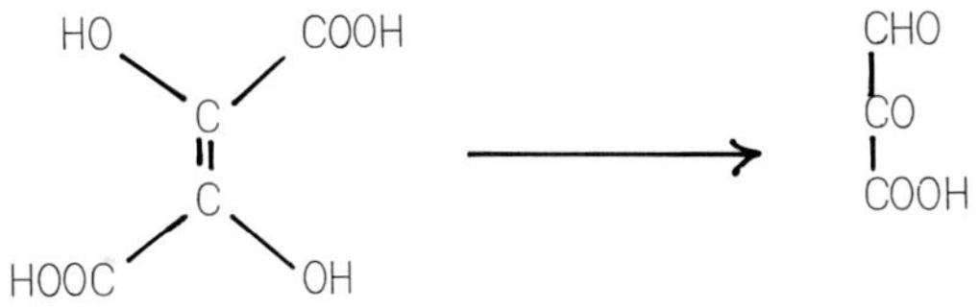

Hydroxymalondialdehyde, tartronaldehyde, HOCH(CHO)$_2$, can be synthesised from 2-chloro-2-ethoxy-ethanoyl chloride as shown by thw following scheme of reactions (B.Horst and H.J.May, Ber., 1966, 99, 3771).

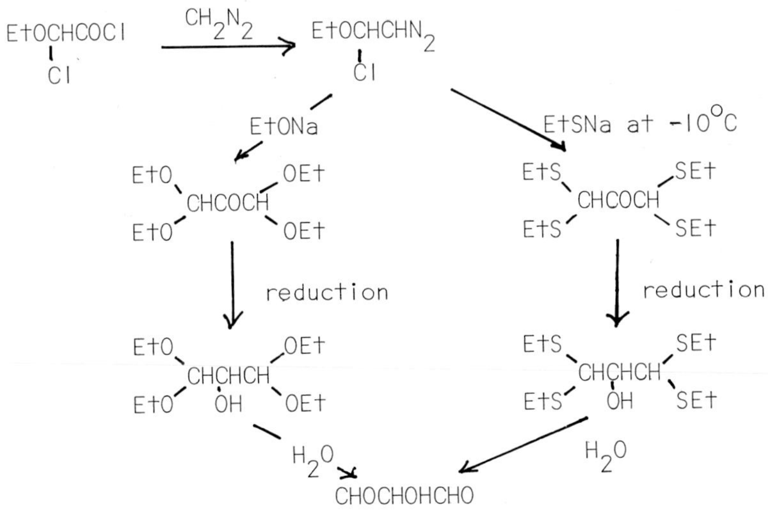

Malondialdehyde can be generated from 1,1,3,3-tetraethoxy-propane (F.Mashio and Y.Kimura, Nippon Kagaku Zasshi, 1962, 83, 303) and then can be coupled with aryldiazonium tetrafluoroborates to provide substituted arylazomalon-dialdehydes (mesoxaldialdehyde-2-phenylhydrazones) (C.Reichardt and W.Grahn, Ber., 1970, 103, 1065).

Spectroscopic measurements of ^{15}N labelled compounds indicate that in solution arylazomalondialdehydes exist as hydrogen-bonded mesoxaldialdehyde-2-phenylhydrazones.

The synthesis of malondialdehyde from 1,1,3,3-tetraethoxypropane can be achieved by treatment with aqueous hydrochloric acid at 55°C and the product has b.p. 120-122°C/0.25mm., Its reaction with diethyl oxomalonate gives the diethyl α-(diformyl) tartrate tautomer with m.p. 128-130.5°(R.B.Woodward, Fr. Pat., 1,495,047).

Chapter 19

TRIHYDRIC ALCOHOLS AND THEIR OXIDATION PRODUCTS
(continued)

Dihydroxycarboxylic and Trihydroxycarboxylic Acids and
Related Compounds

B.J.Coffin

I *Dihydroxycarboxylic Acids.*

Mevalonic acid, $HOCH_2CH_2C(CH_3)(OH)CH_2COOH$, is an important
intermediate in the study of many biosynthetic processes.
A number of methods are available for the synthesis of
labeled mevalonates (J.W.Cornforth and R.H.Cornforth,
Biochem. Soc. Symp., No.29, 5, 1969). Many of these involve
a Reformatsky reaction between acetoacetaldehyde dimethyl
acetal and ethyl bromoacetate. It has been demonstrated that
ester enolates can be generated quantitatively by use of
hindered amines (M.W.Rathke, J.Amer. chem. Soc., 1970, 92,
3222; M.W.Rathke and A.Lindert, *ibid.*,1971, 93, 2318). This
procedure is ideally suited to the synthesis of $[3-^{13}C]-$
labelled mevalonates by the following sequence of reactions
(D.E.Cane and R.H.Levin, J. Amer. chem. Soc., 1976, 98, 1183).

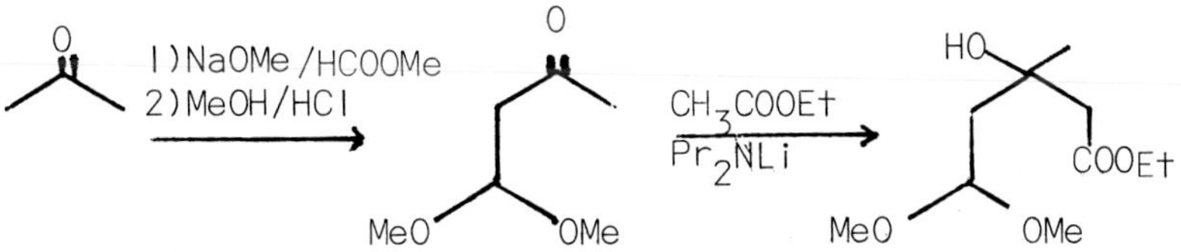

This sequence of reactions can also be adapted to the synthesis of $[3,4-^{13}C_2]$ -mevalonate. Acetoacetaldehyde-2-^{13}C dimethyl acetal is made by reaction of acetone-2-^{13}C with sodium methoxide and methyl formate, followed by methanolic hydrogen chloride (E.E.Royals and K.C.Brannock, J. Amer. chem. Soc., 1953, 75, 2050).

Mevalonolactone can be made by the reaction of $MeCOCH_2CH_2OH$ as its tetrahydropyranyl ether with acetic acid and naphthyl lithium containing an excess lithium, followed by hydrolysis (Fr. Pat., 2,098,755).

Labelled mevalonolactone can be prepared from labelled acetic acid by the following sequence of reactions (J.A. Lawson et al.,Synthesis, 1975, 11, 729)

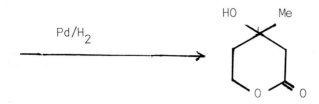

Biochemical Reactions involving Mevalonic acid.

The use of substrates doubly labelled with stable isotopes is ideally suited to the study of molecular rearrangements. The study of methyl migrations during lanosterol biosynthesis has used mass spectrometric analysis of samples of labelled acetic acid obtained by degradation (R.K.Maudgal *et al.*, J. Amer. chem. Soc., 1958, 80, 2589; J.W.Cornforth *et al.*, Tetrahedron, 1959, 5, 311). The combination of double labeling techniques and ^{13}C nmr using vicinal carbon-carbon couplings has been used as a probe for both bond-making and bond-scission processes. The use of $[1,2-^{13}C_2]$- acetate to study terpenoid and polyketide biosynthesis results in pairs of doublets in the ^{13}C-nmr spectrum of the product arising from vicinal carbons which originate from a common molecule of acetate (H.Seto *et al.*,J. Amer. chem. Soc., 1973, 95, 8461; A.G.McInnes *et al.*,Chem. Comm., 1974, 282). Any process which breaks a bond between paired acetate carbons will result in ^{13}C-nmr signals which are enhanced but unsplit. The use of $[1,2-^{13}C_2]$ - acetate leads to the biosynthetic formation of mevalonate containing three groups of paired atoms: C-1 and C-2, C-3 and C-3', and C-4 and C-5.

Since C-1 is subsequently lost as carbon dioxide in the format
formation of isopentenyl pyrophosphate, all carbons derived

from C-2 of the mevalonate give rise to singlets. As long as
the C-3,3' pair or the C-4,5 pair remain intact the
corresponding ^{13}C-nmr signals in any derived metabolite
appears as coupled doublets. If at any time either of these
bonds is broken the spectrum resulting from the product
exhibits enhanced singlets. This method cannot be used
to detect directly the breaking of bonds derived from
either C-2,3 or C-3,4 bonds of mevalonate. These two
processes may be observed by using mevalonate doubly labelled
at the appropriate sites. The use of labelled mevalonate has
the further advantage of providing direct evidence for the
intermediacy of mevalonate in the biosynthetic sequences.

There are a number of reports on the biosynthesis of
trans, trans-farnesol from mevalonate in various organisms
and tissues (J.B.Richards and F.W.Hemming, Biochem. J., 1972
128, 1345; P.Benvenite *et al.*,Phytochemistry, 1970, 9, 1673).
Enzyme systems can be obtained for the formation of
farnesol pyrophosphate and *trans, trans*-farnesyl triphosphate
(I.Shechter, Biochem. Biophys. Acta, 1973, 316, 222).   Work
with *S. Dulleta* can lead to the formation of *trans, trans*-
farnesol and *trans*-nerolidol (R.D.Goodfellow, Y.-S. Huang
and H.E.Radtka, Insect Biochem., 1972, 2, 467).

trans, trans-farnesol

nerolidol

geranyl geraniol

Farnesyl pyrophosphate synthetase as well as catalysing the reaction of natural substrates, catalyses the reactions of the substrates: dimethyl allyl pyrophosphate, geranyl pyrophosphate and isopentenyl pyrophosphate.

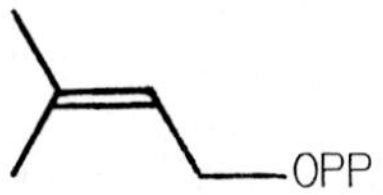

dimethyl allyl
pyrophosphate

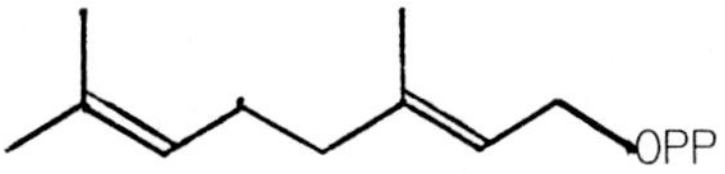

geranyl pyrophosphate

isopentenyl pyrophosphate

It can also produce enzymatically 10,11-dihydrofarnesyl pyrophosphate from isopentenyl pyrophosphate and 6,7-dihydrogeranyl pyrophosphate.

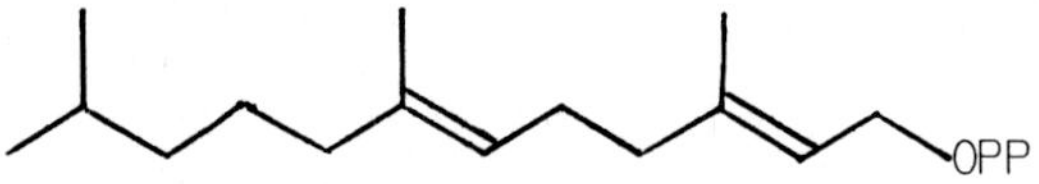

10,11-dihydrofarnesyl pyrophosphate

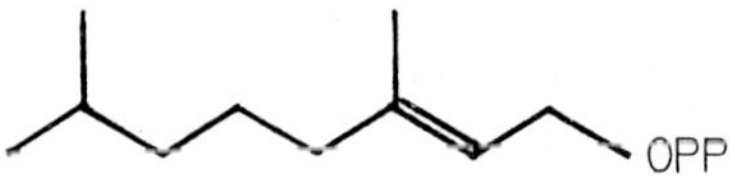

6,7-dihydrogeranyl pyrophosphate

Since the pyrophosphate unit of 6,7-dihydrogeranyl pyrophosphate seems to be bound to the enzyme it is probable that many other compounds can act as a substrate for the enzyme (P.W.Holloway and G.Popjak, Biochem. J., 1967, 104, 57; G.Popjak *et al., ibid.,* 1969, 111, 333).

Using an enzyme extract from pumpkin, the higher homologues of isopentenyl pyrophosphate (K.Ogura *et al.,* Chem. Comm. 1972,881), geranyl pyrophosphate (T.Nishino *et al.,* J. Amer. chem.Soc., 1972, 94, 6849), geranylgeranyl pyrophosphate (*ibid. idem.,* 1971, 93, 794) and squalene pyrophosphate (K.Ogura *et al., ibid.,*1972, 94, 307) can be obtained from mevalonate.

 cis,trans-Farnesol,

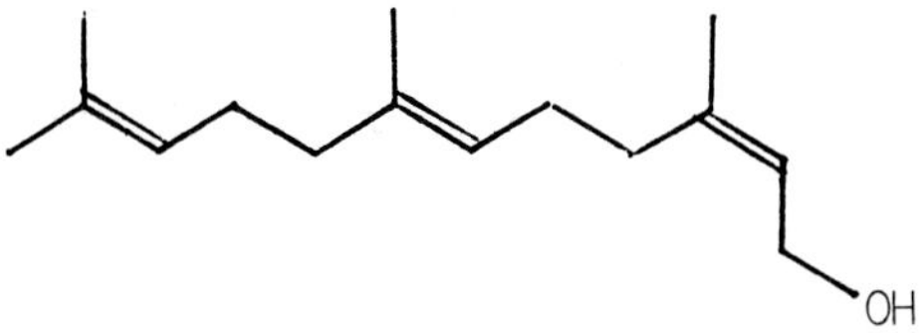

is an important intermediate in the formation of many polycyclic sesquiterpenes and is found in *Pinus radiata* seedlings (G.Jacob *et al.,*Phytochemistry, 1972, 11, 1683) and may be formed from $\left[1-^{3}H\right]$ — geranyl pyrophosphate. Geranyl pyrophosphate, derived from $\left[4-^{14}C\right]$ — isopentenyl pyrophosphate, is a precursor of *trans.trans*-farnesol, *cis,trans*-farnesol and the corresponding aldehydes (C.George Nascimenta and O.Cori,Phytochemistry, 1971, 10, 1803).

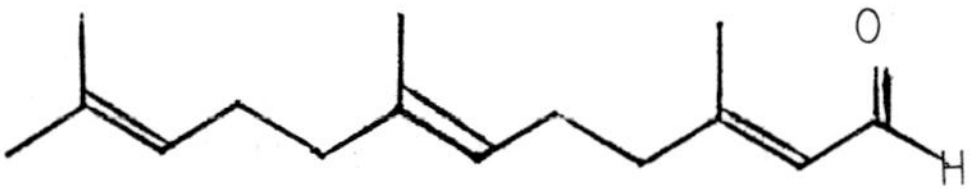

  trans,trans-farnesal

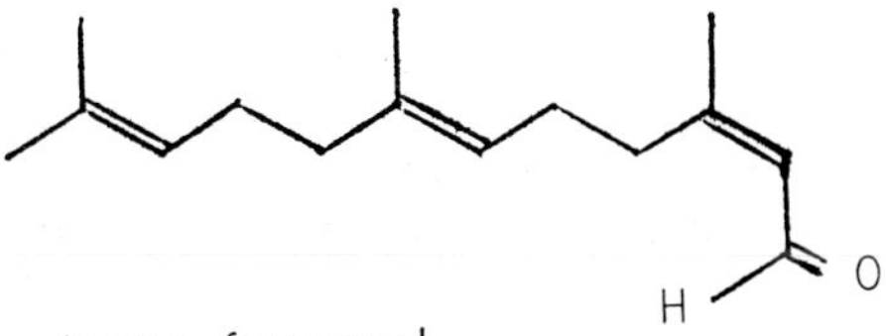

  cis,trans-farnesal

Using the farnesyl pyrophosphate synthetase system of *Pinus radiata, Citrus sinensis* or *Andrographis paniculata,* the formation of both the 2-*cis* and the 2-*trans* double bond

ocurs with the pro-4S proton of mevalonate being specifically
eliminated (K.H.Overton and F.M.Roberts, Chem. Comm., 1973,
378; E.Jedlicki et al., Arch. Biochem. Biophys., 1972, 152,
590).

The enzyme system has a low specificity with regard to the
long-chain part of the allylic substrate and this may permit
rotation about the 2,3-bond in any intermediate carbonium ion.

trans-   cis-

However, when $\left[5-^3H_2, 2-^{14}C\right]$—mevalonate is used,
trans,trans-farnesol retains all the tritium activity, but
with cis,trans-farnesol one-sixth of the total tritium
activity is lost, indicating loss of tritium from C-I
aldehydes as intermediates in the 2,3-isomerisation (K.H.
Overton and F.M.Roberts, Chem. Comm., 1973, 378). Using
sequential labeling, the apparent order of synthesis is
trans,trans-farnesol, trans,trans-farnesal, cis,trans-
farnesal and cis,trans-farnesol (L.Chayet et al., Phytochemistry
1973, 12, 95).   In addition, the pro-IS hydrogen is lost in
the isomerisation of trans,trans-farnesol into cis,trans-
farnesol and in the reverse reaction, the stereochemistry
is opposite in that the pro-IR hydrogen is specifically
lost (K.H.Overton and K.M.Roberts, Chem. Comm., 1974, 385).

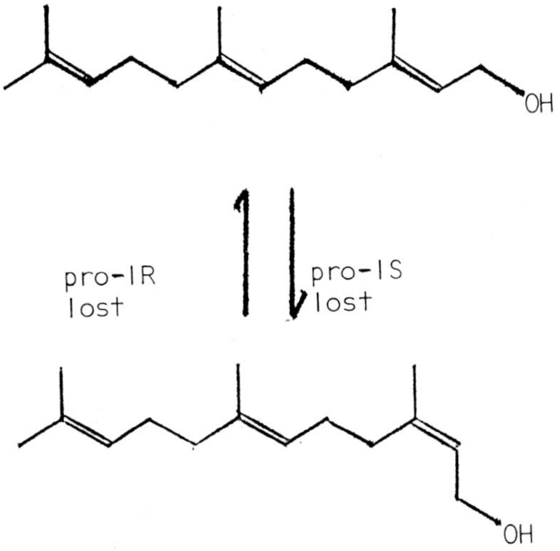

pro-IR
lost

pro-IS
lost

Closely related to farnesol, is methyl 10,11-epoxy-farnesoate, an insect juvenile hormone.

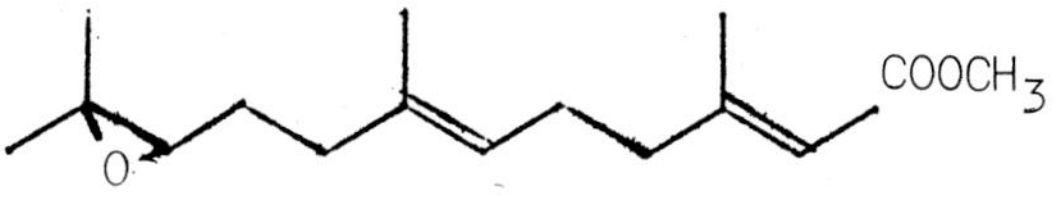

It can be isolated from the tobacco hornworm moth, *Manduca sexta* (K.J.Judy *et al.*, Proc. natl. Acad. Sci., U.S.A., 1973,70, 1509). Using labelled mevalonate, it has been shown that the biosynthesis of the juvenile hormone skeleton follows an isoprenoid pathway with specific incorporation of mevalonate into the compounds:

These compounds are thought to be derived from one homoisoprenoid unit and two isoprenoid units. The homoisoprenoid
unit, homomevalonic acid,

Homomevalonic acid

is produced from propanoate and two acetate fragments
rather than from three acetate units. The probable
synthetic route is shown in the following sequence of
reactions.

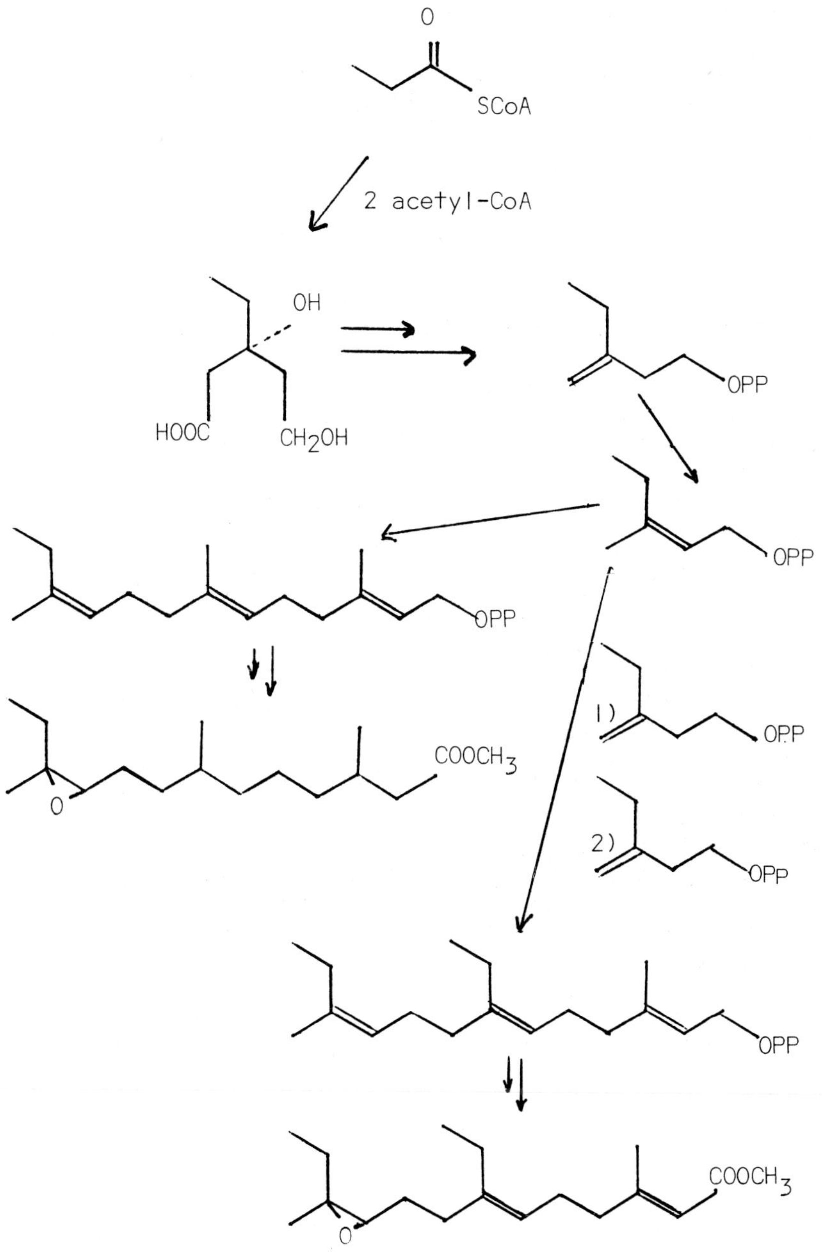
SCoA
2 acetyl-CoA
OH
HOOC
CH2OH
OPP
1)
2)
COOCH3

Abscisic acid has the structure

Abscisic acid

(R.S.Burden and H.R.Taylor, Tetrahedron Letters, 1970, 4071;
S.Isoe *et al.*, *ibid.*, 1972, 2517; T.Oritani and K.Yamashiti,
ibid.,1972, 2521; G.Rybeck, Chem. Comm., 1972, 1190).
It is a widely occurring plant growth regulato (B.V.Milbarrow,
Peanta, 1967, 76, 96). Abscisic acid is found in a number
of ripening fruits, such as avocado, strawberry, banana and
tomato (R.C.Noddle and D.R.Robinson, Biochem. J., 1969, 112,
547).

    Labelled abscisic acid can be obtained by feeding these
plants with $[2-^{14}C]$ —mevalonate. Since abscisic acid
contains a chain *cis* double bond, there are two possible
pathways for its formation which may be distinquished by the
use of pro-4R and pro-4S labelled mevalonates. Initial
formation of the chain *cis* − double bond can take place with
loss of the pro-4R proton in the isomerisation step.
    When the $[4R-^3H]$ −   and $[4S-^3H]$ —mevalonates are
fed to the avocado plant, the results show that the chain
cis double is produced indirectly at some stage from a
trans double bond with the retention of tritium during
the isomerisation. The 4R-proton of mevalonate retained
at C-5 in the formation offarnesol is lost upon cyclisation to
abscisic acid.

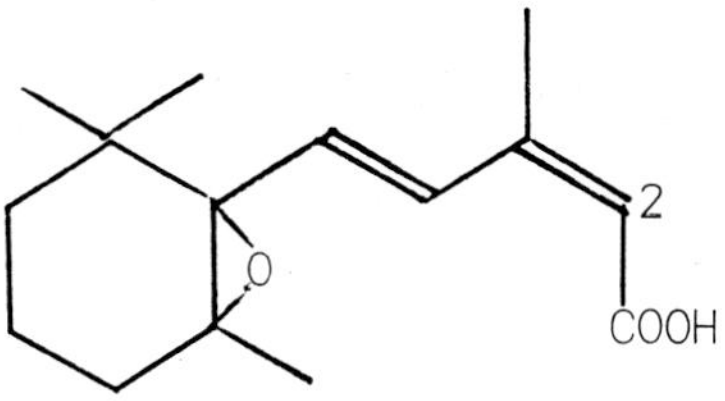

The epoxy compound shown above also acts as a plant growth inhibitor. When the labelled 2-*cis* isomer is used as a precursor of abscisic acid in green tomato fruit, a 1.8% incorporation of activity is seen after 48 hours, whereas the 2-*trans* isomer gives only 0.04% incorporation (B.V.Milborrow and R.C.Noddle, Biochem. J., 1970, 119, 727). When the [2-^{14}C] — compound is used, 1',2'-epi-2-*cis*-xanthoxin acid and another product formed at the same time abscisic acid. In this product the C-1' stereochemistry is opposite to that in abscisic acid. This compound, however, is not a precursor to abscisic acid. However, the (+)-2-*cis*-xanthoxin acid

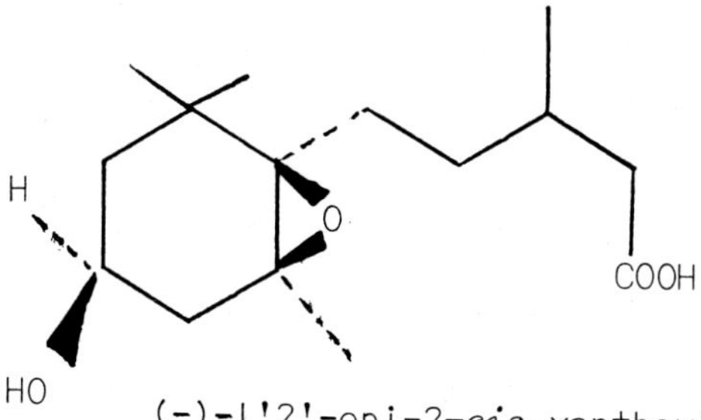

(-)-1'2'-epi-2-*cis*-xanthoxin

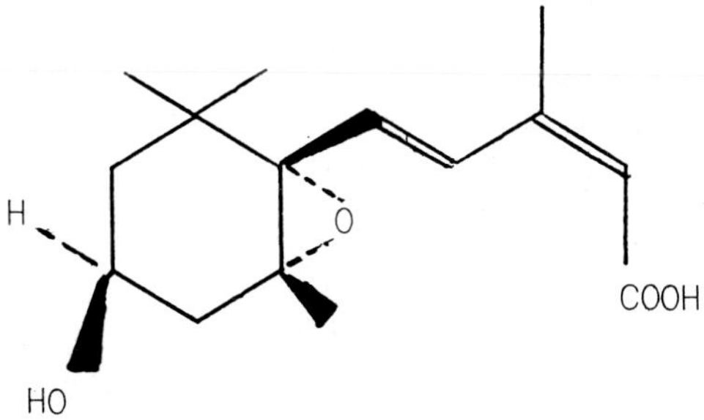

(+)-2-*cis*-xanthoxin

can act as a precursor to abscisic acid in tomato shoots
(B.V.Milborrow and M.Garston, Phytochemistry, 1973, 12, 1597).

The following structure indicates the sites of labeling
and their stereochemistry when abscisic acid is formed from
stereospecifically labelled 2- and 5-tritiated mevalonate:

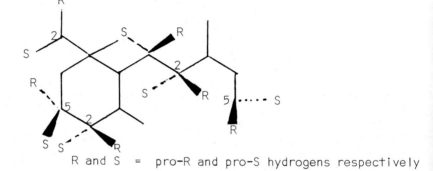

R and S = pro-R and pro-S hydrogens respectively

Clearly in abscisic acid a number of these labels are
completely lost, e.g. at C-1 and C-4';   indeed only one
of the six C-5 hydrogens of mevalonate should be retained in
abscisic acid.

The retention of labels from stereospecifically labelled
mevalonate at C-2 depends not only upon the stereospecificity
of elimination at C-4,5 and C-2',3', but also upon the
presence of isopentenyl isomerase which tends to isomerise
any stereospecifically labelled mevalonate, resulting in
an apparent loss of the 2R-tritium atom and the retention
of the 2S tritium atom.

From [2R-^{3}H]—mevalonate, the three tritium atoms are
retained in abscisic acid at the expected position at C-4
and C-3'.   From [2S-^{3}H] —mevalonate, only one tritium is
retained in abscisic acid.   When [5S-^{3}H] —mevalonate is
used as a precursor, one tritium is retained at C-5.   Thus,
there is an overall *trans* elimination of hydrogen at C-4,5
in the formation of the C-4,5 double bond   (B.V.Milborrow,

Biochem. J., 1972, <u>128</u>, 1135).

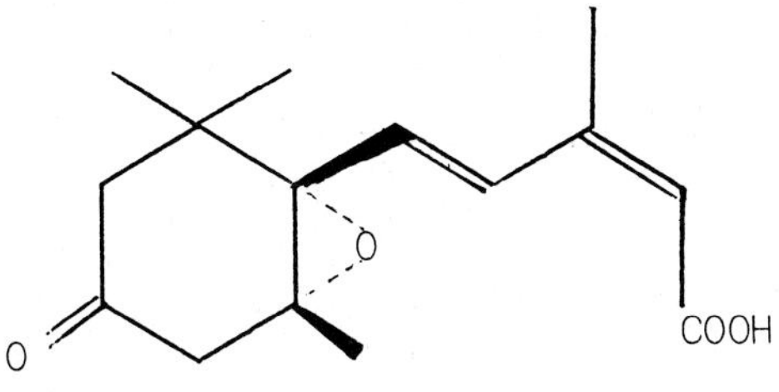

2-*cis*-4'-Oxoxanthoxin acid, shown above, is a very labile compound and rearranges to abscisic acid, suggesting that possibly dehydrogenation at C-4' occurs prior to ring cleavage of the epoxide (H.F.Taylor and R.S.Burden, Proc. R. Soc. London, Ser B, 1972, <u>180</u>, 317).

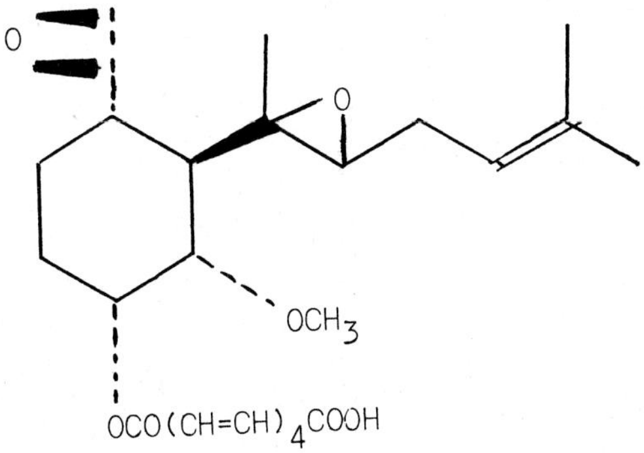

Fumagillin

Fumagillin is an antibiotic extracted from *Aspergillus fumigat us* (D.S.Tarbell *et al.*, J. Amer. chem. Soc., 1961, <u>83</u>, 3096). When $[2-^{14}C]$ —mevalonate is used as a precursor to fumagillin, incorporation of the activity is found to be specifically in the terminal dimethyl group. A mechanism has been suggested for the biosynthesis of fumagillin. Cyclisation of *cis,trans*-farnesol in a manner similar to monoterpene cyclisation and subsequent formation of a four membered ring intermediate. Introduction of a

second double bond into this intermediate in the side chain
and subsequent rearrangement leads to a further intermediate
which is at the same oxidation state as fumagillin   .
(W.B.Turner, 'Fungal Metabolites', Academic Press, New
York, 1972).

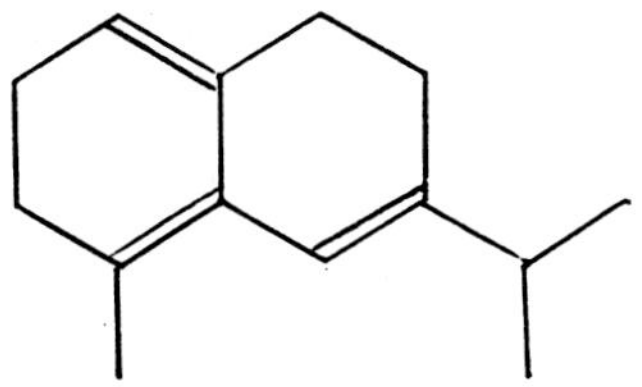

Germacrene C

Germacrene C is the main constituent of the oil of the seed of *Kadsura japonica* (K.Morikawa and Y.Hirosi, Tetrahedron Letters, 1969, 1799). Mevalonate acts as a precursor to germacrene C, catalysed by *K. japonica.*

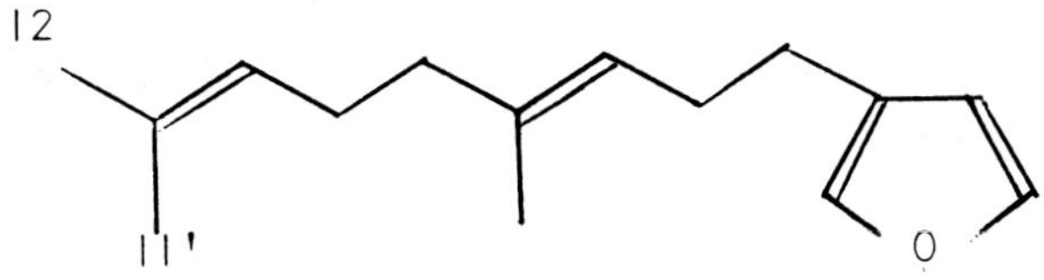

Dendrolasin

Dendrolasin occurs in the ant,*Lasius filiginosus* (A.Quitilco *et al.*, Tetrahedron, 1957, 1, 177). Incorporation of [2-^{14}C] —mevalonate into dendrolasin is poor at C-11, C-11' and C-12 (E.E.Waldner *et al.*, 1969, 52, 15).

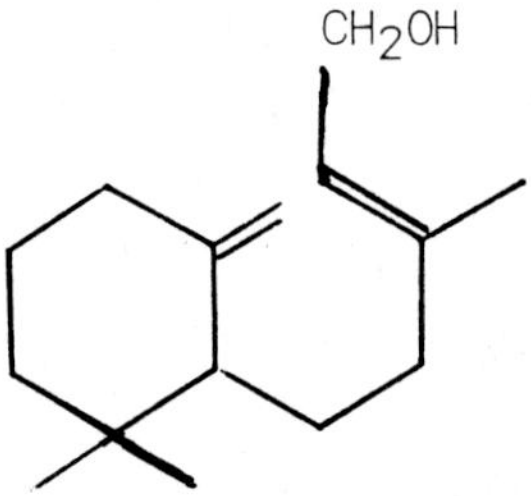

CH$_2$OH

*trans-*γ -Monocyclofarnesol

trans- γ -Monocyclofarnesol can be extracted from the disrupted cells of the leaf-spot fungus, *Helminthosporium siccans* (K.T.Suzuki *et al.*, Chem. Comm., 1971,527). Using [2-^{14}C] —mevalonate a high degree of incorporation of activity can be found in the *trans-*γ - monocyclofarnesol.

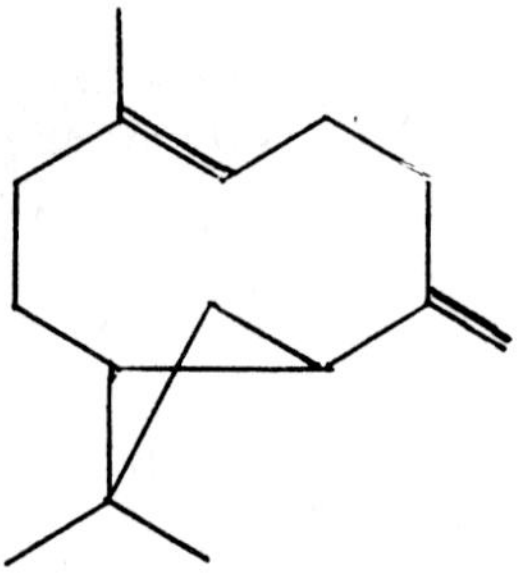

Caryophyllene

Caryophyllene is a major sesquiterpene constituent of peppermint, *Mentha piperita* (L.H.Zalkow *et al.*, J. Amer. chem. Soc., 1960, 82, 6354; R.Croteau and W.Loomis, Phytochemistry, 1972, 11, 1055). When [2-^{14}C] —mevalonate is fed to peppermint plants through the cut stem, the activity is incorporated into the caryophyllene and its distribution is shown in the diagram below (F.E.Regnier *et al.*, Abstr. 150th. National meeting of the American Chemical Society, Atlantic City, N.J., Sept. 1965, p.1160).

This distribution shows that the isopentenyl pyrophosphate derived portion of caryophyllene incorporates the label more extensively than that derived from the dimethyl allyl pyrophosphate. There is also some loss of specificity from the label originally at C-2 of mevalonate in the C-4 methyl and the C-8 methylene in caryophyllene, indicating the presence of an isopentenyl pyrophosphate isomerase.

Petasin is derived from *Petasites hybidus* (A.Aebi and C.Djerrasi, Helv., 1959, 42, 1785), and is a member of the eremorphilane series. Using $[2\text{-}^{14}C]$ —mevalonate as a precursor in this plant, specific incorporation occurs in petasin (J.A.Zabkiewiaz, R.A.B.Keates and C.J.W.Brooks, Phytochemistry, 1969, 8, 2087; C.J.W.Brooks and R.A.B. Keates, *ibid*, 1972, 11, 3235). Degradation of the product

shows that the label is incorporated into the isopropylidene
group and specifically at C-3 and C-9. It has been suggested
that the eremorphilane group of sesquiterpenes arises from
the eudesmane skeleton by a shift of the methyl group at
C-10 to C-5 with the sequential loss of the C-1 (or C-9)
proton. When [4R-^{3}H] -mevalonate is used specific
labeling at C-4 occurs, supporting the concept of the
methyl migration taking place with the simultaneous
1,2-hydride shift. The reduction of petasin with lithium
aluminium hydride gives the diol which contains only two
tritium atoms due to the substituents at positions 1 and 7
being base labile, leading to partial exchange.

OCOC(CH$_3$)=CH(CH$_3$)

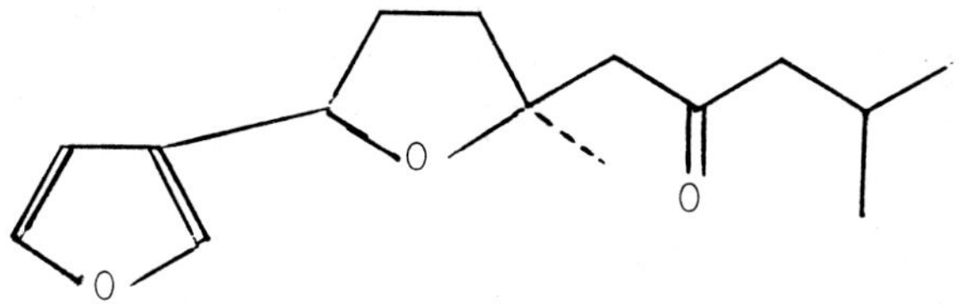

Ipomeamarone

Ipomeamarone is found in the root of the sweet potato, *Ipomoea batatas,* after infection with the black rot fungus, *Ceratocystis fimbriata* (I.Uritani, M.Uritani and H.Yamada, Phytopathology, 1960, 50, 30). It can act as a potential antifungal agent against the invading fungus (I.Uritani and T.Akazawa, Science, 1955, 121, 216). Using $[2-^{14}C]$ - mevalonate precursor, degradation of the product to ipomeanic acid indicates that all of the activity from the mevalonate is retained. This establishes ipomeamarone as a sesquiterpene.

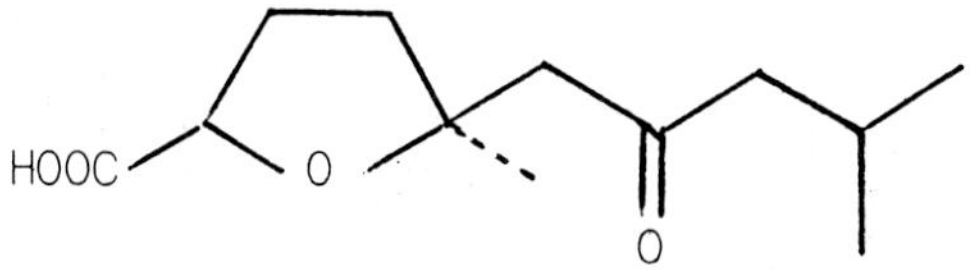

Ipomeanic acid

Caratol is a principal sesquiterpene alcohol of the seeds of the carrot, *Daucus carota* (V.Kykora *et al.,* Coll. Czech. chem. Comm., 1961, 26, 788). Its probable biogenesis from *cis,trans-*farnesol is shown below (M.Soucek, *ibid.,* 1962, 27, 2929).

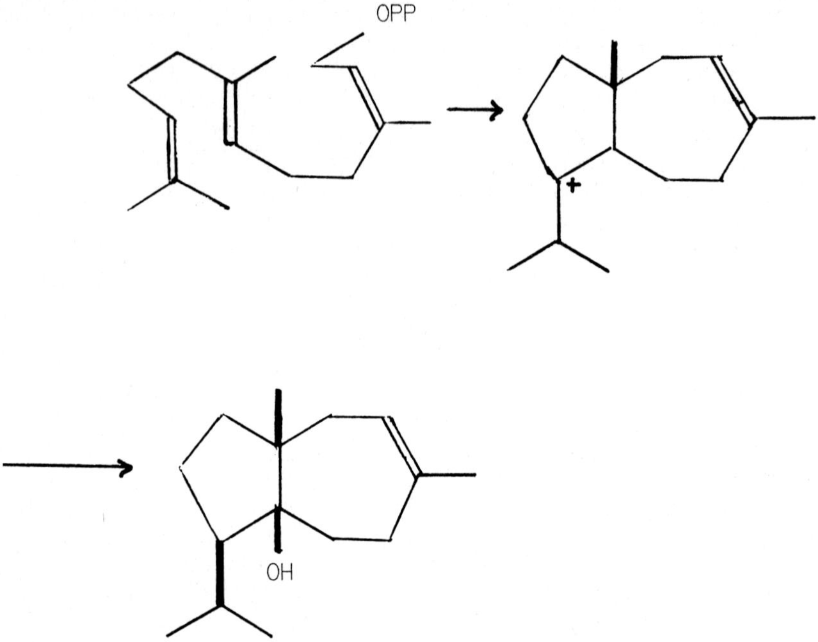

Caratol

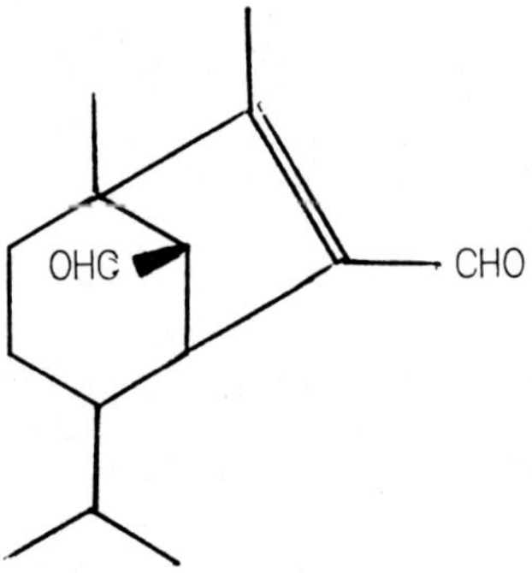

Helminthosporal

Helminthosporal is obtained from the fungus, *Helminthosporium sativum*, which is responsible for the seeding blight, root rot and leaf spot of a number of cereals.

The biosynthesis of helminthosporal from $[2\text{-}^{14}\text{C}]$ -
mevalonate gives low incorporation of activity, but
oxidation and subsequent ozonolysis produces a lactone
retaining a large percentage of the original activity of the
helminthosporal. However, in this degradation one carbon
atom, the exocyclic carbon of the α,β -unsaturated
aldehyde is specifically lost (P. de Mayo et al.,
Experientia, 1962, 18, 359). A biosynthetic route for the
formation of helminthosporal involves a germcrene
intermediate which subsequently cyclises to an intermediate
tricyclic compound, Sativene

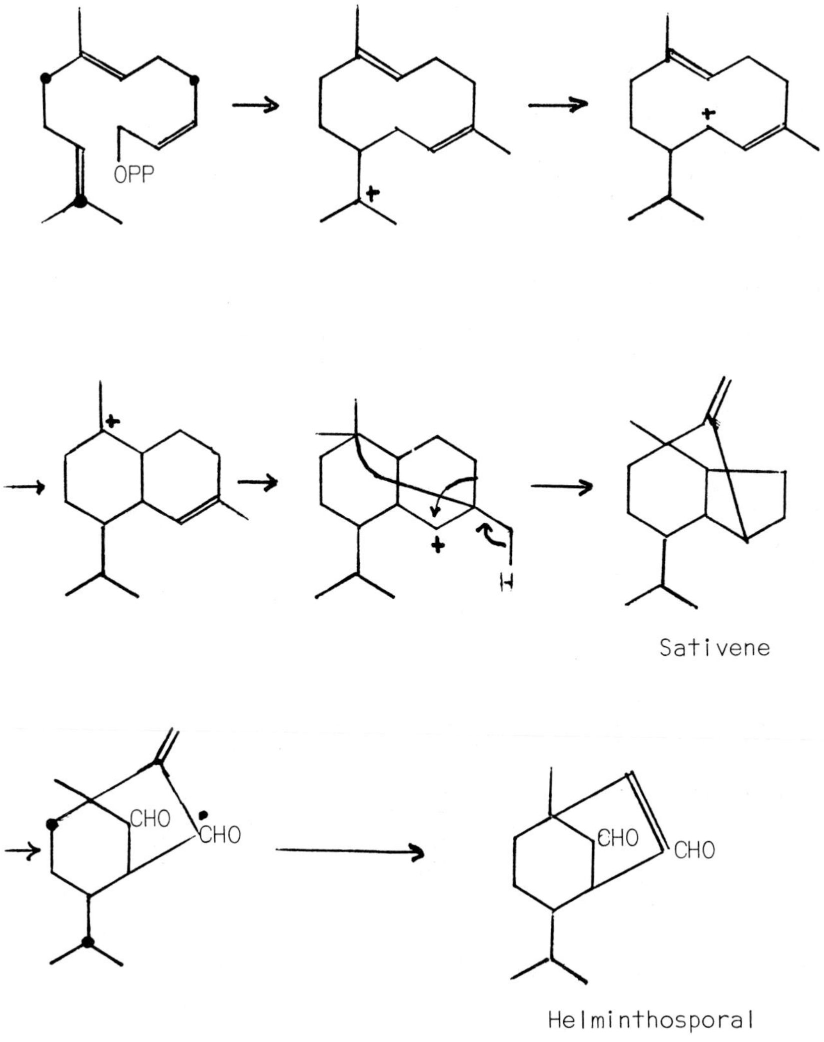

Sativene

Helminthosporal

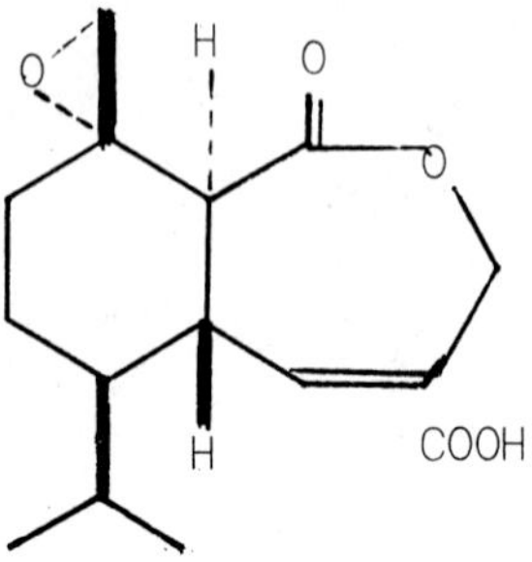

Avocettin

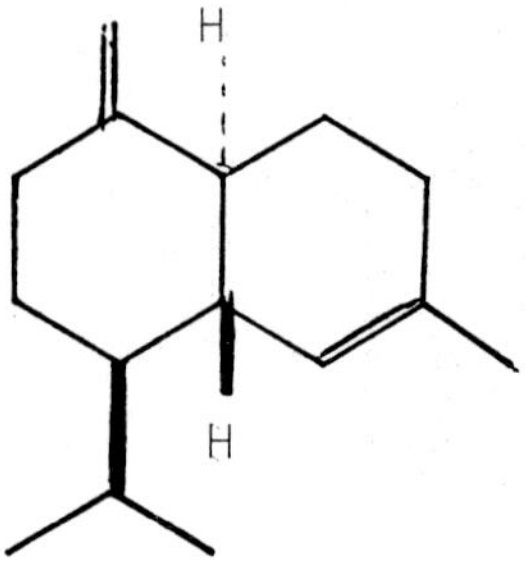

ent-γ-Cadinene

Avocettin can be isolated from *Anthostoma avocetta* and may be thought of as the end product of the extensive oxidation of ent-γ-cadinene.

[2-^{14}C]-Mevalonate is incorporated into avocettin at C-3, C-9 and in the isopropyl side chain. Using [4R-^{3}H] mevalonate in admixture with [2-^{14}C]-mevalonate intact incorporation occurs with one tritium attached to the carbon bearing the isopropyl group. The mechanism of the formation of the isopropyl group of avocettin has been studied using [5R-^{3}H]- and [5S-^{3}H]-mevalonates. Degradation of the product shows that the formation of the isopropyl group involves the stereospecific migration of the pro-5S proton of mevalonate (D.Arigoni, Pure Appl. Chem., 1975, 41, 219).

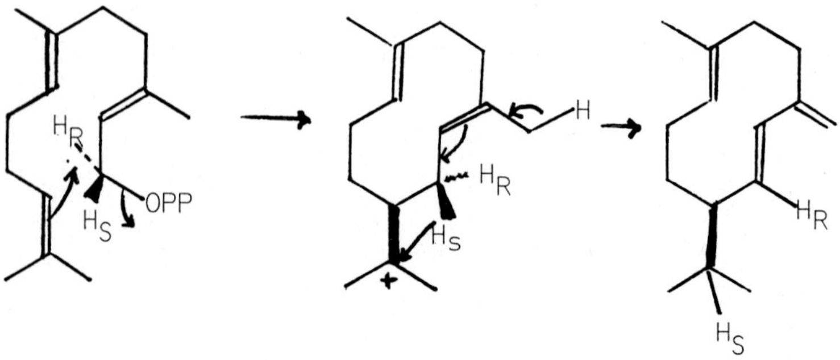

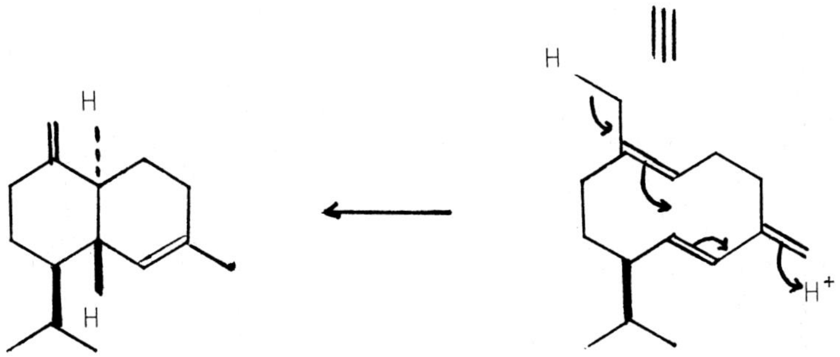

Alkaloids of Nuphar

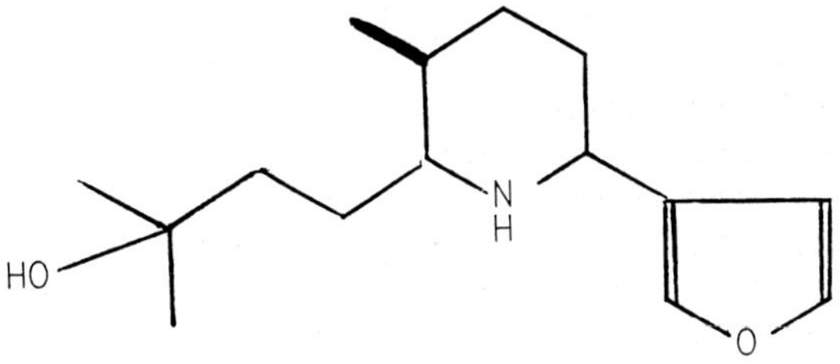

Nupharamine

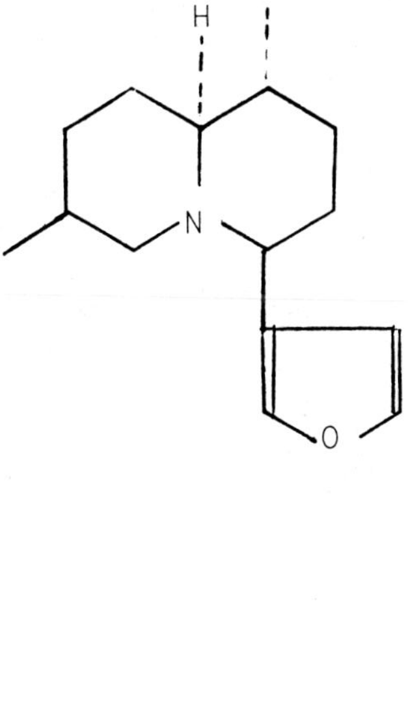

Deoxynapharidine

Nupharamine and deoxynapharidine are typical examples alkaloids found in the genus of aquatic plants, *Nuphar* (O.E.Edwards in 'Cyclopentanoid Terpene Derivatives', ed. W.I.Taylor and A.R.Battersby, Marcel Dekker, New York, 1969, p. 357). Both [2-^{14}C] - and [3,4-^{14}C$_2$] - mevalonates are each specifically incorporated, showing that a sesquiterpene biosynthetic pathway is involved.

Mycophenolic acid

Mycophenolic acid is isolated from *Penicillium brevi-compactum* (S.H.Birkinshaw, H.Raistrick and D.J.Ross, Biochem. J., 1952, 50, 630). [2-^{14}C] -mevalonate is incorporated only into the side chain of mycophenolic acid (A.J.Birch *et al.*, Proc. Chem. Soc. London, 1957, 33). The seven carbon side chain is considered to be derived from a C$_{15}$ farnesyl unit, and oxidation of the side chain probably precedes *O*-methylation. A reasonable biosynthetic scheme involves the following sequence of reactions.

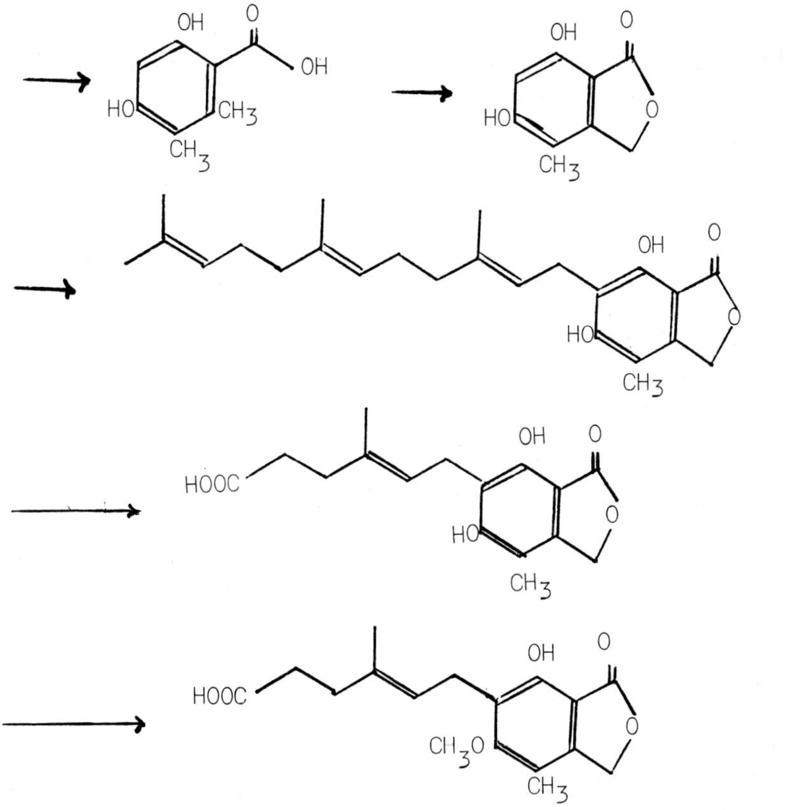

Illudins

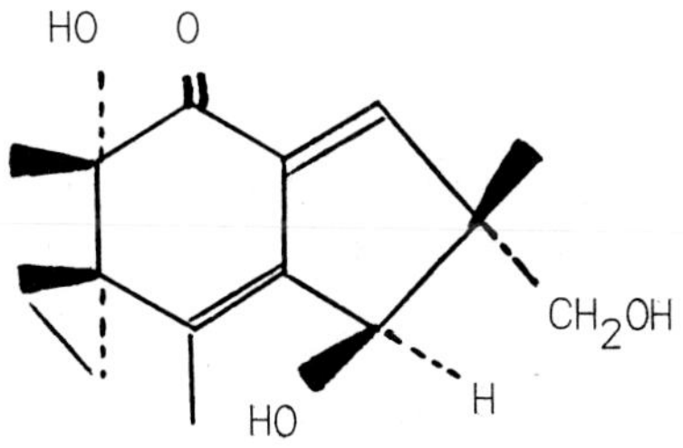

Illudin S

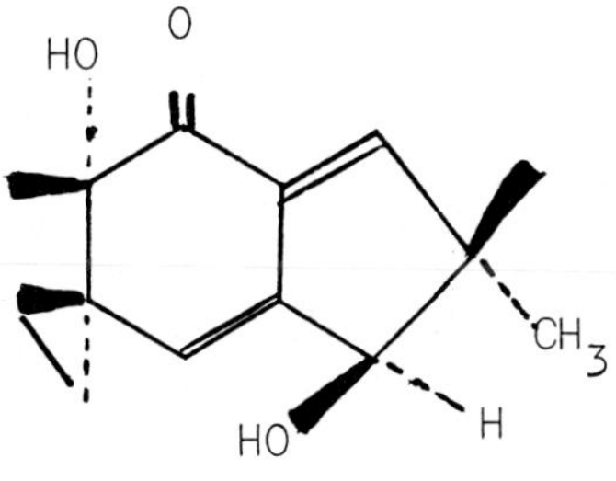

Illudin M

The fungal metabolites, illudin S and illudin M, can be obtained from the *Basidomycetes Clitocybe illudus* and *Lampteromyces japonicus*. They exhibit antibacterial and antitumour activity (T.McMorris and M.Anchel, J. Amer. chem. Soc., 1963, 85, 831).

Using [4R-³H, 2-¹⁴C] -mevalonate, one tritium atom is specifically retained in the formation of illudin M, and degradation of the illudin M after feeding with [2-³H₂, 2-¹⁴C] -mevalonate indicates retention of three (rather than four) tritium atoms (J.R.Hanson and T.Marten, J. Amer. chem. Soc., 1973, 95, 171). The mechanism of the biosynthesis of the illudins involves a bicyclic intermediate which is formed by the loss of a proton derived from C-2 of the mevalonate precursor.

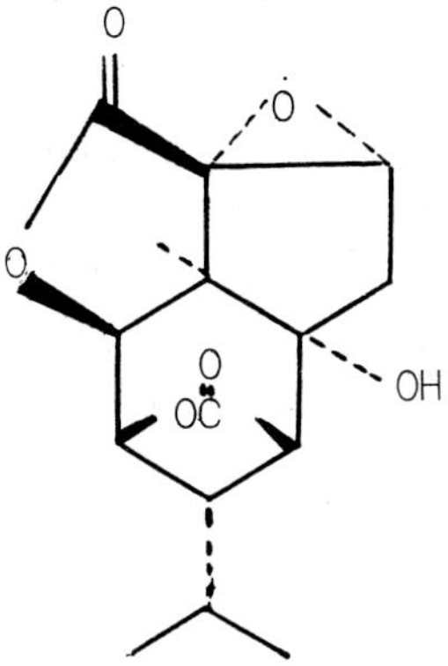

Picrotoxinin

Picrotoxinin is part of a complex picrotoxin found in several species of the *Menispermaceae* (L.A.Porter, Chem. Rev., 1967, 67, 440) together with the closely related coriamyrtin which occurs in a number of Coriarie species (T.Okuda and T.Yoshida, Tetrahedron Letters, 1964, 694)

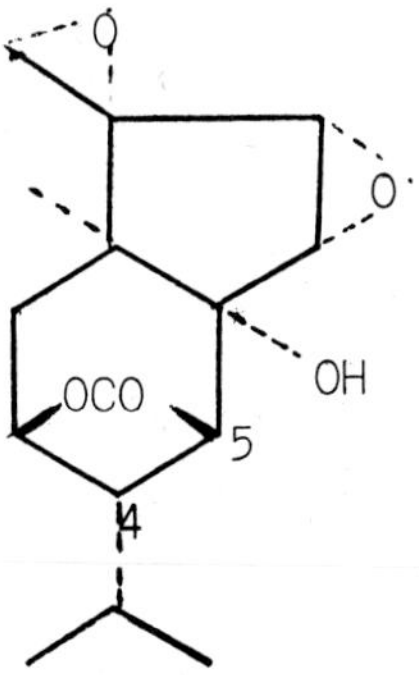

Coriamyrtin

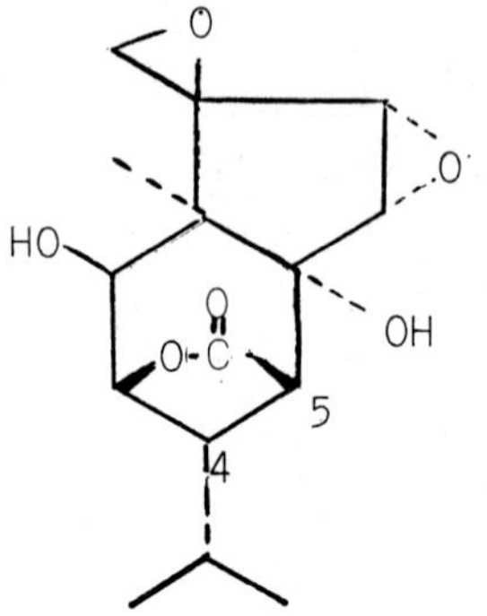

Tutin

Tutin can be considered chemically with coriamyrtin and can be isolated from *Coriaria japonica* (T.Okuda and T.Yoshida, Tetrahedron Letters, 1965, 2137, 4191).

Picrotoxinin can be considered to be derived biogenetically by the oxidative partial removal of the A and B rings of 17-methylandrostane derivatives (H.Conroy, J. Amer. chem. Soc., 1952, 74, 3046) and this group of compounds can be considered to be derived from farnesol by cyclisation and two 1,2-methyl migrations (A.D.Cross, Quart. Rev. chem., 1960, 14, 317).

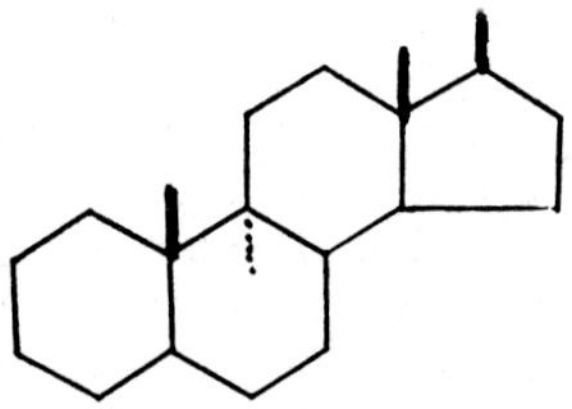

17- methylandrostane

Dendrobine

Dendrobine is also closely related to picrotoxinin and can be isolated from *Dendrobium nobile*. It totally incorporates [2-^{14}C] -mevalonate and this pathway is specific (M.Yamazaki *et al.*, Chem. pharm. Bull. Japan, 1966, 14, 1058). It is thought that dendrobine is derived from a cadalene intermediate in which cyclisation and ring cleavage subsequently occur to give the picrotoxinin skeleton which is then aminated.

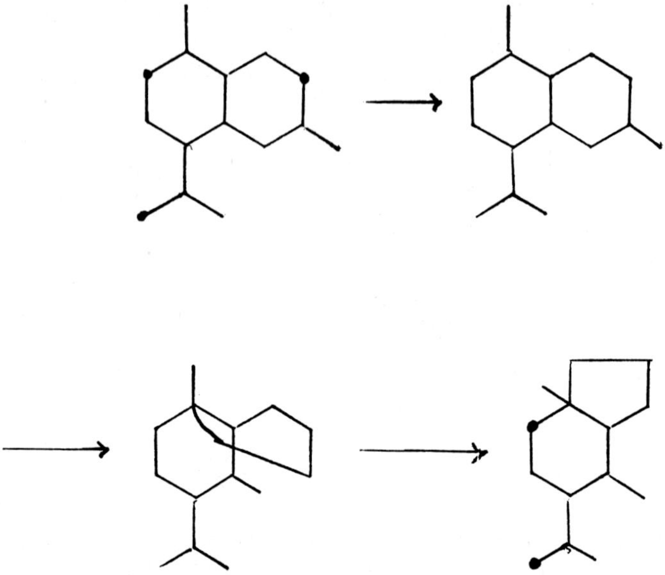

There is low incorporation of activity into both coriamyrtin and tutin in *C. japonica* using $[2\text{-}^{14}C]$ – and $[4\text{-}^{14}C]$ -mevalonates. Degradation of coriamyrtin gives the labeling as shown below (A) (M.Biollez and D.Arigoni, Chem. Comm., 1969, 633). Double labeling studies of the synthesis of tutin in *C. japonica* gives specific incorporation for tritium (Diagram B) (A.Corella *et al.*, Chem. Comm., 1969, 684).

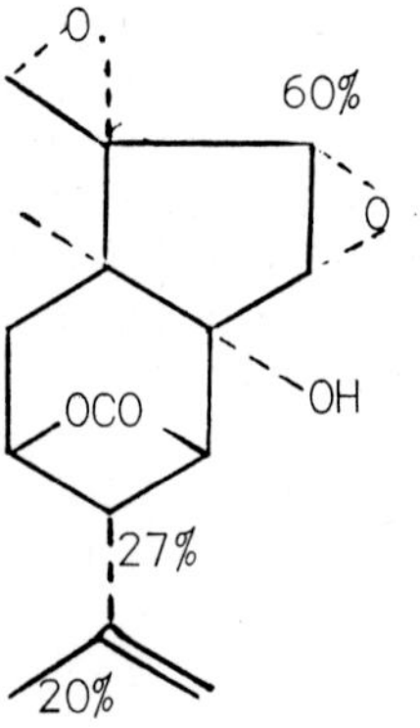

Diagram A

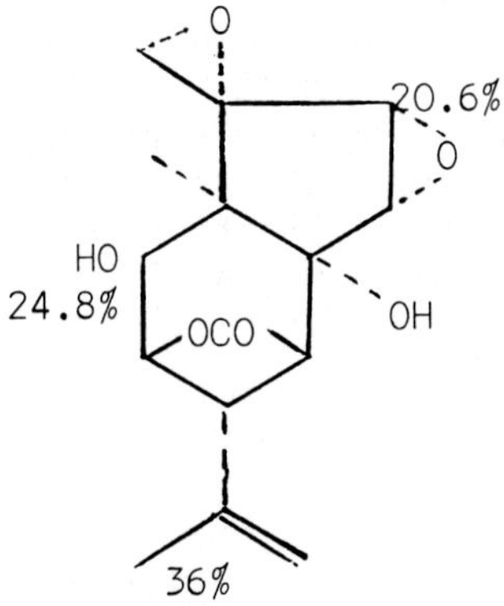

Diagram B

Copaborneol

Copaborneol is a major constituent of *Pinus sylvestris*
oil (M.Kolbe and L.Westfelt, Acta Chem. Scand., 1967, 21,
585; 1970, 24, 1623) and may take part in the biosynthesis
of tutin. Copaborneol labeled with tritium is found to be
a precursor of tutin (K.W.Turnbull *et al.*, Chem. Comm.,
1972, 598). In turn, it is related to the biosynthesis
of the other compounds of this group.

   cis, trans-Farnesol is a key intermediate in the
formation of dendrobine as shown by the use of $[2-^{14}C]$ -
mevalonate as a precursor (O.S.Edwards, J.L.Douglas and
B.Mootoo, Canad. J. Chem., 1970, 48, 2517). The initial
cyclisation of *cis,trans*-farnesol produces a compound with
a positive charge at C-11 which rearranges to give a positive
charge at C-1 of a muurolane derivative. The use of $[4R-^{3}H]$
mevalonate distinquishes some of the mechanisms of hydrogen
migration involved in this transformation. That both
hydrogens from C-1 of *trans,trans*-farnesol are retained in
dendrobine indicates a non-oxidative isomerisation to
cis,trans-farnesol. The pro-5R proton of mevalonate
specifically migrates from the carbon derived from C-1 of
farnesol to C-8 of dendrobine.

Sativene

Sativene is a sesquiterpene and degradation of a number of closely related metabolites show that all of the C-5 protons of mevalonate are retained in sativene. It is the pro-5R proton which migrates to the isopropyl side chain (D.Arigoni, Pure Appl. Chem., 1975, 41, 219). The isopropyl side chain retains the biogenetic identity of the methyl group and is specifically derived from C-2 of mevalonate. The suggested biosynthetic scheme for sativene is shown below.

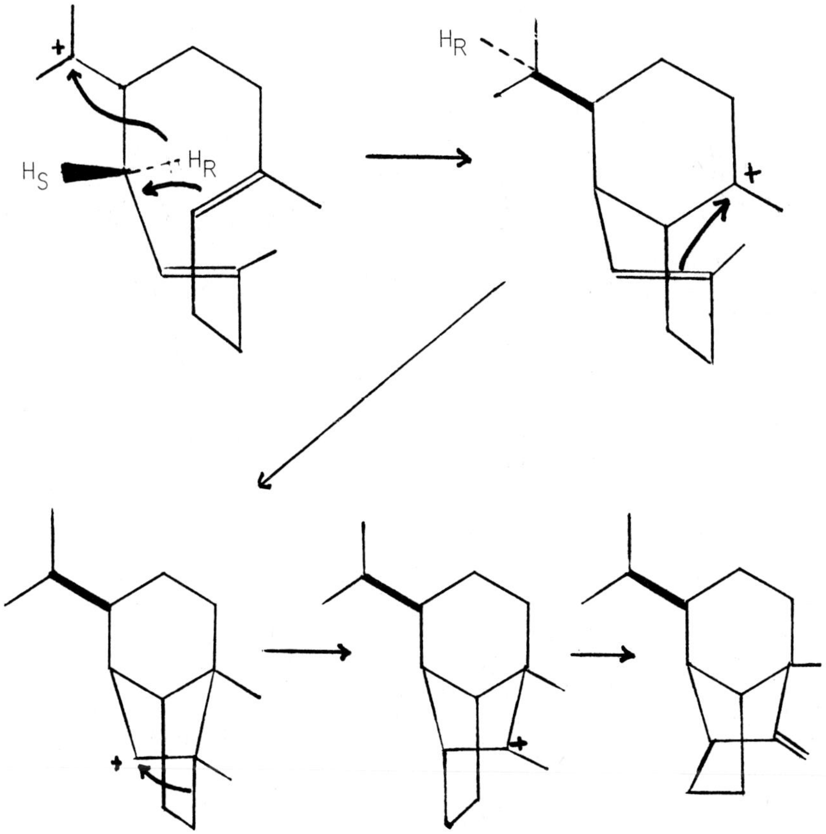

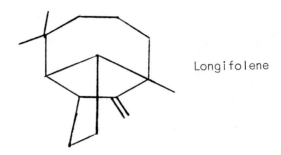

Longifolene

Longifolene occurs in a number of Pinus species
(J.L.Simonsen and D.H.R.Barton, Ed., "The Terpenes', Vol. III,
Cambridge, 1952, p. 92). There are two possible biosynthetic
pathways for the formation of (-)-longifolene depending on
the enantiomeric nature of the initially formed 11-membered
ring compound. Since the two carbon bridge of (+)-
longifolene is specifically derived in *Pinus ponderosa,* the
formation is direct and does not involve a process which
could equilibrate the label in the two carbon bridges. The
stereospecific nature of a 1,3-hydrogen shift to C-3 of
(+)-longifolene is established. The migrating hydrogen can be
shown to be derived from the pro-5R position in mevalonate and
to be located in the *exo* position. With (-)-longifolene
the migrating hydrogen is thus derived from the pro-5S proton
in mevalonate.

The suggested biosynthetic scheme for the formation of
longifolene starts with *cis,trans*-farnesol and involves
a 1,2-shift of the 3,4-bond to the 2,4-bond, possibly with
the simultaneous formation of the 3,7-bond (C.Ourisson, Bull.
Soc. Chim. Fr., 1955, 895; D.Arigoni, Pure Appl. Chem.,
1975, 41, 219).

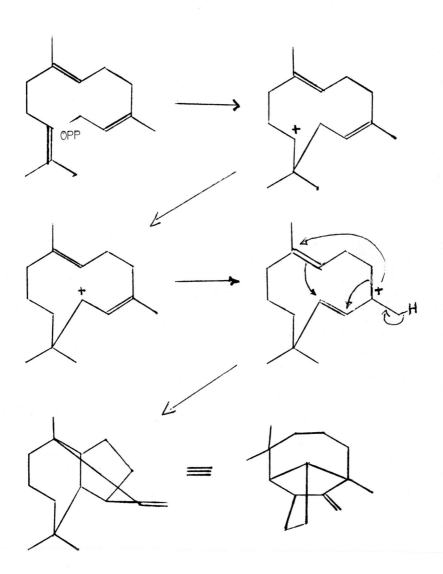

272

The formation of *cis,trans*-farnesol from *trans,trans*-farnesol does not involvean oxidation-reduction reaction sequence but the formation of nerolidyl pyrophosphate intermediate where an inversion of the C-I protons of farnesol can occur. This can be a nonenzymic process (C.George-Nascimento, R.Pont-Lezica and O.Cori, Biochem. Biophys. Res. Comm., 1971, 45, 119).

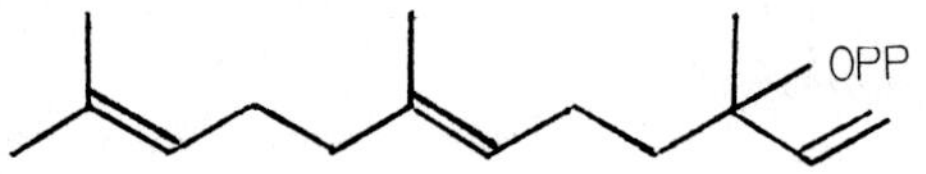

Nerolidyl pyrophosphate

There are three parameters which are important in rationalising the observed stereochemical preferences for hydrogen migration.

a)   The face of attack on the isopropylidene double bond (re or si).

b)   The conformation of the cyclising chain to the olefinic hydrogen of the isopropylidene group (*syn* or*anti*).

c)   The intermediacy of *cis* or *trans* farnesyl pyrophosphate.

In the biosynthesis of both (-)-sativene and (-)-longifolene *cis,trans*-farnesyl pyrophosphate is involved. The initial cyclisation takes place on the *re*-face of the isopropylidene group from an *anti*-conformation.

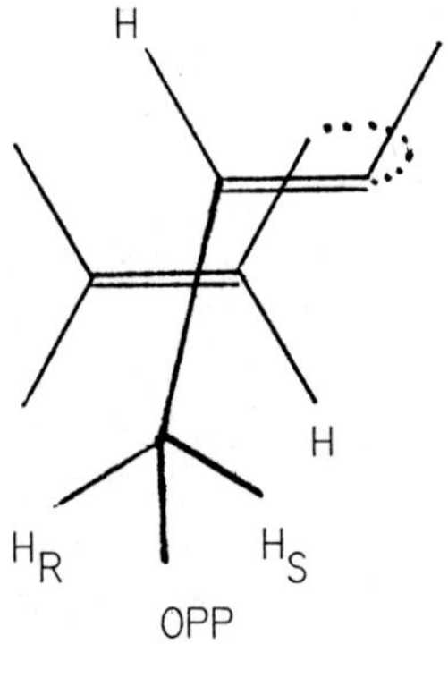

re, anti, cis

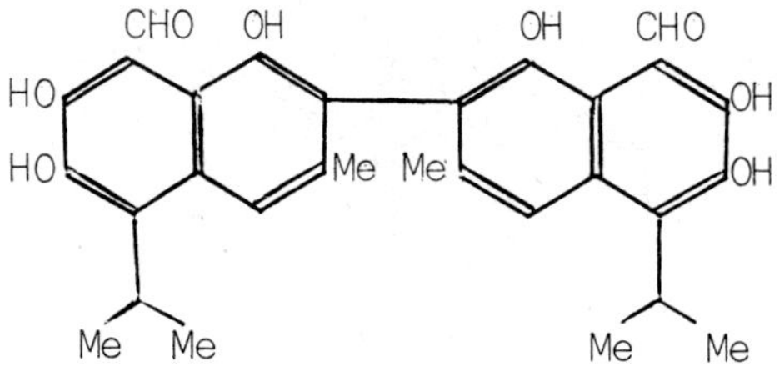

Gossypol

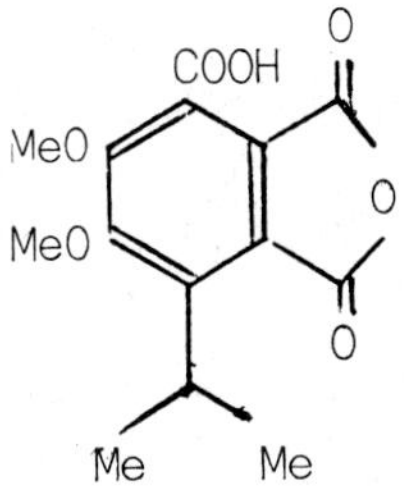

Gossic acid

Gossypol is a toxic compound and occurs in the resin glands of the seeds, tap-root bark and roots of the cotton plant, *Gossypium hirsutum* (H.D.Boyce *et al.*, Oil Soap (Chicago), 1941, 18, 27; F.H.Smith, Nature, 1961, 192, 888).

Since [2-^{14}C] -mevalonate is incorporated into gossypol, its formation follows an isoprenoid pathway (P.F.Heinstein *et al.*, J. biol. Chem., 1962, 237, 2643).

Labelled gossic acid retains all the activity of the starting material and thus the initial cyclisation does not involve *cis,trans*-farnesol which would lead to a loss of one-third of the label in gossic acid. If the mode of cyclisation were as in gossic acid a key compound in the degradation of labelled gossypol is a gossypol hexamethyl ether in which both C-15 formyl groups are removed. This compound exhibits the same activity as gossypol and confirms the specific incorporation of [2-^{14}C] -mevalonate. These results confirm the intact incorporation of the farnesyl unit.

If the mode of cyclisation is as in diagram A:

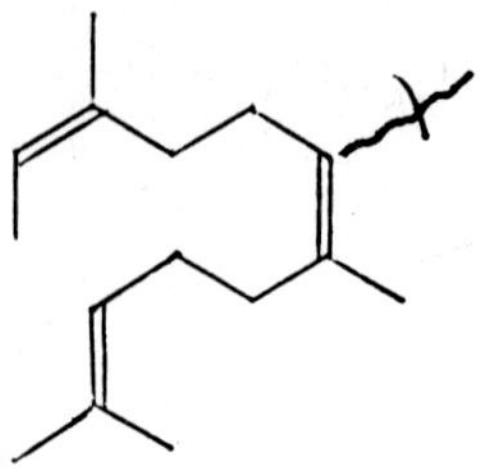

Diagram A

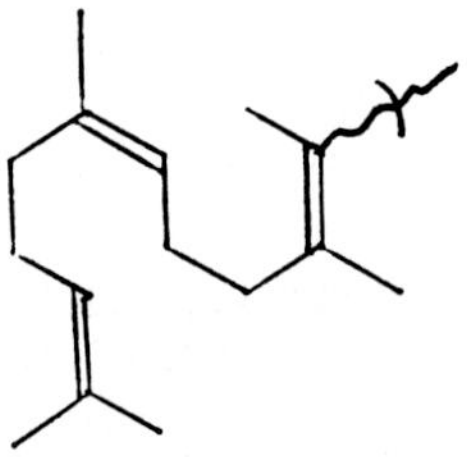

Diagram B

$[2-^{14}C]$ -mevalonate should lead to unlabelled gossic acid and $|2-^{14}C|$ -farnesol should give labelled gossic acid. If the cyclisation is as in diagram B, then the opposite result should occur. In practice gossic acid is labelled from $[2-^{14}C]$ -geraniol and not from $[2-^{14}C]$ -farnesol, thereby supporting the initial cyclisation of B.

All four isomers of farnesol can be used as precursors. Only the two *cis* isomers are incorporated and *cis,cis*-farnesol is the most effective precursor which agrees with the cyclisation of the above intermediate which requires *cis,cis*-farnesol.

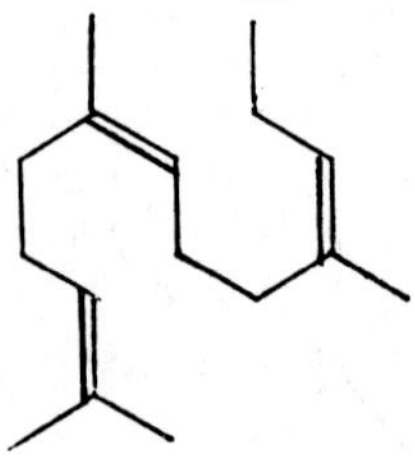

cis,cis-farnesol

Marasmic acid

Marasmic acid is an antibacterial compound from the mold,
Marasmus conigenus (J.J.Dugan *et al.*, J. Amer. chem. Soc.,
1966, 88, 2838).
[2-^{14}C] -mevalonate is incorporated into marasmic acid
with one-third of the label at the geminal dimethyl group
which suggests the following biosynthetic scheme.

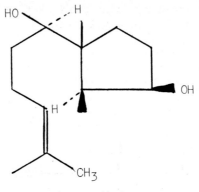

Cyclonerodiol

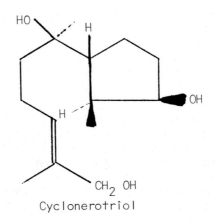

Cyclonerotriol

Cyclonerodiol and the related cyclonerotriol can be isolated from *Fusarium culmorium* (J.R.Hanson *et al.*, J. chem. Soc. Perkin I, 1975, 1586; R.Evans *et al.*, Chem. Comm., 1975, 814).

Using [4,5-$^{13}C_2$] -mevalonate the folding of the farnesyl chain can be shown by all three ^{13}C-^{13}C couplings remaining. With [2-$^3H_2$, 2-^{14}C] -mevalonate considerable tritium is removed by the action of prenyl isomerase, but [4R-^3H, 2-^{14}C] - and [5-$^3H2$, 2-^{14}C] -mevalonates are incorporated without loss of the label.

trans-Farnesyl pyrophosphate is a precursor of cyclonerodiol which itself acts as a precursor of the triol. Nerolidiol is not a precursor indicating that an enzyme bound intermediate may be involved.

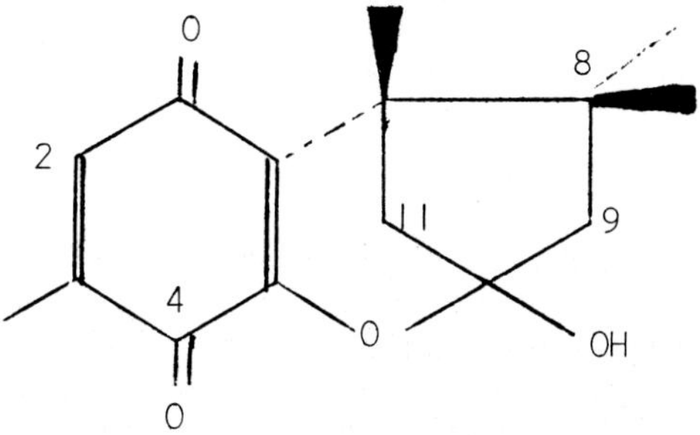

Lagopodin B

Lagopodin B is a quinone antibiotic from *Coprinus lagopus* (P.Bollinger, Dissertation No. 3595, ETH, Zurich, 1965). [2-^{14}C] - Mevalonate is specifically incorporated and degradation shows that one-third of the activity is at C-11, one-third is at the C-8methyl and one-sixth each is at C-2 and C-14, suggesting that cuparene is an intermediate (D.Arigoni and N.Buzzolini, unpublished work). Cuparene, cuparenol and cuparenone are all efficient precursors.

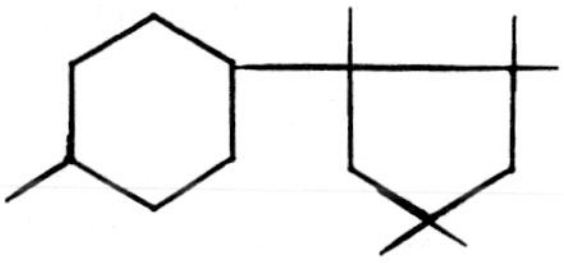

Cuparene

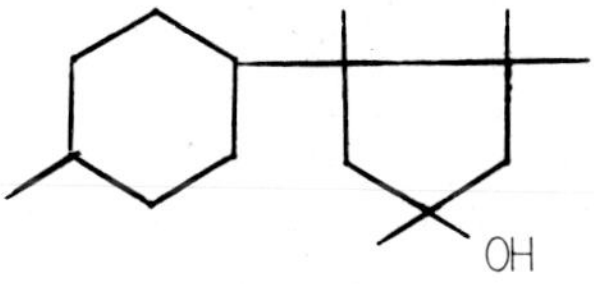

Cuparenol

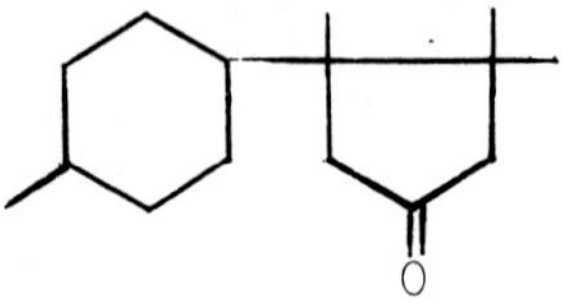

278

Cuparenone

The mechanism of the second cyclisation from bisabolene type compound can be investigated using [4R-^{3}H, 2-^{14}C] -mevalonate which is specifically incorporated into lagopodin B.   Degradation shows that half the tritium is at C-2 and the remaining tritium is found at C-9.   The cyclisation thus involves a 1,4-hydride shift from C-6 to C-9 and does not involve a bisabolene intermediate.

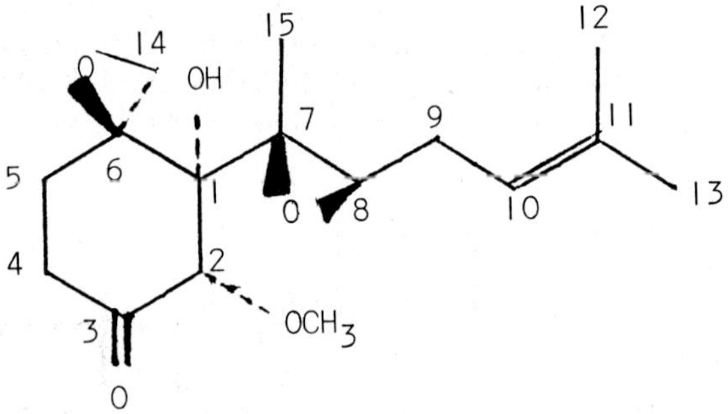

Ovalicin

Ovalicin can be isolated from the culture filtrate of the fungus, *Pseudorotium ovalis* and exhibits antibiotic, immunosuppressive and antitumour activity (P.Bollinger *et al.*, Helv., 1973, 56, 819).   Ovalicin is closely related to fumagillin (E.Corey and B.Snider, J. Amer. chem. Soc.,

1972, 94, 2549).

The observed ^{13}C nmr spectrum of ovalicin derived from the incorporation of $[3,4-^{13}C_2]$ -mevalonate is consistent with the biosynthetic pathway involving the intermediacy of β -bergaotene (D.E.Cane and R.H.Levin, J. Amer. chem. Soc., 1976, 98, 1183).

β-Bergamotene

C-10 and C-11 of ovalicin are derived from C-4 and C-3 respectively of mevalonate. C-1 and C-6 of ovalicin, derived from C-4 and C-3 respectively of a common molecule of mevalonate give rise to the observed pair of doublets, J = 5.3Hz. By contrast, C-1 and C-7, although adjacent, are derived from distinct mevalonate units.

erythro-DL-2,3-dihydroxybutanoic acid, $CH_3CH(OH)CH(OH)COOH$, can be made by oxidation of *trans*-crotonic acid. This, on reaction with diazomethane followed by hydrolysis, gives L-4-methyl-erythrulose (M.Viscontini *et al.*, Helv., 1972, 55, 570).

Homoviridifloric acid, 3-carboxy-4-methylhexane-2,3-diol, can be isolated as an ester of trachelanthamidine which is curassavine, the major alkaloid of *Heliotropium curassavicum* (C.C.J.Culvenor *et al.*, Chem. Comm., 1978, 423). This plant also contains heliotrine and lasiocarpine, both esters of heliotridine (T.R.Rajagopalan and V.Batra, Indian J. Chem., 1977, 15B, 494). In addition, this plant can also contain the minor alkaloids, coramandalin, and heliovicine which are esters of trachelamidine with (+)-viridifloric acid and (-)-trachelanthic acid respectively.

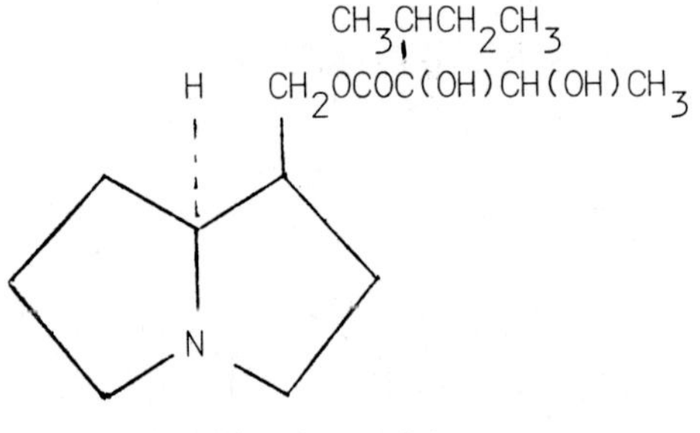

Homoviridifloric acid

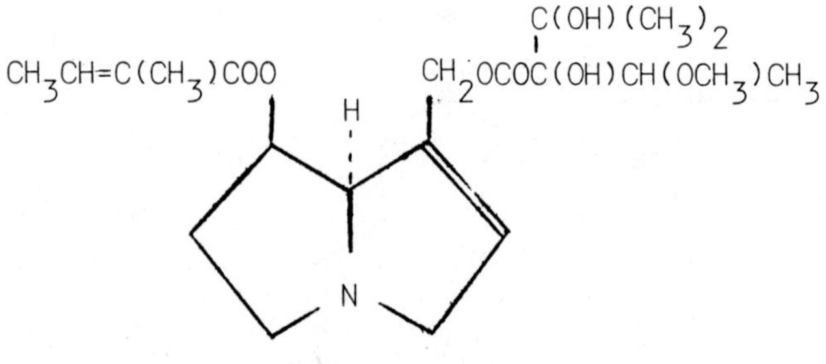

Lasiocarpine

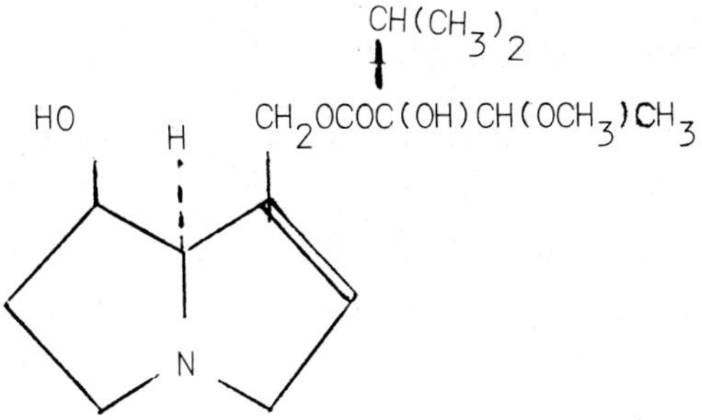

Heliotrine

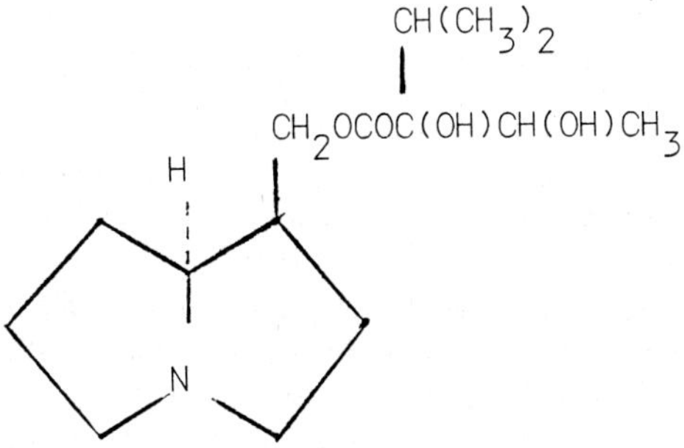

(+)-viridifloric acid

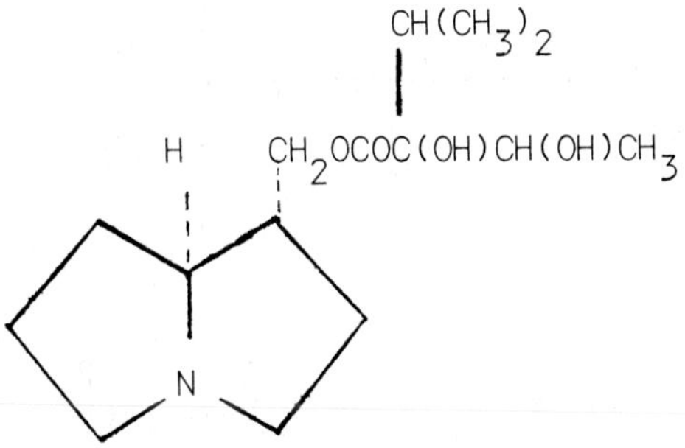

(-)-trachelanthic acid

2 *Amino Hydroxy Carboxylic acids*

α-Hydroxy and α-alkoxyamino acids are of interest because of their relation to 6-methoxypenicillins and cephalosporin analogues β-Hydroxy-α -amino acids can be made by the reaction of a glycine derivative with an aldehyde or a ketone usually in alkaline solution (M.Suzuki *et al.*, Chem. and Ind., 1973, 228).

The anion of ethyl N,N-bis(trimethylsilyl)glycinate reacts with non-enolisable aldehydes to provide hydroxy amino acids (K.Ruhlmann *et al.*, 1970, 10, 393).

$$(Me_3Si)_2NCH_2 \;COOEt \qquad + \qquad RCHO \longrightarrow (Me_3Si)_2NCH \overset{\displaystyle COOEt}{\underset{\displaystyle \underset{R}{\overset{|}{CHOH}}}{}}$$

$$\longrightarrow RCH(OH)CH(NH_2)COOH$$

The base treatment of the copper complex of the glycine/pyruvate Schiff base and its reaction with aldehydes provides β-hydroxy-α -amino acids.   The free acid can be liberated by precipitating the copper with hydrogen sulphide (T.Ichikawa *et al.*, J. Amer. chem. Soc., 1970, 92, 5514).

$$\xrightarrow{\quad H_2S \quad} R\,CH(OH)CH(NH_2)COOH$$

β- Hydroxy-α -amino acids which are made by the reaction of aldehydes with glycine derivatives at a high pH are enriched in one enantiomer if the *N*-substituted glycine is part of a resolved cobalt(III) complex

(Y.N.Belokon *et al.*, Izvest. Akad. Nauk. SSSR., Ser. Khim., 1973, 156). The (+)-enantimer of potassium bis(*N*-salicylideneglycinato) cobalt(III) can be made by resolution of the racemate with brucine and this, on reaction with acetaldehyde at pH 11.2, gives a mixture of D-threonine and D-allothreonine.

β-Hydroxy-α-amino acids can be synthesised from methyl α-alkyl isocyanoacetates by reaction with aldehydes to provide oxazolines. The latter give hydroxyamino acids by ring-opening with aqueous hydrochloric acid (M.Suzuki *et al.*, Chem. and Ind., 1973, 228; J. org. Chem., 1973, 38, 3572).

Alternatively, methyl α-alkyl isocyanacetates can be reacted with an acid chloride to provide a product which, on hydrogenation in the presence of palladium, provides the β-hydroxy-α-amino acid.

The above methods favour the formation of the *threo* and

erythro isomer respectively.

β-Hydroxy-α-amino acids can also be prepared from oxazolin-2-one-4-carboxylates by base cleavage. *N*-Acetyl oxazolin-2-thiono-4-carboxylates can be cleaved by a base such as potassium t-butoxide to give the derivatives of α-aminoacrylic acids (D.Hoppe, Angew. Chem. internat. Edn., 1973, <u>12</u>, 656).

These can then be converted into a number of β-functional -α-amino acids. A similar route starts with α-isothiocyanatoacetates but proceeds via 2-benzylthiooxazolin-4-carboxylates to yield α-amino acrylates.

The reaction of *p*-nitrobenzenesulphonoxyurethane with ethyl ββ-dimethyl acrylate in the presence of a base yields the corresponding aziridine which, on acetolysis and hydrolysis, gives β-hydroxyvaline (C.Berse and P.Bessette, Canad. J. Chem., 1971, <u>49</u>, 2610)

γ-Hydroxy- α -amino acids can be prepared by the photochlorination of α-amino acids possessing a γ-hydrogen atom. This gives the corresponding amino γ-lactone after hydrolysis (H.Faulstich *et al.*, Ann., 1973, 560).

Lactones can also be made from diethyl allyl acetamidomalonate which, on treatment with bromine followed by hydrobromic acid, gives diastereoisomeric 2-amino-5-bromo-4-valerolactone hydrobromide. Hydrolysis of this provides ($\pm$)- γ -hydroxy-proline (Y.K.Lee and T.Kaneko, Bull. chem. Soc. Japan, 1973, 46, 4924).

α-Amino and α-hydroxy amino acids possessing an L-configuration exhibit, in addition to the strong positive c.d. maximum at 210-215 nm., used for configurational assignments, a weak negative long wavelength band at 235-240 nm. (C.Toniolo, J.phys. Chem., 1970, 74, 1390). The bands 210-215 nm. are due to **n-π^*** transitions of the carboxy group and the 235-240 nm. bands are due to coupling of the non-bonding heteroatom with the chromophoric transition of the carbonyl group (J.C.Craig and W.E.Pereira, Tetrahedron, 1970, 76, 3457; G.Barth *et al.*, Chem. Comm., 1969, 355).

4-Amino-2-hydroxybutanoic acid, $NH_2CH_2CH_2CH(OH)COOH$, and its 3-methyl derivative can be made from the isooxazolidones, derived from the nitrone, $CH_2 = N(OEt)$, and acrylates, $RCH = CHCOOMe$, by a 1,3-dipolar addition (H.Sato *et al.*, Bull. chem. Soc. Japan, 1976, 49, 2815).

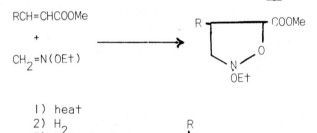

$$R = H \text{ or } Me$$

L-(-)-4-Amino-2-hydroxybutanoic acid can be obtained from an aqueous medium containing DL-4-amino-2-hydroxybutanoic acid by culturing acetobacter (T.Miyaki and K.Matsumoto, Jap. Pat., 75 19,988).

4-Chloro-3-oxobutanoic acid can be converted into 4-amino-3-hydroxybutanoic acid (M.Kurono *et al.*, Jap. Pat., 76 39,634).

$$ClCH_2COCH_2CONHPh \xrightarrow{NaBH_4} ClCH_2CH(OH)CH_2CONHPh$$

$$\xrightarrow{\text{NH}_3} \quad NH_2CH_2CH(OH)CH_2CONHPh$$

$$\xrightarrow{\text{hydrolysis}} \quad NH_2CH_2CH(OH)CH_2COOH$$

L-(-)-4-Amino-2-hydroxybutanoic acid can be formed by the following scheme (T.Naito and S.Nakagawo, U.S. Pat., 1970, 3,823,187).

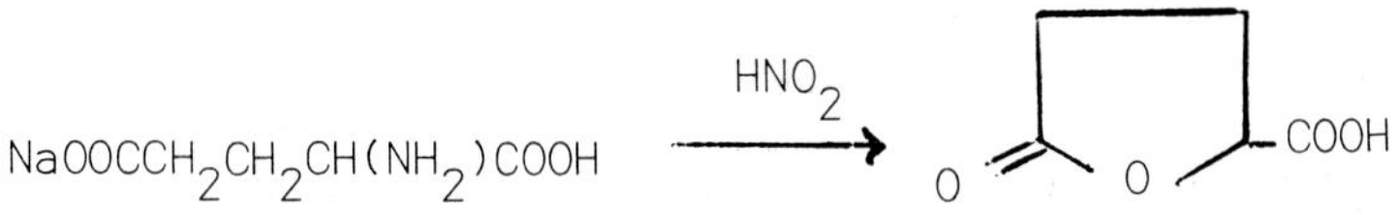

$$NaOOCCH_2CH_2CH(NH_2)COOH \xrightarrow{\text{HNO}_2}$$

α-carboxy- **γ** -butyrolactone

$$\xrightarrow{\text{NH}_3/\text{EtOH}} \quad H_2NOCCH_2CH_2CH(OH)COOH$$

L-2-hydroxyglutamic acid

$$\xrightarrow{\text{NaOCl}} \quad H_2NCH_2CH_2CH(OH)COOH$$

L-(-)-4-amino-
-2-hydroxybutanoic acid

(S)-4-Amino-2-hydroxybutanoic acid can be formed from the corresponding diamino acid (Y.Horiuchi *et al.*, Agric. biol. Chem., 1976, <u>40</u>, 1437) or from 3-acetoxy-2-methoxy-1-pyrrollidine.

$$\longrightarrow \quad H_2NCH_2CH_2CH(OH)COOH$$

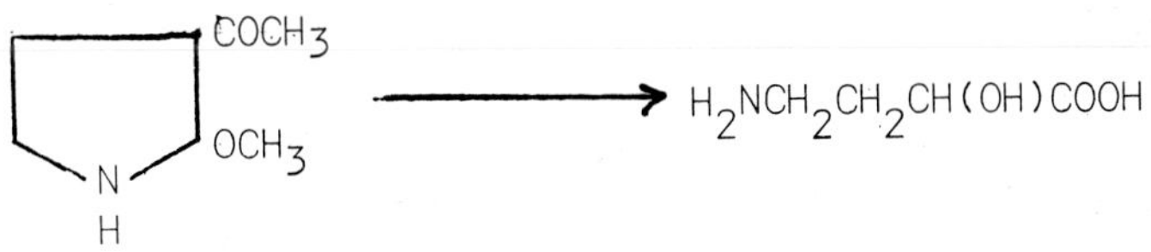

4-Amino-3-hydroxybutanoic acid can be *O*-silylated and then reacted with a haloalkanoate ester to provide a product which, on hydrolysis, yields 4-hydroxypyrrolidin-2-one-1-yl alkyl carboxylic acid esters (R.Monguzzi *et al.*, Ger. Pat., 2,759,033).

$H_2NCH_2CH(OH)CH_2COOH$ $\xrightarrow{\begin{array}{c}1)\ (Me_3Si)_2NH\\2)\ BrCH_2COOEt\end{array}}$

$\xrightarrow{\text{hydrolysis}}$

N-Acyl-3-hydroxy-2-amino acids cleave, on oxidation with lead tetraacetate, to give *N*-acylhydroxyglycine derivatives (W.Oettmeier, Ber., 1970, 103, 2314).

$$\begin{array}{c}RCONHCHCOOEt\\|\\R'CHOH\end{array} \xrightarrow{Pb(OAc)_4} RCONHCH(OH)COOEt$$

General Methods for the synthesis of amino acids

a) By the alkylation of α -hydroxy and methoxy glycines (D.Ben-Ishai and Z.Bernstein, Tetrahedron, 1977, 33, 3261).

$$H_2NCH_2CH_2CH(OH)COOH \xrightarrow{RX} RNHCH_2CH_2CH(OH)COOH$$

b) By the reduction of **α**-ketoacid oximes (M.Jubault *et al.*, Chem. Comm., 1977, 250; J.Pospisek and K.Blaha, Coll. Czech. chem Comm., 1977, _42_, 1068).

$$RCH_2C(=NOH)COOH \xrightarrow{\text{reduction}} RCH_2CH(NH_2)COOH$$

c) By the reduction of phenylhydrazones or Schiff bases of. **α**-keto acids (T.Tabokovic *et al.*, Croat. Chem. Acta., 1977, _49_, 497; K.Nakamura, A.Ohno and S.Oka, Tetrahedron Letters, 1977, 4593; N.H.Khan and A.R.Kidiveri, J. org. Chem., 1973, _38_, 822).

$$RCH_2C(=NNHPh)COOH \xrightarrow{\text{reduction}} RCH_2CH(NH_2)COOH$$

d) By the alkylation of diethyl acetamidomalonate, followed by hydrolysis (R.L.N.Harris and T.Tectic, Austral. J. Chem., 1977, _30_, 649; J.L.Kelieg *et al.*, J. med. Chem., 1977, _20_, 721; G.P.Gardini and G.Palle, J. labelled Comp. Radiopharm., 1977, _13_, 339; Y.K.Lee and T.Kaneko, Bull. chem Soc. Japan, 1973, _46_, 2924).

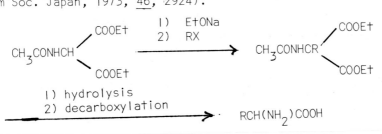

A modification of the above method includes an half-ester stage, permitting decarboxylation in refluxing dioxan, to the ester of the N-acylamino acid so that an enzymic hydrolysis step to give a mixture of N-acyl-L-**α** -amino acid and N-acyl-D-**α** -amino acid ester may be incorporated (A.Berger *et al.*, J. org. Chem., 1973, _38_, 457).

e) By use of the synthesis and subsequent hydrolysis
of hydantoins (P.M.Hardy *et al.*, Chem. Comm., 1977, 759).

$$HOCH_2(CH_2)_4CHO \xrightarrow{HCN} HOCH_2(CH_2)_4CH(OH)CN$$

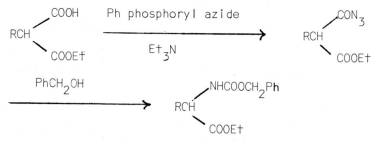

$$\xrightarrow{base} H_2N(CH_2)_5CH(NH_2)COOH$$

f) By a modified Curtius reaction with malonic acid
half-ester using phenyl phosphoryl azide with triethylamine,
followed by the addition of benzyl alcohol to give
N-benzyloxycarbonyl amino acid esters (S.Yamada *et al.*,
Tetrahedron Letters, 1973, 2343).

$$RCH\genfrac{}{}{0pt}{}{COOH}{COOEt} \xrightarrow[Et_3N]{Ph\ phosphoryl\ azide} RCH\genfrac{}{}{0pt}{}{CON_3}{COOEt}$$

$$\xrightarrow{PhCH_2OH} RCH\genfrac{}{}{0pt}{}{NHCOOCH_2Ph}{COOEt}$$

g) By the alkylation of methyl nitroacetate, followed
by reduction and hydrolysis (E.Kiji and S.Za, Bull.chem.
Soc. Japan, 1973, **46**, 337)

$$NO_2CH_2COOCH_3 \xrightarrow{RX} NO_2\overset{R}{\underset{}{C}}HCOOCH_3 \xrightarrow[\substack{2)\ hydrolysis}]{1)\ reduction} NH_2\overset{R}{\underset{}{C}}HCOOH$$

h) By the alkylation of N,N-bis(trimethylsilyl)glycine trimethyl silyl ester, $(Me_3Si)_2NCH_2COOSiMe_3$, as its sodio derivative (E.Bayer and K.Schmidt, Tetrahedron Letters, 1973, 2051).

Routes to amino acids, other than **α**-amino acids, are not easily generalised since particular synthetic strategy must be chosen to provide the required location of the functional groups.

Dehydro-amino acids can be prepared as follows.
a) By an ene reaction between *N*-benzylidene amino acid esters and diethyl azodicarboxylate (R.Grigg *et al.*, Chem. Comm., 1978, 109).

b) By the dehydrochlorination of *N*-chloro-*N*-acylamino acid esters (A.Kolar and R.K.Olsen, Synthesis, 1977, 457; H.Poisel, Ber., 1977, 110, 942, 948).

c)   By the thermolysis of β-alkyl sulphinyl amino acids in the presence of a phosphine as a sulphinic acid receptor (D.H.Rich and J.P.Tam, J. org. Chem., 1977, 42, 3815).

d)   By a base catalysed elimination from a β-chloro-alkyl amino acid derivative (A.Srinivasen *et al*., J. org. Chem., 1977, 42, 2256).

e) By the rearrangement of acylimines formed by the treatment of o-chloranil-oxazol-5(4H)-one adducts with base (J.M.Riordan *et al.*, J. org. Chem., 1977, 42, 236).

$$\text{HCONHCH(Ph)COOH}$$

Serine, $HOCH_2CH(NH_2)COOH$, can be prepared by the 18-crown-6-catalysed reaction of azide anions with methyl 3-hydroxy-2-bromopropanoate, followed by reduction of the product. Isoserine, $NH_2CH_2CH(OH)COOH$, is formed by the same reaction if no catalyst is used (K.Nakamura *et al.*, Tetrahedron Letters, 1977, 4593).

$$HOCH_2CHBrCOOMe \xrightarrow{\ N_3^-\ } HOCH_2CHN_3COOMe$$

$$\xrightarrow{\text{reduction}} HOCH_2CH(NH_2)COOMe$$

Derivatives of O-methylserine, $CH_3OCH_2CH(NH_2)COOH$, and

S-methylserine, $CH_3SCH_2CH(NH_2)COOH$, can be prepared from glycine via the *N*-benzyloxycarbonyl-aziridine-2-carboxylate (Z.Bernstein and D.Ben-Ishai, Tetrahedron, 1977, **33**, 881).

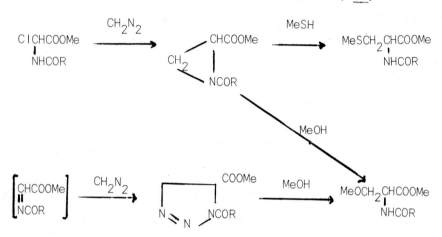

(βR) [β-^{2}H] -L-Homoserine and the (β S)-isomer are synthesised by a route involving the *cis*-addition of hydroxylamine to cinnamic acid; (γR) [γ-^{2}H] -DL-homoserine and the (γ S)-isomer are obtained from the stereospecifically labelled 3-phenyl [1-^{2}H] propanol (D.Coggiola *et al.*, Chem. Comm., 1976, 143).

The optically active seleno amino acids can be made by
the nucleophilic displacement of a tosyl group by either
benzyl selenoate or the selenide anion from a suitably
protected *O*-tosylserine derivative   (J.Roy *et al.*, J. org.
Chem., 1970, <u>35</u>, 510;   C.S.Panda, J.Rudick and R.Walter,
ibid.,1970, <u>35</u>,1440).

$$CH_2OTs \atop RCONHCHCOOR \quad \xrightarrow{R'Se^-} \quad CH_2SeR' \atop RCONHCHCOOR$$

N-Benzoylserine methyl ester, $HOCH_2CH(NHCOPh)COOMe$, can
be converted into the oxazolidine derivative by reaction
with phosgene   (T.Inui *et al.*, Bull. chem. Soc. Japan,
1970, <u>43</u>, 1582).

$$HOCH_2CHCOOMe \atop NHCOPh \quad \xrightarrow{COCl_2} \quad PhCON—COOMe$$

Both the *erythro* and the *threo* form of β-arylserine
with thionyl chloride undergo substitution by chlorine
with the retention of configuration. The chlorination of the
threo form proceeds directly whereas chlorination of the
erythro form proceeds *via* an intermediate *trans*oxazoline
(S.H.Pines *et al.*,J. org. Chem., 1969, <u>34</u>, 1621).

The bis(L-serinato)copper(II) complex reacts with excess formaldehyde at pH 7 to 9 to provide a bis(amino acidato) copper (II) complex from which β -hydroxymethylserine can be isolated by decomposition of the complex with hydrogen sulphide (J.R.Brush *et al.*, J. Amer. chem. Soc., 1973, 95, 2034).

bis(L-serinato)copper(II) complex   (as copper complex)

Both L-serine-*O*-phosphate, $(HO)_2OPOCH_2CH(NH_2)COOH$, and L-threonine-*O*-sulphate can be deaminated in the presence of pyridoxal-5-phosphate to the corresponding keto-acid (J.H.Thomas, K.S.Dodgson and N.Tudrall, Biochem. J., 1968, 110, 687).

$$(HO)_2P(O)OCH_2CH(NH_2)COOH \longrightarrow (HO)_2P(O)OCH_2COCOOH$$

L-serine reacts in the presence of pyridoxal with indole and with hydrogen sulphide to give DL-tryptophan and DL-cystine respectively. These reactions proceed through an intermediate pyridoxylidene amino acrylic acid, derived by the β-elimination from the Schiff base. (K.Korte and Y.Izumi, Bull. chem.Soc. Japan, 1971,44, 2287).

$$\text{HOCH}_2\text{CH(NH}_2\text{)COOH} \quad \xrightarrow{\text{H}_2\text{S}} \quad \begin{array}{l}\text{SCH}_2\text{CH(NH}_2\text{COOH}\\ |\\ \text{SCH}_2\text{CH(NH}_2\text{)COOH}\end{array}$$

It is possible to convert optically active serine into the racemic form of cysteine by the following methods.

   a) The reaction of ethyl serinate hydrochloride with thionyl chloride to generate β-chloroalanate hydrochloride which, after protection of the amino group, reacts with thioacetic acid to give a product which may be hydrolysed to cysteine (T.Furata and T.Ishimaru, Nippon Kagaku Zasshi, 1968, 89, 716).

$$\text{HOCH}_2\text{CH(NH}_2\text{)COOEt.HCl} \quad \xrightarrow{\text{SOCl}_2} \quad \text{ClCH}_2\text{CH(NH}_2\text{)COOEt.HCl}$$

$$\xrightarrow{\text{Ac}_2\text{O}} \quad \text{ClCH}_2\text{CH(NHCOCH}_3\text{)COOEt} \quad \xrightarrow[\text{2) hydrolysis}]{\text{1) CH}_3\text{COSH}} \quad \text{HSCH}_2\text{CH(NH}_2\text{)COOH}$$

   b) The direct reaction of thioacetic acid and acetic anhydride with serine to give the azlactone intermediate which can then be hydrolysed to cysteine (P.Rambacher, Ber., 1968, 101, 2595).

$$\text{HOCH}_2\text{CH(NH}_2\text{)COOH} \quad \xrightarrow{\text{CH}_3\text{COSH/Ac}_2\text{O}} \quad$$

$$\xrightarrow{\text{H}_2\text{O}} \quad \text{HSCH}_2\text{CH(NH}_2\text{)COOH}$$

Threonine, $CH_3CH(OH)CH(NH_2)COOH$, can be made by the base-catalysed reaction of ethanolamine with acetaldehyde in the presence of copper (II) oxide. After the initial oxidation of ethanolamine to glycine, two molecules of acetaldehyde condense with the copper/glycine complex to generate the oxazolidine which can be isolated and hydrolysed to yield threonine and allo-threonine (P.Maldonado *et al.*, Bull. soc. Chim. Fr., 1971. 2933).

$HOCH_2CH_2NH_2$

$+$

CH_3CHO

hydrolysis

$\xrightarrow{\hspace{2cm}}$

$\xrightarrow{\text{CuO}}$

$CH_3CH(OH)CH(NH_2)COOH$ $+$ *allo*threonine

Threonine and allo-threonine can be obtained by the resolution of potassium bis(N-salicylidene-glycinato)cobalt (III) using brucine. The free enantiomers are liberated with aqueous acetic acid.

L-threonine undergoes ammonolysis via a double inversion reaction involving aziridine ring formation giving L-*threo*-2,3-diaminobutanoic acid which can be isolated as its N^α -tosyl derivative. Treatment of this with sodium in liquid ammonia leads to the formation of L-*threo*- and D-*threo*-2,3-diaminobutanoic acid. D- and L-*erythro*-2,3-diamino-butanoic acid can be made similarly from threonine (E.Atherton and J.Meinhofer, J.Antibiot., 1972, <u>25</u>, 539).

Threonine can be converted into *erythro*-3-chloro and 3-bromo-2-aminobutanoic acid using respectively phosphorus trichloride and the carbon tetrabromide/triphenylphosphine adduct (A.Srinavasen *et al.*, J. org. Chem., 1977, <u>42</u>, 2256; T.Wieland *et al.*, Ann., 1977, 806).

$CH_3CH(OH)CH(NH_2)COOH$ $\xrightarrow{Ph_3P.CBr_4}$ $CH_3CHBrCH(NH_2)COOH$

Methyllanthione, HOOCCH(NH$_2$)CH(CH$_3$)SCH$_2$CH(NH$_2$)COOH, can be made by converting DL-threonine into DL-β-methylcysteine which is then added to 2-acetamidoacrylic acid (H.D.Belitz, Tetrahedron Letters, 1974, 749).

CH$_3$CH(OH)CH(NH$_2$)COOH $\xrightarrow{\hspace{3cm}}$ HSCH(CH$_3$)CH(NH$_2$)COOH

CH$_2$=CH(NHCOCH$_3$)COOH

$\xrightarrow{\hspace{3cm}}$ HOOCCH(NH$_2$)CH(CH$_3$)SCH$_2$CH(NHCOCH$_3$)COOH

hydrolysis

$\xrightarrow{\hspace{4cm}}$ HOOCCH(NH$_2$)CH(CH$_3$)SCH$_2$CH(NH$_2$)COOH

The stereospecific synthesis of (2S,3R)-2-amino-3-mercaptobutanoic acid can be achieved according to the following scheme (J.L.Morell, P.Fleckenstein and E.Gross, J. org. Chem., 1977, <u>42</u>, 355).

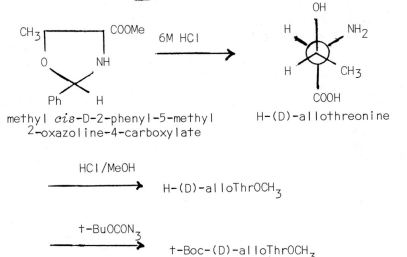

methyl *cis*-D-2-phenyl-5-methyl 2-oxazoline-4-carboxylate

H-(D)-allothreonine

$\xrightarrow{\text{HCl/MeOH}}$ H-(D)-alloThrOCH$_3$

$\xrightarrow{\text{t-BuOCON}_3}$ t-Boc-(D)-alloThrOCH$_3$

$$\xrightarrow{\text{TsCl}/\text{C}_5\text{H}_5\text{N}}$$

(Newman projection: H, OR″, NHR′, H, CH₃, COOMe)

t-Boc-(D)-alloThr(Ts)OCH₃

$$\xrightarrow{\text{CH}_3\text{COSK}}$$

(Newman projection: H, SR, NHR′, CH₃, H, COOCH₃)

$$\xrightarrow{\text{CF}_3\text{COOH}}$$

(Newman projection: H, SH, NH₂, CH₃, H, COOH)

Hydroxy groups may be protected as t-butyl ethers for the use of hydroxy amino acids in peptide synthesis. This can be done by converting the N-benzoyloxycarbonyl acid into its p-nitrobenzyl ester, followed by treatment with isobutene to provide the t-butyl ether. The amino and carboxylic acid protecting groups can be cleaved by catalytic hydrogenation (E.Wuensch and J.Jentsch, Ger. Pat., 1973, 1,493,658).

3 *Amino mercapto carboxylic acids*

Cysteine, $HSCH_2CH(NH_2)COOH$, is present as a number of derivatives in human urine (K.Yao *et al.*, Physiol. Chem. Phys., 1970, 10, 195).   Methyl selenocysteine, $CH_3SeCH_2CH(NH_2)COOH$, is a constituent of *Astragalus bisculatus* (C.M.Chow *et al.*, Phytochemistry, 1971, 10, 2693). Biosynthetic studies suggest that this compound can be made by the same enzymes which are responsible for the synthesis of *S*-methylcysteine, $CH_3SCH_2CH(NH_2)COOH$.

Optically active cysteine derivatives can be prepared by the addition of a thiol to either methyl 2-phthalimido-acrylate or 4-methyleneoxazolone in the presence of cinchone alkaloid (H.Pracejus *et al.*, J. prakt. Chem., 1977, 329, 219).

Cysteine can be prepared by the addition of benzyl mercaptan to 2-acetamidoacrylic acid to give a racemic product, *N*-acetyl-*S*-benzylcysteine. This can be resolved by its reaction with aniline in the presence of papain when the L-isomer forms an anilide which can be readily separated. The D-isomer is converted into D-cysteine (A.Schoeberl *et al.*, Ber., 1969,102, 1767).

$$CH_2=CCOOH \quad \xrightarrow{\quad} \quad PhCH_2SCH_2CHCOOH \quad \xrightarrow[\text{2)Na/NH}_3]{\text{1)hydrolysis}} \quad HSCH_2CH(NH_2)COOH$$
$$\underset{NHCOMe}{\big|} \qquad\qquad \underset{NHCOMe}{\big|}$$

The alkylation of homocysteine lactone to S-alkyl-($\pm$)-homocysteine can be achieved by using primary alkyl halides in the presence of sodium methoxide (H.M.Kolenbrander, Canad. J. Chem., 1969, 43, 3271)

$$\underset{\substack{| \\ S \quad\text{---}\quad CO}}{CH_2CH_2CHNH_2} \quad \xrightarrow{\text{RX/MeONa}} \quad RSCH_2CH_2CH(NH_2)COOH$$

The dehydro amino acid derivatives can be made via 4-alkylidene-2-chloromethyloxazol-5-ones (H.Kurita *et al.*, Bull. chem. Soc. Japan, 1968, 41, 2758) or by the elimination of the thiol group from cysteine derivatives (L.Kisfaludy *et al.*, Acta. Chim. Acad. Sci. Hung., 1969, 59, 159).

ββ-Disubstituted-*N*-formyl-*S*-benzylcysteine can be made from ethyl isocyanacetate by reaction with ketones to give isocyanoacrylates followed by treatment with benzyl mercaptan (U.Schollkopf and D.Hoppe, Ann., 1973, 799).

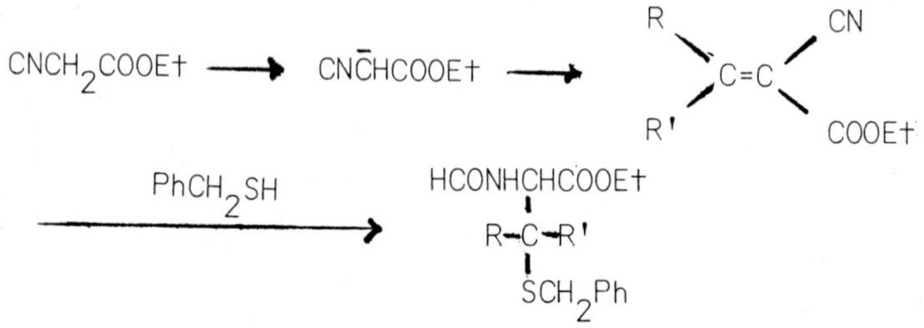

The following scheme illustrates the preparation of β -functionalised α -amino acids (D.Hoppe, Angew. Chem. internat. Edn., 1973, 12, 656).

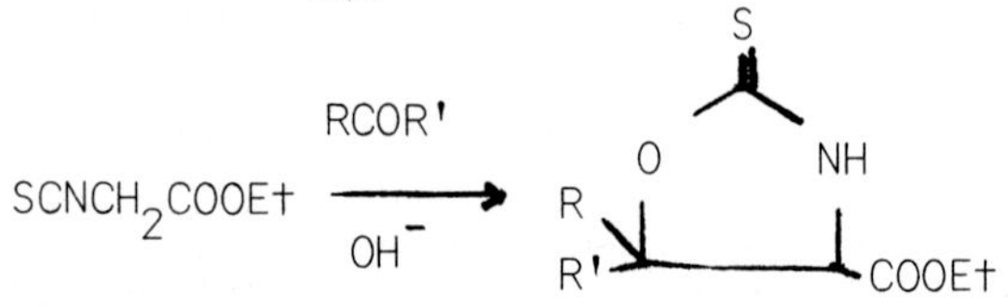

ethyl 2-thioxo-1,3-oxazolidine-4-carboxylate

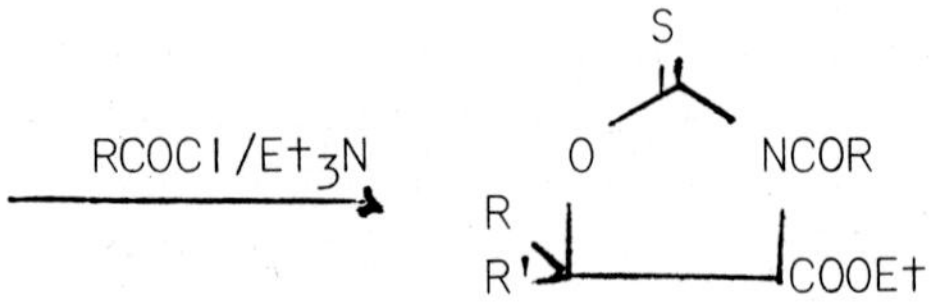

ethyl 3-acyl-2-thioxo-1,3-oxazolidine-4-carboxylate

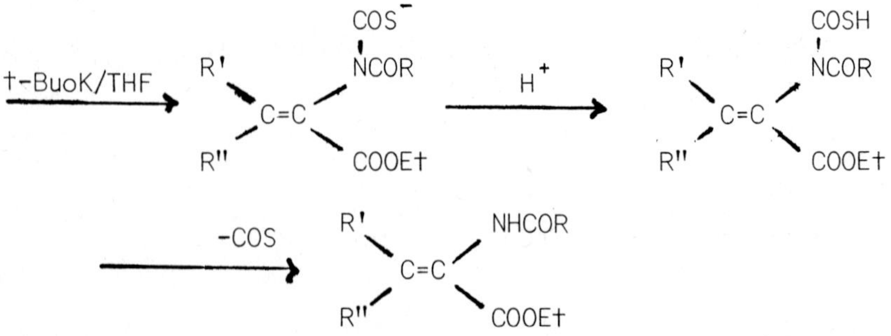

ethyl α-(N-acylamino)acrylates

β-Substituted-*N*-formyl-*S*-benzylcysteines can be used as cysteine precursors since the formyl and the benzyl group can be easily removed (U.Schollköpf and D.Hoppé, Angew. Chem. internat. Edn., 1970, *3*, 236).

$$NC\bar{C}COOEt \quad + \quad R'COR'' \longrightarrow \left[\begin{array}{c} R' \\ \diagdown \\ R'' \end{array} C=C \begin{array}{c} COOEt \\ \diagup \\ \diagdown \\ NHCHO \end{array} \right]$$

$$\xrightarrow[\text{2)OH}^-]{\text{1)PhCH}_2\text{SH}} \quad \begin{array}{c} R' \;\; SCH_2Ph \\ \diagdown \;\; | \\ CCHCOOH \\ \diagup \;\; | \\ R'' \;\; NHCHO \end{array} \quad \xrightarrow{H_2/PdC} \quad \begin{array}{c} R' \\ \diagdown \\ R'' \end{array} CHCH(NH_2)COOH$$

Selenocysteine can be made by the sodium borohydride reduction of selenocystine (A.Shrift *et al.*, Plant Physiol., 1976, *58*, 248).

α -^{2}H-S-Benzyl-($\pm$)-cysteine can be made from benzyl chloromethyl sulphide and diethyl acetamidomalonate using ^{2}HCl/^{2}H$_2$O for the hydrolysis-decarboxylation stage (D.A.Upson and V.J.Hruby, J. org. Chem., 1976, *41*, 1353)

$$PhCH_2SCH_2Cl \quad + \quad \begin{array}{c} COOEt \\ | \\ ^-CNHCOMe \\ | \\ COOEt \end{array} \longrightarrow \quad PhCH_2SCH_2 \begin{array}{c} COOEt \\ | \\ CNHCOMe \\ | \\ COOEt \end{array}$$

$$\xrightarrow{^2HCl/^2H_2O} \quad PhCH_2SCH_2C^2H(NH_2)COOH$$

$$Me_2NH + {}^2H_2C(COOMe)_2 + \begin{array}{c} COOEt \\ | \\ HCNHCOMe \\ | \\ COOEt \end{array} \longrightarrow Me_2N\overset{{}^2H}{\underset{{}^2H}{C}} - \begin{array}{c} COOEt \\ | \\ CNHCOMe \\ | \\ COOEt \end{array}$$

$$\xrightarrow{MeI} \quad Me_3\overset{+}{\underset{{}^2H}{N}}\overset{{}^2H}{C} - \begin{array}{c} COOEt \\ | \\ CNHCOMe \\ | \\ COOEt \end{array} \xrightarrow{PhCH_2SNa} \quad PhCH_2S\,\overset{{}^2H}{\underset{{}^2H}{C}} - \begin{array}{c} COOEt \\ | \\ CNHCOMe \\ | \\ COOEt \end{array}$$

$$\xrightarrow{{}^2HCl/{}^2H_2O} \quad PhCH_2\overset{{}^2H}{\underset{{}^2H}{S}C} - \begin{array}{c} {}^2H \\ | \\ CNH_2 \\ | \\ COOH \end{array}$$

Cysteine reacts with ninhydrin to provide a product which has a spiro structure (G.Prota and E.Ponsiglione, Tetrahedron, 1973, 29, 4271).

Both cysteine and cystine are completely desulphurised under relatively mild conditions using Raney nickel. Methionine is not desulphurised under the same conditions (M.T.Perlstein et al., Biochem. Biophys. Acta., 1971, 236, 174).

Cysteine is oxidised to cystine by diethyl azodicarboxylate in a few minutes (C.R.Axen et al., J. org. Chem., 1967, 32, 4117). However, iodine oxidises L-cysteine to cysteic acid (O.G.Lowe, J. org. Chem., 1977, 42, 2524) and aqueous chlorine oxidises cysteine to a mixture of cysteic acid and cysteinylcysteic acid (P.G.Gordon, Austral. J. Chem., 1973,

$\underline{26}$,1771).

diethyl azodicarboxylate

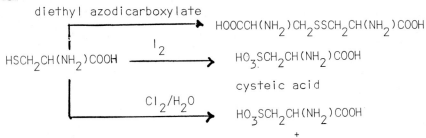

HOOCCH(NH$_2$)CH$_2$SSCH$_2$CH(NH$_2$)COOH

HSCH$_2$CH(NH$_2$)COOH $\xrightarrow{\text{I}_2}$ HO$_3$SCH$_2$CH(NH$_2$)COOH

cysteic acid

$\xrightarrow{\text{Cl}_2/\text{H}_2\text{O}}$ HO$_3$SCH$_2$CH(NH$_2$)COOH

+

HO$_3$SCH$_2$CH(NH$_2$)CONHCH(NH$_2$)CH$_2$SO$_3$H
cysteicylcysteic acid

Cysteine reacts with linoleic acid hydroperoxide in ethanol to give the following products (H.W.Gardiner *et al.*, Lipids, 1977, $\underline{12}$, 655).

Me(CH$_2$)CH(OOH)CH=CHCH(OOH)CH$_2$(CH$_2$)$_7$COOH + HSCH$_2$CH(NH$_2$)COOH

$\downarrow$

Me(CH$_2$)$_4$CH(OH)CH=CHCH(OH)CH(CH$_2$)$_7$COOH
                          SCH$_2$CH(NH$_2$)COOH

+

HOOC—N ... S ... (CH$_2$)$_7$ COOH

The oxidation of the cysteine/phenylacetylene adduct provides two diastereoisomeric *cis*-S-(β -styryl)-L-cysteine ⌐S⌐oxides (J.F.Carson and L.E.Boggs, J. org. Chem., 1967, $\underline{32}$, 673)

PhC=CH
+
HSCH$_2$CH(NH$_2$)COOH

$\rightarrow$

Ph   SCH$_2$CH(NH$_2$)COOH
    C=C
H   H

$\longrightarrow$

Ph   SOCH$_2$CH(NH$_2$)COOH
    C=C
H   H

Cysteine reacts with *o*-quinones through the thiol group, whereas lysine undergoes addition through the amino group, followed by oxidation (W.S.Pierpoint, Biochem. J., 1969, $\underline{112}$, 609)..

L‑ cysteine reacts with formaldehyde and other carbonyl compounds including monosaccharides to give thiazolidine carboxylic acid derivatives (H.Hellstrom *et al.*, J. chem. Soc., (B), 1969,1103; R.G.Kallen, J. Amer.chem. Soc., 1971, 93, 6227; R.Bogner *et al.*, Ann., 1970, 738, 68; J.P. Fourneau *et al.*,Compt. Rend., 1971, 272, 1515).

R = R' = H; R = H, R' = COOH

Cysteine reacts with pyridoxal to give a product containing a thiazolidine ring whereas other amino acids with hydroxyl groups form cyclic carbinolamines (Y.Matsushima, Chem. pharm. Bull. Japan, 1968, 16, 2046; E.H.Abbott and A.C.Martell, J. Amer. chem. Soc., 1970, 92, 1745). Cystine reacts with pyridoxal phosphate to give thiocysteine, ammonia and pyruvic acid (A.Rinaldi and C. De Marco, Arch. Biochem. Biophys., 1970, 138, 697).

Cysteine and *N*-methyl-*N*-nitroso-toluene-*p*-sulphonamide react together to give cystine, possibly via the S-nitroso derivative of cysteine (V.Schulz and D.R.McCalla, Canad. J. Chem., 1969, 47, 2021).

Felinine, $HOCH_2CMe_2SCH_2CH(NH_2)COOH$, can be made by the addition of cysteine to 2-methyl-4-hydroxybut-1-ene (A.Scholberl *et al.* Ber., 1968, 101, 373).

$HSCH_2CH(NH_2)COOH$

   + $\longrightarrow$ $HOCH_2C(CH_3)_2SCH_2CH(NH_2)COOH$

$HOCH_2CH_2C(CH_3)=CH_2$

An analogue of cysteic acid, α-sulpho-β-alanine can be prepared from the reaction of β-alanine with oleum. The reaction possibly involves the addition of sulphur trioxide to the enol form of the acid (D.Wagner *et al.*, Tetrahedron Letters, 1968, 4479).

$1-(\pm)$-Amino-7-hydroxy-4,5-dithiaheptane-1-carboxylic acid, $HOOCCH(NH_2)CH_2CH_2SSCH_2CH_2OH$, m.p. 216-217.5°C, can be made by the addition of cyanogen bromide to a solution of $(\pm)$-homocystine in concentrated ammonia containing 2-mercaptoethanol (C.Abe and C.Ressler, J. org. Chem., 1974, 39, 253).

$HOOCCH(NH_2)CH_2S$ $HOOCCH(NH_2)CH_2SSCH_2CH_2OH$
                               | + $HSCH_2CH_2OH \longrightarrow$ +
$HOOCCH(NH_2)CH_2S$ $HOOCCH(NH_2)CH_2SH$

I-L-Amino-7-hydroxy-4,5-dithiaheptane-l-carboxylic acid, m.p. 229°C, and $[\alpha]_D^{24.5}$ +42.6°, can be made by the addition of cyanogen bromide to a solution of L-cysteine in dilute hydrochloric acid containing 2-mercaptoethanol.

I-L-Amino-6-hydroxy-3,4-dithiahexane-l-carboxylic acid, $HOOCCH(NH_2)CH_2SSCH_2CH_2OH$, m.p. 182.5-183.5° and $[\alpha]_D^{24.5}$ -242° (in IM acetic acid) can be made by the addition of cyanogen bromide to a solution of L-cysteine hydrochloride in dilute hydrochloric acid containing 2-mercaptoethanol *(idem. ibid.)*

$$HOOCCH(NH_2)CH_2SH \quad + \quad HSCH_2CH_2OH \xrightarrow{CNBr} HOOCCH(NH_2CH_2S$$
$$HOCH_2CH_2S$$

I-L-Amino-4,5-dithiahexane-l,6-dicarboxylic acid, $HOOCCH(NH_2)CH_2CH_2SS\ CH_2COOH$, m.p. 183-185°C, and $[\alpha]_D^{24.5}$ +6.1° can be made by the reaction of cyanogen bromide with L-cysteine L-cysteine in hydrochloric acid containing 2-mercaptoethanol *(idem. ibid.)*

$$HOOCCH(NH_2)CH_2SH \quad + \quad HSCH_2COOH \xrightarrow{CNBr} HOOCCH(NH_2)CH_2S$$
$$HOOCCH_2S$$

I-L-Amino-6-hydroxy-3,4-dithiahexane-l-carboxylic acid-L-diamino-4,5-dithiaheptane-l,7-dicarboxylic acid, $HOOCCH(NH_2)CH_2CH_2SSCH_2CH(NH_2)COOH$, can be made by the reaction of cyanogen bromide with equimolar amounts of L-homocysteine and L-cysteine in dilute hydrochloric acid *(idem. ibid.)*

$$HOOCCH(NH_2)CH_2SH \ + \ HOOCCH(NH_2)CH_2CH_2SH \ \xrightarrow{CNBr} \ \begin{array}{l} HOOCCH(NH_2)CH_2S \\ | \\ HOOCCH(NH_2)CH_2CH_2S \end{array}$$

Symmetrical disulphides can be prepared by treating the mercaptan in dilute hydrochloric acid with approximately two equivalents of cyanogen bromide. Thus, L-cysteine gives L-cystine.

$$2 \ HSCH_2CH(NH_2)COOH \ \xrightarrow{CNBr} \ \begin{array}{l} SCH_2CH(NH_2)COOH \\ | \\ SCH_2CH(NH_2)COOH \end{array}$$

Cyanogen bromide can be considered to be acting as a mild oxidising agent, and can be useful for the synthesis of mixed symmetrical disulphides from mercaptans.

The mixed disulphide of cysteine and penicillamine reacts with cyanide ions to provide 2-amino-2-thiazoline-4-carboxylic acid (T.R.C.Boyde, J. chem. Soc. (C), 1968, 2751).

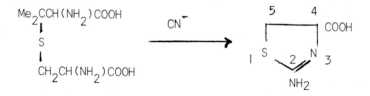

Vinylene-L-cysteine sulphones cyclise by reaction with base to yield the two epimeric cyclic sulphones (J.F.Carson, L.E.Boggs and R.Lundin, J. org. Chem., 1968, 33, 3739)

S-Alkylation with cysteine can occur and leads to
products as shown by the synthesis of tryptathione from
cysteine and 3a-hydroxy-1,2,3,3e,8,8a-hexahydropyrrolo
[2,3-b] indole-2-carboxylic acid (W.E.Savige and A.Fontana,
Chem. Comm., 1976, 600).

Alkenyl cysteine sulphoxides can be cyclised to thiazine
-S-oxides (J.F.Carson and R.E.Lundin, jnr., J. chem. Soc.
Perkin I, 1976, 1195).

Alkynyl cysteine-S-oxide and the corresponding dioxide in the presence of base undergo initial addition of the amino group to the triple bond to give the cyclic sulphoxide and the cyclic sulphone respectively (J.F.Carson and L.E.Boggs, J. org. Chem., 1971, $\underline{36}$, 611).

2-Acetamido-S-methylcysteine undergoes oxidation to cystine derivatives using iodine. This method can also be applied to the preparation of cyclo-L-cystine (B.Kamber, Helv., 1971, $\underline{54}$, 927).

MeSCH$_2$CH(NHCOMe)COOH $\xrightarrow{\text{I}_2}$

The radiolysis of oxygenated cysteine solutions (J.P.Barton and E.J.Packer, Internat. J. Radiation, 1970, $\underline{2}$, 159) and cysteine formate solutions (M.Morita $et~al.$, Bull. chem. Soc. Japan, 1971, $\underline{44}$, 2257) proceeds by pathways involving sulphur radicals of cysteine. An aqueous solution cysteine on radiolysis breaks down to give sulphur radicals of cysteine (C.J.Dixon and D.W.Grant, J. Phys. Chem., 1970, $\underline{74}$, 941).

HSCH$_2$CH(NH$_2$)COOH· $\xrightarrow{h\nu}$ ·SCH$_2$CH(NH$_2$)COOH

Phaeomelanins can be formed $in~vivo$ by the 1,6-addition reaction of cysteine to dopaquinone, produced by the enzymic oxidation of tyrosine. The first step involves the formation

of 2-S-cysteinyl dopa and 5-S-cysteinyl dopa which can then
undergo cyclisation and oxidation   (G.Mizuraea *et al.*,
Experientia, 1969, 25, 920;   G.Prota   *et al.*, *ibid.*, 1970,
26, 1058;   L.Minle *et al.*, Gazzetta, 1970, 100, 46L).

where   R = $CH_2CH(NH_2)COOH$

   Cysteine reacts with levulinic acid to give a bicyclic
compound, a substituted pyrroldinothiazolidine derivative
(G.L.Oliver *et al.*,J. Amer. chem. Soc., 1958, 80, 702).

   α-Angelica lactone with cysteine gives a stable lactam
(D.K.Black, J. chem. Soc. (C), 1966, 1123).

$\Delta^{\alpha,\beta}$ Butenolides react with cysteine to form β-thio-adducts (S.M.Kupchan *et al.*, J. org. Chem., 1970, 35, 3539).

$$CH_2CH(NH_2)COOH$$

$$+ \ HSCH_2CH(NH_2)COOH \longrightarrow$$

R = Me; R' = R" = H
R' = Me; R = R" = H
R" = Me; R = R' = H

Methionine, $CH_3SCH_2CH_2CH(NH_2)COOH$

N- Formyl methionine, $CH_3SCH_2CH_2CH(NHCHO)COOH$, is found in proteins of *Clostridium pasteurianum rubredoxin* (K.F.McCarthy and W.Levenberg, Biochem. biophys. Res. Comm., 1970, 40, 1053), whilst N-acetyl methionine is found in honey bee proteins. *L-* Methionine can be extracted from a *Streptomycete* fermentation (J.P.Scannel *et al.*, J. Antibiot., 1971, 24, 239).

*N-*Homomethionine, $CH_3SCH_2CH_2CH_2CH(NH_2)COOH$, is a constituent of cabbage (Y.Suketa *et al.*, Chem. pharm. Bull. Japan, 1970,18, 249). Methods have been described for the synthesis of methionine-^{14}C and methionine-[35]S (K.Samachocka and J.Kowalezyk, Radiochem. Radioanalyst Letters, 1970, 4, 131). Iodomethane-^{11}C can be used to synthesise methyl labelled methionine (C.H.Langstrom and H.Lundqvist, Internat. Appl. Radiat. Isotopes, 1976, 27, 357).

L- Selenomethionine and L-selenoethionine can be made from L-2-amino-4-bromobutanoic acid (H.D.Jakubka *et al.*, Coll. Czech. chem. Comm., 1968, 13, 3910).

$$\underset{\substack{\text{Br} \\ | \\ CH_2 \\ | \\ CH_2 \\ | \\ NH_2CHCOOH}}{} \xrightarrow{\ PhCH_2SeNa\ } \underset{\substack{PhCH_2Se \\ | \\ CH_2 \\ | \\ CH_2 \\ | \\ NH_2CHCOOH}}{} \xrightarrow[\substack{2)\ \ RI}]{1)\ \ Na/NH_3} \underset{\substack{RSe \\ | \\ CH_2 \\ | \\ CH_2 \\ | \\ NH_2CHCOOH}}{}$$

R = Me or Et

Dehyro-L-methionine, (+)-(S)-E-2-amino-4-methylthiobut-3-enoic acid

can be made from 3-acetamido-3,3-bis(ethoxycarbonyl) propionaldehyde, AcNHC(COOEt)$_2$CH$_2$CHO. The mercaptal can be formed by reaction with methyl mercaptan and then can be saponified, decarboxylated and converted into the methyl ester, AcNHCH(COOMe)CH$_2$CH(SMe)$_2$. This mercaptal gives the monosulphoxide with sodium metaperiodate which, on pyrolysis, gives dehyro-L-methionine as its N-acetyl methyl ester as a mixture of the E and Z-isomers (K.Balenovic and A.Deljac, Rev. Trav. chim., 1973, 92, 117).

Methionine can be oxidised to its sulphone, CH$_3$SO$_2$CH$_2$CH$_2$CH(NH$_2$)COOH, by use of trichloroisocyanuric acid (M.Z.Afassi, Tetrahedron Letters, 1973, 4893).

L-methionine can be oxidised to L-(S)-methionine sulphoxide, CH$_3$SOCH$_2$CH$_2$CH(NH$_2$)COOH, using HAuCl$_4$ (E. Bordignon et al.,Chem. Comm., 1973, 878).

Aerial oxidation of methionine to its S-oxide is catalysed by the presence of bisulphite ions (M.Ionoue and H.Hikoya, Chem. pharm. Bull. Japan, 1971, 19, 1216).

The conversion of methionine sulphoxide into methionine sulphimine proceeds with the retention of configuration (B.W.Christensen and AKjaer, Chem. Comm., 1969,934).

S- Methyl methionine undergoes electrochemical reduction to a mixture of methionine and 2-aminobutanoic acid (T.Iwasoki *et al.*, Chem. and Ind., 1973, 1163).

2-Amino-3-mercaptobutanoic acid, $CH_3CH(SH)CH(NH_2)COOH$, can be prepared as follows.

The condensation of acetaldehyde with hippuric acid in acetic anhydride in the presence of sodium acetate gives a mixture of the *cis* and the *trans*-2-phenyl-4-ethylidene-5-oxazolone, m.p. 90-92°C. The addition of benzyl mercaptan, followed by hydrolysis, gives the two diastereoisomers of 2-amino-3-benzylmercaptobutanoic acid. The benzyl group can be removed by the action of sodium in liquid ammonia (J.Hoogmastens, P.J.Claes and H.Vanderhaeghe, J.org. Chem., 1974, 39, 425).

N-Formyl-2,2,5-trimethyl-4-carboxythiazolidine, m.p. 163-167°C, can be made from 2-amino-3-mercaptobutanoic acid by first reacting it with acetone in the presence of concentrated hydrochloric acid, and then with formic acid and sodium formate (*idem., ibid., loc. cit.*).

erythro-($\pm$)-N-Formyl-2,2,5-trimethyl-4-carboxythiazolidine
m.p. 199-203°C, can be similarly be prepared from *erythro*-
DL-2-amino-3-mercaptobutanoic acid (*idem.*, *ibid.*, *loc. cit.*).

Hydrolysis of optically active N-formyl-2,2,5-trimethyl-
4-carboxythiazolidine gives optically active 2-amino-3-
mercaptobutanoic acid as hydrochlorides, *erythro*-isomers,
m.p. 206-208°C, · *threo*-isomers, m.p. 147-153°C.

threo-2-Amino-3-mercaptobutanoic acid can be S-benzylated
with benzyl chloride in a sodium in liquid ammonia. This
yields 2-amino-3-benzylmercaptobutanoic acid, m.p. 165-166.5°,
which can be converted into the *threo*-DL-2-acetamido-3-
benzylmercaptobutanoic acid, m.p. 104.5-107°C, by reaction
with acetic anhydride.
    erythro-DL-2-acetamido-3-benzylmercaptobutanoic acid,
m.p. 140-141.5°C, can be similarly prepared from *erythro*-
DL-2-amino-3-mercaptobutanoic acid (*idem.*, *ibid.*, *loc. cit.*).

4   *Diamino Carboxylic Acids*

Ornithine, $NH_2CH_2CH_2CH_2CH(NH_2)COOH$, is found to generate the
naturally occurring 2-amino-4-oxo-pentanoic acid,
$CH_3COCH_2CH(NH_2)COOH$, in *Clostridium stricklandii*
(C.Y.Tsuda and H.C.Friedmann, J. biol. Chem., 1970, 245, 5914).

The large scale preparation of ornithine can be achieved starting with glutamic acid (V.Gut and K.Poduska, Coll. Czech. chem. Comm., 1971, 36, 3470).

L-ornithine can be formed from L-arginine (J.S.Bland and J.E.W.Kean, Chem. Comm., 1971, 1024).

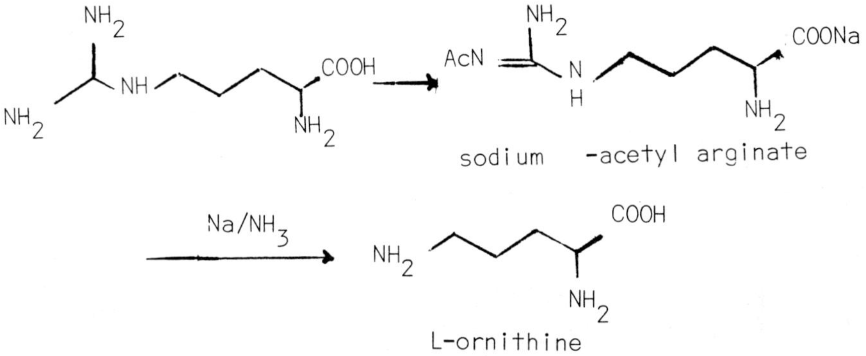

sodium -acetyl arginate

L-ornithine

The mechanism of the mild reductive cleavage in the above scheme is shown below.

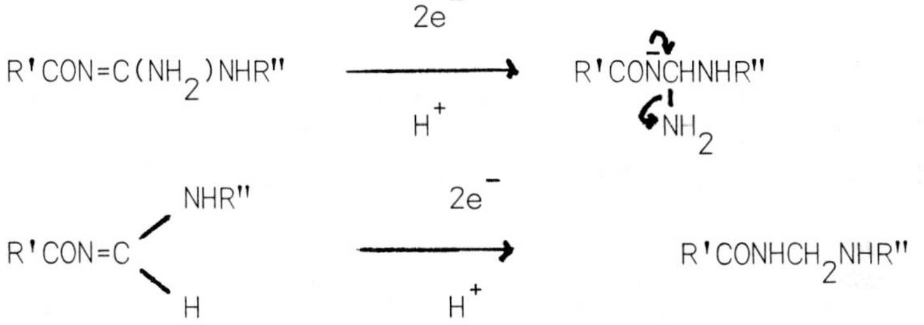

$$R'CON=C(NH_2)NHR'' \xrightarrow[H^+]{2e^-} R'CONCHNHR''$$

$$R'CON=C \begin{array}{c} NHR'' \\ \diagdown \\ H \end{array} \xrightarrow[H^+]{2e^-} R'CONHCH_2NHR''$$

Labelled L-ornithinine can be prepared from S-ethyl thiouronium bromide labelled with [14]C (K.V.Viswanathan et al., Radiochem. Radioanlyst. Letters, 1976, 26, 301), whilst L-ornithine-5-^{3}H can be made by the catalytic tritiation of 4-cyano-2-aminobutanoic acid (M.Havranek and I.Mezo, Acta. Chem. Acad. Sci. Hung., 1973, 77, 341).

N^δ-Hydroxyornithine can be prepared from the reaction of diethyl acetamidomalonate with 3-(N-tosyl-N-benzyloxy) aminopropyl bromide. This method provides the diastereoiso- meric D- and L-isomers which can be resolve enzymatically.

$$\underset{\substack{| \\ \text{COOEt}}}{\overset{\substack{\text{COOEt} \\ |}}{CH_3CONHCH}} \quad + \quad \underset{PhCH_2O}{\overset{Ts}{>}}NCH_2CH_2CH_2Br \quad \longrightarrow$$

$$\underset{\substack{| \\ \text{COOEt}}}{\overset{\substack{\text{COOEt} \\ |}}{CH_3CONHCCH_2CH_2CH_2N}}\underset{OCH_2Ph}{\overset{Ts}{<}} \quad \longrightarrow \quad NH_2CH(OH)CH_2CH_2CH(NH_2)COOH$$

Arginine reacts with nitromalondialdehyde in aqueous alkaline solution to give Nδ-(5-nitro-2-pyrimidyl)ornithine (A.Signor et al., Biochemistry, 1971, <u>10</u>, 2734).

$$\underset{\substack{| \\ \text{CHO}}}{\overset{\substack{\text{CHO} \\ |}}{NO_2CH}} \qquad \underset{NH_2}{\overset{NH}{\underset{\|}{C}}}NHCH_2CH_2CH_2CH(NH_2)COOH$$

$$\longrightarrow \qquad \underset{NO_2}{\text{[pyrimidine ring]}} NHCH_2CH_2CH_2CH(NH_2)COOH$$

Arginine, $NH_2C(=NH)NHCH_2CH_2CH_2CH(NH_2)COOH$, can be made as its N-methyl-L-form via the ornithine analogue by using the hydantoin route, starting with the following compound.

$$\underset{NHCO}{\overset{CONH}{\diagdown}}\underset{CH_2CH_2CH_2N_3}{\overset{CH_3}{C\diagup}}$$

This can be converted enzymatically into α-methyl-L-arginine and α-methyl-L-ornithinine (H.Marhr *et al.*, J. Antibiotics, 1976,29, 821).

$$\begin{matrix} CONH \\ | \\ NHCO \end{matrix} C \begin{matrix} CH_3 \\ \\ CH_2CH_2CH_2N_3 \end{matrix} \longrightarrow NH_2C(=NH)NHCH_2CH_2CH_2 \begin{matrix} NH_2 \\ | \\ CCOOH \\ | \\ CH_3 \end{matrix}$$

The synthesis of side chain mono-, di- and trimethyl arginines can be achieved by using ornithine and the corresponding N-methylated S-methyl isothiouronium iodide (A.Patthy *et al.*,Acta Biochem. Biophys. Acad. Sci. Hung., 1977, 12, 191).

The condensation of L-arginine with 3-oxoglutaric acid provides a mixture of diastereisomers which can be reduced to the natural products, nopaline and its isomer, isonopaline (D.Cooper and J.L.Firmin, Org. Prep, Proceed. Internat., 1972, 9, 99; R.E.Jensen *et al.*, Biochem. Biophys. Res. Comm., 1977, 75, 1066).

All four isomers of octopine, $NH_2C(=NH)NHCH_2CH_2CH_2CH-$ (NHCHCH$_3$COOH)COOH, can be formed from the reaction of D- and L-arginine with pyruvic acid, followed by sodium borohydride reduction (J.F.Biellmann, G.Braulaut and L.Walien, Bio. org. Chem., 1977, 6, 89).

$$NH_2C(=NH)NHCH_2CH_2CH_2CH(NH_2)COOH$$

$$\xrightarrow{CH_3COCOOH} NH_2C(=NH)NHCH_2CH_2CH_2 \begin{matrix} HOOCCCH_3 \\ \| \\ N \\ | \\ CHCOOH \end{matrix}$$

$$\xrightarrow{NaBH_4} NH_2C(=NH)NHCH_2CH_2CH_2 \begin{matrix} HOOCCHCH_3 \\ | \\ NH \\ | \\ CHCOOH \end{matrix}$$

Lysine, $NH_2CH_2CH_2CH_2CH_2CH(NH_2)COOH$

Mono-, di- and tri-substituted lysines are present in human urine (Y.Kamimoto and J.Akazawa, J. biol. Chem., 1970, 245, 5751). N^{ε} -trimethyl lysine, $(CH_3)_3\overset{+}{N}(CH_2)_4CH(NH_2)COO^{-}$, is found in histones (K.Hempel *et al.*, Naturwissen schaftan, 1968, 55, 37). Many methylated derivatives of lysine are found in animal protein (W.M.Kuell and R.S.Adelstein, Biochem. Biophys. Res. Comm., 1969, 37, 59; M.Hardy *et al.*, Biochem. J., 1970, 117, 44; R.R.Weihung and E.D.Horn, Nature, 1970, 227, 1263). L-γ-oxalysine is a lysine antagonist in *E. Coli* found in strptomycetes (E.O.Stapley *et al.*, An. Inst. Farmacol. Espan., 1970, 1966-1967 (publ. 1970), 15-16, 185).

Lysine can be prepared from 2-cyanopiperidine via a hydantoin intermediate (R.Kikumoto *et al.*, Ger. Pat., 2,256,731; T.I.Samolovich *et al.*, Zhur. Org. Khim., 1969, 5, 579).

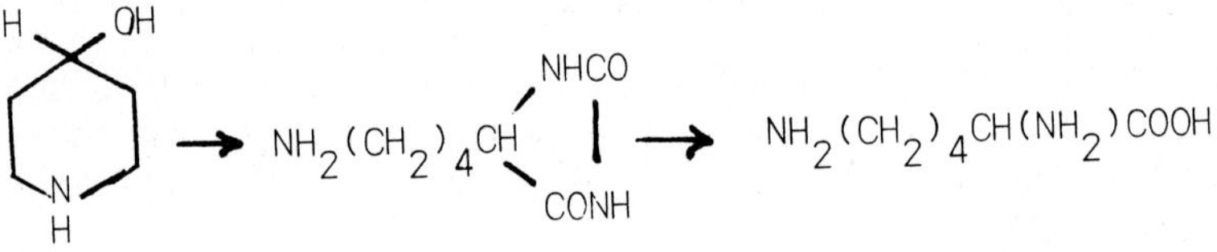

DL-lysine-6-^{14}C can be made as illustrated below (J.Pichat, J.Tostainand and C.Baret, Bull. Soc. Chim. Fr., 1970, 1837).

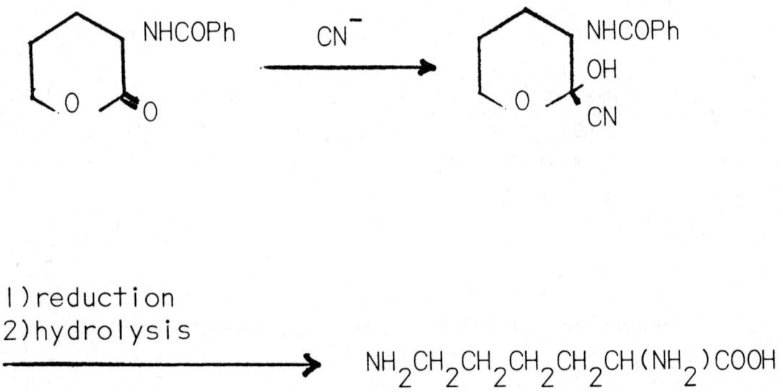

1) reduction
2) hydrolysis

$NH_2CH_2CH_2CH_2CH_2CH(NH_2)COOH$

The synthesis of 4,5-dideuterated or tritiated (+)- and (-)-lysine can be achieved as shown below (A.C.AJansen, K.E.T.Kerling and E.Havinga, Rec. trav. Chim. 1970, **89**, 861).

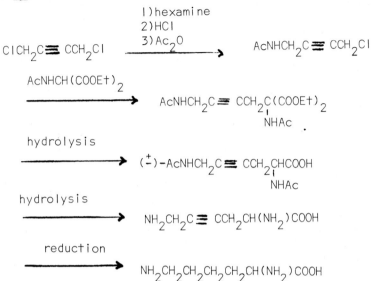

When either the D- or the L-form of 2,3-diaminopropanoic acid undergoes cyanomethylation, the product can be reduced to provide the D- and the L-form of 4-azalysine (J.Kola, Coll. Czech. chem. Comm., 1969, **34**, 630).

β-Chloro-L-alanine can beutilised in the syntheses of L-4-selenalysine, $NH_2CH_2CH_2SeCH_2CH(NH_2)COOH$, and 4-selena-homolysine, $NH_2CH_2CH_2CH_2SeCH_2CH(NH_2)COOH$, (T.Sadeh *et al.*, J. Pharm. Sci., 1976, **65**, 623; C. De Marco *et al.*, Gazzetta, 1976, **106**, 211).

N^{ε} -monomethyllysine is the major product from the reaction of L-lysine with formaldehyde in dilute neutral solutions (Y.Kakimoto and S.Akazawa, J. biol. Chem., 1970, **245**, 5751; J.Puskas and E.Tyihak, Periodici Polytech., 1969, **13**, 261).

Methyl glyoxal reacts with L-lysine to give a yellow polymeric material which is made up of a series of 3-hydroxypyrrole moieties bridged by vinylene or 1,2-hydroxy-

-2-oxo-butylene groups (A.Bonsignore *et al.*, Ital. J. Biochem., 1977, <u>25</u>, 162).

5 *Hydroxy Oxo Carboxylic Acids*

Tetronic acids

The chemistry of the tetronic acids has been reviewed (L.J.Haynes and J.R.Plimmer, Quart. Rev., 1960, <u>14</u>, 292). There are two basic types of tetronic acids with the following formulae.

Tetronic acid α-Tetronic ac

The term "butenolide" was used to describe buteno or crotonolactones (T.Klobb, Compt. Rend., 1900, 130, 1254). Chemical Abstracts has adopted the furanone system for naming these compounds. Thus, $\Delta^{\beta\gamma}$-butenolides are the 2(3H)-furanones and $\Delta^{\alpha\beta}$-butenolides are referred to as the 2(5H)-furanones. Thus, tetronic acid is referred to as 4-hydroxy-2(5H)-furanone and α-tetronic acid as 3-hydroxy-2(5H)-furanone.

2(3H)-furanone   2(5H)-furanone
$\Delta^{\beta\gamma}$ -butenolide $\Delta^{\alpha\beta}$ -butenolide

Pulvinic,acid can now be called 3-hydroxy-5-oxo- α 4-diphenyl- 2(5H) -furan-acetic acid; pulvinic acid dilactone is 3,6-diphenylfuro [3,2 b] furan-2,5-dione and pulvinone becomes 3-phenyl-4-hydroxy-5-phenylmethylene-2(5H)-furanone.

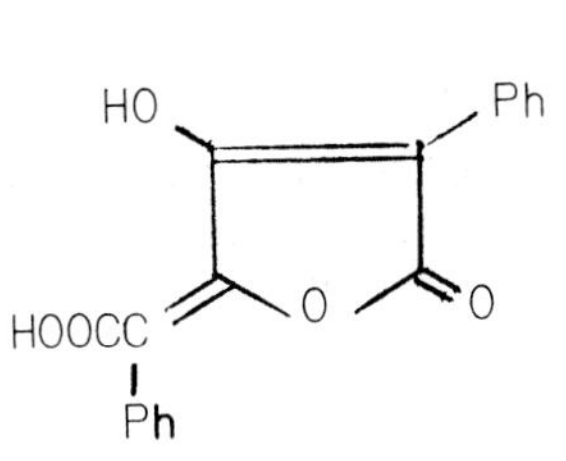

HO   Ph

HOOCC
|
Ph

Pulvinic acid

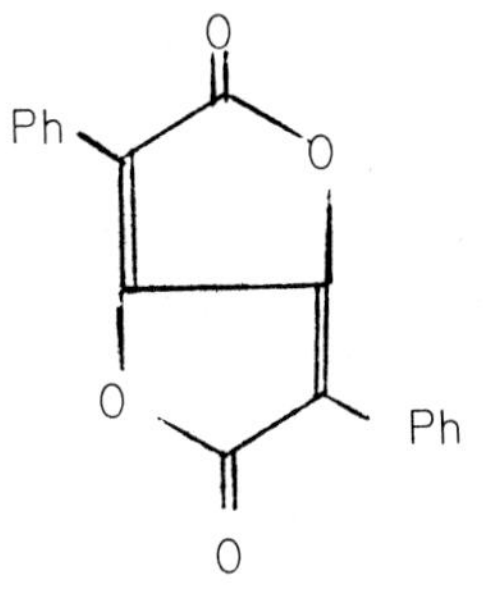

Pulvinic acid dilactone

HO   Ph

CH
|
Ph

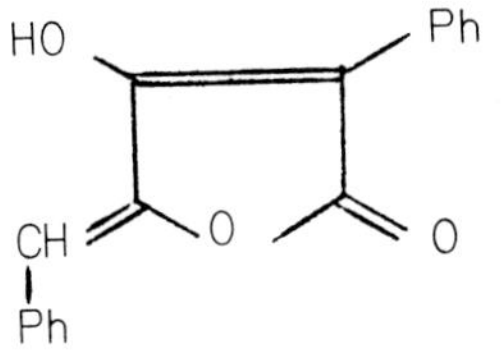

Pulvinone

The following are some of the methods available for the synthesis of tetronic acids.

a)   From 2-benzyl-4-oxazolylglyoxamide with sodium hydroxide (J.W.Cornforth and R.H.Cornforth, J. chem. Soc., 1957, 158).

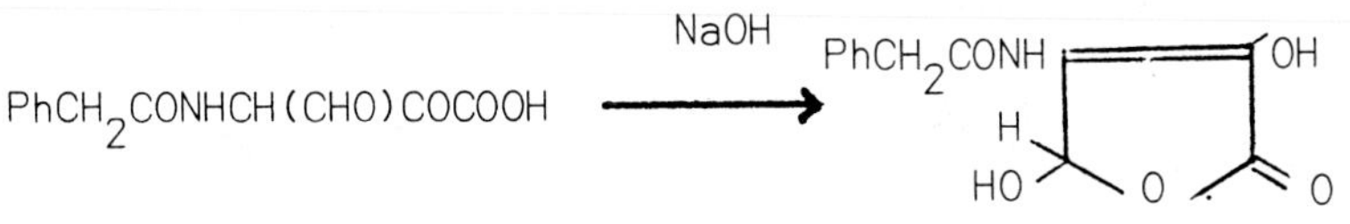

$PhCH_2CONHCH(CHO)COCOOH$ $\xrightarrow{NaOH}$ $PhCH_2CONH$ ... OH
H
HO   O   O

b) From acetyltetronic acids by the reaction with aromatic aldehydes in the presence of hydrogen chloride gas (R.W. Turner, Ger. Pat., 2,101,639, 1971).

c) From 2,4-dioxoalkanoic acids by reaction with aromatic aldehydes in the presence of piperidine to provide, after reaction with acid anhydrides or acyl chlorides, β-acyl-γ-aryl-α-tetronic acids (K.Thomas, Brit. Pat., 1963, 933,509).

d) By the condensation of 2-oxobutanoic acid with aldehydes in the presence of concentrated sulphuric acid (W.Roedel and U.Hempel, Chem. Abs., 1974, 11804r).

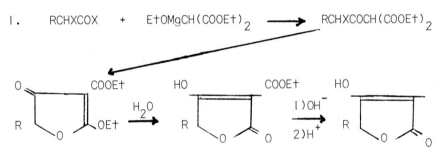

$$CH_3CH_2COCOOH \quad + \quad RCHO \xrightarrow{H_2SO_4}$$

e) From halogenoacyl malonates (T.P.C.Mulholland et al., J. chem. Soc. Perkin I, 1972, 1225).

1. $RCHXCOX \quad + \quad EtOMgCH(COOEt)_2 \longrightarrow RCHXCOCH(COOEt)_2$

$\xrightarrow{H_2O}$ $\xrightarrow[2)H^+]{1)OH^-}$

2. $CH_3CHBrCHBrCOCH(COOEt)_2$

$\xrightarrow{NaOH}$ $\xrightarrow[2) HCl]{1) NaOH}$

$\xrightarrow{H_2/catalyst}$

This method has been used to prepare optically active tetronic acids and $\gamma\gamma$-diphenyl tetronic acid. The O-acyl compounds are made in the presence of TlOEt and the products undergo a Fries migration in the presence of $TiCl_4$ in nitrobenzene to provide α-acyltetronic acids (J.Bloomer *et al.*, Chem. Comm., 1972, 243; Tetrahedron Letters, 1973, 163).

3.

$$CH_3CH_2COCH(COOEt)_2 \xrightarrow{Br_2} CH_3CH_2COCBr(COOEt)_2$$

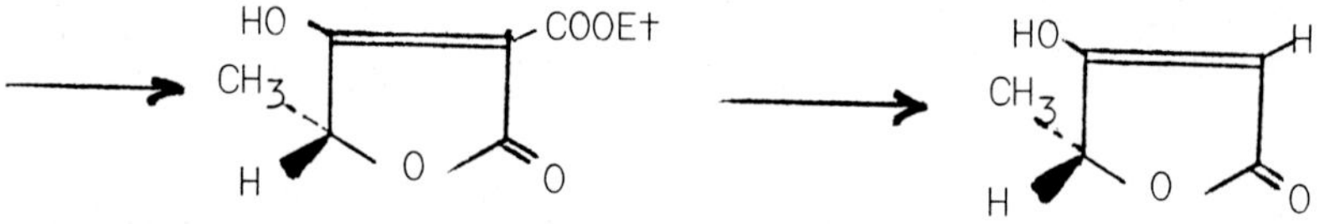

This method has been used to make (S)-carolic acid (F.H. Andersen *et al.*, Acta. Chem. Scand., 1974, 28, 130; J.L. Bloomer and W.F.Hoffmann, Tetrahedron Letters, 1969, 4339; J.P. Jacobsen, T. Reffstrup and P.M. Boll, Acta. Chem. Scand., Ser. B, 1977, 31B, 505).

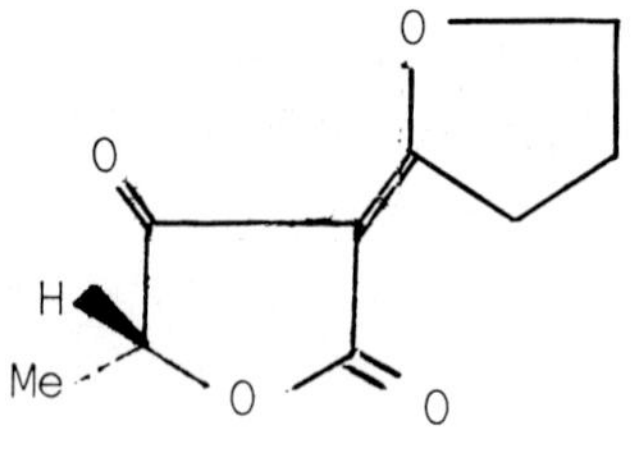

Carolic acid

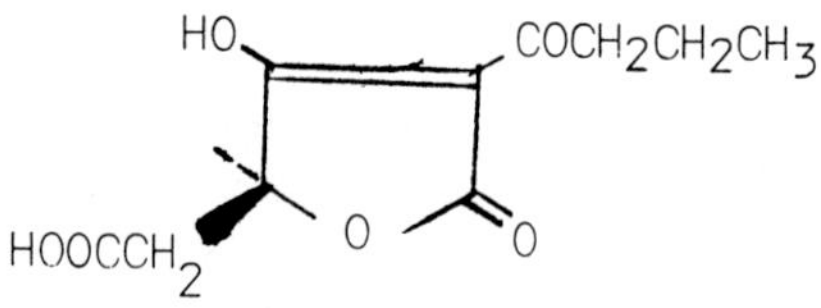

Carlosic acid

In the study of the biosynthesis of carolic acid it is apparent that carlosic acid is the penultimate precursor of carolic acid. The relative stereochemistries of carolic and carlosic acids is of interest. Carolic acid has the *R*-configuration (P.M.Boll *et al.*, Acta. Chem. Scand., 1968, 22, 3251) and an *R*-configuration for carlosic acid would be consistent with a simple decarboxylation process where the chiral centre is not affected. An *S*-configuration for carlosic acid would require a considerably more complex biosynthetic conversion into the *R*-form of carolic acid. Carlosic acid has the *S*-configuration (J.L.Bloomer and F.E.Kappler, Chem. Comm., 1972, 1047). The assignment of the *S*-configuration depends on the following information. The methyl ester of *S*-malic acid (H.Arakawa, Naturwiss, 1963, 50, 441) can be acetoacetylated with diketen and the resultant ester is cyclised and hydrolysed to a product identical to that derived from (-)-carlosic acid.

f) By the bromination of keten acetals to dialkoxy bromo-esters which, on pyrolysis, give α-methyl (or phenyl) β-ethyl tetronates (S.M.McElvain and W.R.Davie, J. Amer. chem. Soc., 1952, 74, 1816).

$$RCH=C(OR')_2 \xrightarrow{\quad Br_2 \quad} RCHBrC(OR')_2CHRCOOR'$$

$$\xrightarrow{\quad heat \quad}$$

R' = Et   R = Me
R' = Me   R = Ph

g) From bromoketo esters with an alkyl substituent in the active methylene group to provide alkyl tetronic acids (A.Svensen and P.M.Boll, Tetrahedron, 1973, 29, 4251; B.Reichert and H.Schafer, Arch. Pharm., 1958, 291, 134).

$$R'CHBrCOCHRCOOEt \longrightarrow$$

R' = H or alkyl

h) From 2-propargyloxytetrahydropyran and ethyl carbonate in the presence of sodium ethoxide, followed by heating with zinc chloride at 150°C to give ethyl tetronate (J.F.Gillespie and C.C.Price, J. org. Chem., 1957, 22, 780).

$$ROCH_2C \equiv CH \;+\; OC(OEt)_2 \xrightarrow{\quad NaOEt \quad} ROCH_2C(OEt)_2CH_2COOEt$$

$$\xrightarrow[150°C]{\quad ZnCl_2 \quad}$$

i) From the disodium salt of 3,3'-dioxoadipic acid by alkylation with alkyl halides, followed by treatment with concentrated sulphuric acid (H.G.Raubenheimer and D.H. Dekock, J.S.Afr. Chem. inst., 1972, 25, 70).

$$R = Me, PhCH_2 \text{ or } CH_2C \equiv CH$$

j) From dioxane derivatives by the action of concentrated sulphuric acid (H.Stacel, Arch. Pharm., 1963, 296, 479).

k) By the condensation of pyruvic acid with dimethylamine hydrochloride with formaldehyde (E.E.Galanatey, A.Szabo and J.Fried, Tetrahedron Letters, 1963, 415).

$$CH_3COCOOH + HCHO + Me_2NH \longrightarrow$$

These compounds exist as zwitterions and can react with potassium thioacetate to give the dilactone.

α -Benzamido-$\Delta^{\alpha\beta}$-butenolides can be made by similar methods (D.N.Reinhoudt *et al.*, Rec. trav. Chim., 1968, 87, 1153; H.C.Byerman *et al.*, *ibid.*, 1966, 85, 347).
α -Imino-β-triphenylmethylthiomethyl-$\Delta^{\alpha\beta}$-butenolide can also be made by this method (J.E.Dolfini *et al.*,J. org. Chem., 1963, 34, 1582; A.G.Long and A.F.Turner, Tetrahedron Letters, 1962, 421).

1) By the acid catalysed condensation of benzaldehydes with pyruvic acid (E.D.Stecher *et al.*, J. org. Chem., 1965, 30, 1800).

If pyruvic acid and benzaldehydes are condensed in aqueous solution with aromatic amines, α-aminophenyl derivatives are obtained.

m)   By the ring-opening of 2-ethyl-4-ethoxycarbonyl-5-methyl-3(2H)-furanone with potassium hydroxide, and subsequent ring closure provides α-acetyl- β -ethyl tetronic acid   (D.De Rijke and H.Boelens, Rec. Trav. Chim., 1973, 92, 731)

Similarly, 5-amino-4-phenyl-3(2H)-furanone is converted into α-phenyl tetronic acid   (S.Umio et al.,Chem. Abs., 1969, 71, 101695b).

n)   From diethyl oxalate by reaction with 2-nitrotoluene in the presence of sodium ethoxide   (T.Saken *et al.*, Bull. chem. Soc. Japan, 1964, <u>37</u>, 1166, 1171).

Diethyl oxalate reacts with *o*-hydroxyacetophenones in the presence of sodium ethoxide to yield **β**-benzoyl- **α**-hydroxy- **Δ**$^{\alpha\beta}$ -butenolides   (P.Niviere *et al.*, Bull. Soc. chim. Fr., 1965, 3658;   W.L.Parker and F.Johnson, J. Amer. chem. Soc., 1969, <u>91</u>, 7208).

Ring closure of α-hydroxy- β -3-methoxypropionyl- $\Delta^{\gamma\beta}$ -
butenolide provides the corresponding reduced γ -pyrone
butenolides (A.Caudet *et al.*, Compt. rend., 1970, 270, 1127).

With the corresponding α-hydroxy- β -3-methoxy-2-propenoyl-
$\Delta^{\alpha\beta}$ -butenolide, a γ-pyrone butenolide is produced which,
in turn, can react with thiols and amines (A.Caudet *et al.*,
Compt. rend., 1971, 272, 107).

The γ-pyrone butenolide reaction with hydrazine leads to the ring opening of the γ-pyrone ring rather than the butenolide ring (A.Caudet *et al.*, Bull. Soc. Chim. Fr., 1973, 1707).

This method allows the formation of 2-aroyl- α -hydroxy-butenolides from acetophenones, 2-acetylthiophens and 2-acetylbenzofuran (P.J.Bargnoux *et al.*, Compt. rend., 1973, 276, 1041). This involves condensation with diethyl oxalate, hydroxymethylation of the resulting pyruvic acid derivative, followed by acid catalysed ring closure (G.Durantin *et al.*, Chem. Ther., 1972, 7, 472; J.Paris, Compt. rend., 1974, 278, 1149). The condensation of diethyl ketone with diethyl oxalate in the presence of sodium ethoxide yields the γ-lactone of 3-methyl-2,4-dihydroxy-2,4-hexadienoic acid which can be reduced with Raney nickel to give α-hydroxy- β-methyl- γ -ethyl- $\Delta^{\alpha\beta}$ -butenolide (A.Fabryey, Rocz. Chim., 1966, 40, 1657)

When bis(4-methoxybenzyl)ketone is condensed with diethyl oxalate in the presence of sodium ethoxide, 2,5-dianisylcyclopentane-1,3,5-trione can be made. This, on pyrolysis, gives α -(4-methoxyphenyl)- β -hydroxy- γ - (4-methoxybenzylidene)- $\Delta^{\alpha\beta}$ -butenolide (V.Ojima *et al.*, Phytochemistry, 1973, 12, 25). This , in turn, can be methylated with diazomethane to give the β-methoxy derivative, or on demethylation it yields α-4-hydroxy-benzylidene- β-hydroxy- γ-4-hydroxybenzylidene- $\Delta^{\alpha\beta}$ -butenolide which is identical with the naturally occurring product from the culture filtrate of *Aspergillus terreus*.

o) By reduction of 2-(substituted amino) maleic esters to yield enaminone analogues of tetronic acid which can be hydrolysed to tetronic acid (J.V.Greenhill *et al.*, J. chem. Soc. Perkin I, 1975, 588).

p) 1,3-Dihydroxyacetone reacts with 2-aryl-5-oxazolone in the presence of lead tetraacetate in THF to provide unsaturated oxazolones which can be hydrolysed and recyclised to give butenolides. The addition of ethyl vinyl ether protects the 1,3-dihydroxyacetone *in situ* (H.S.Tan *et al.*, Rec. trav. Chim., 1969, 88, 209; C.Armengaud *et al.*, Compt. rend., 1962, 254, 2181).

Reactions of Tetronic Acids

The results of oxidation of various tetronic acids with chromium trioxide/sulphuric acid are shown below (F.B.Reid *et al.*, J. org. Chem., 1950, 15, 572; 1951, 16, 33).

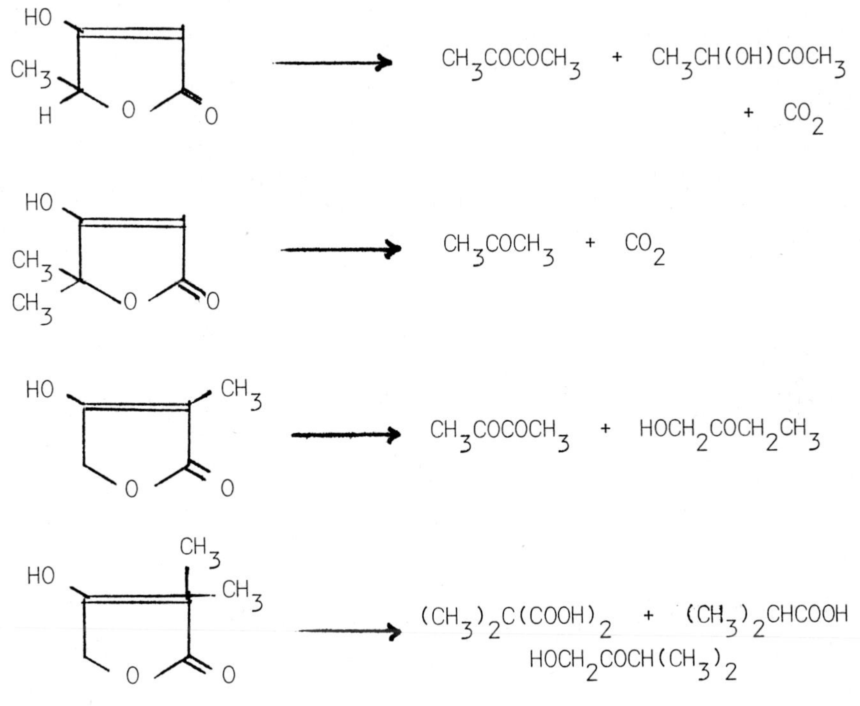

$$CH_3COCOCH_3 \quad + \quad CH_3CH(OH)COCH_3$$
$$+ \quad CO_2$$

$$CH_3COCH_3 \quad + \quad CO_2$$

$$CH_3COCOCH_3 \quad + \quad HOCH_2COCH_2CH_3$$

$$(CH_3)_2C(COOH)_2 \quad + \quad (CH_3)_2CHCOOH$$
$$HOCH_2COCH(CH_3)_2$$

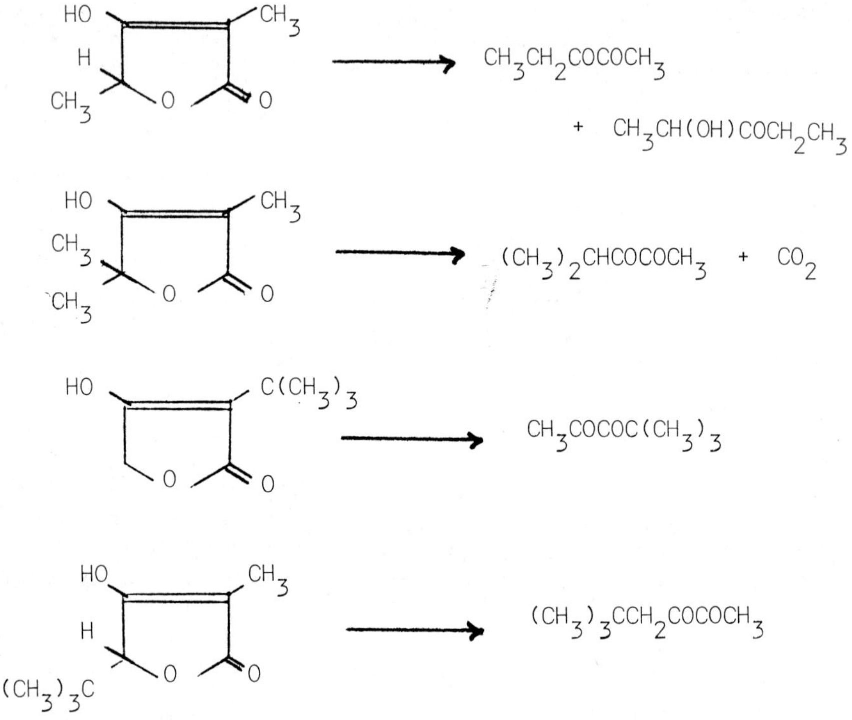
HO
CH3
H
CH3
O
O
CH3CH2COCOCH3
+ CH3CH(OH)COCH2CH3
HO
CH3
CH3
CH3
O
O
(CH3)2CHCOCOCH3 + CO2
HO
C(CH3)3
O
O
CH3COCOC(CH3)3
HO
CH3
H
(CH3)3C
O
O
(CH3)3CCH2COCOCH3

γ-Ethoxy-$\Delta^{\beta\gamma}$-butenolide can be converted into α-ethyl-β-carboxymethyl-$\Delta^{\beta\gamma}$-butenolide as shown below (J.E.DeGraw, Tetrahedron, 1972, 28, 967). The latter compound is a key intermediate in the synthesis of pilocarpine (A.V.Chumachenko *et al.*, Zh. org. Khim., 1972, 8, 1100).

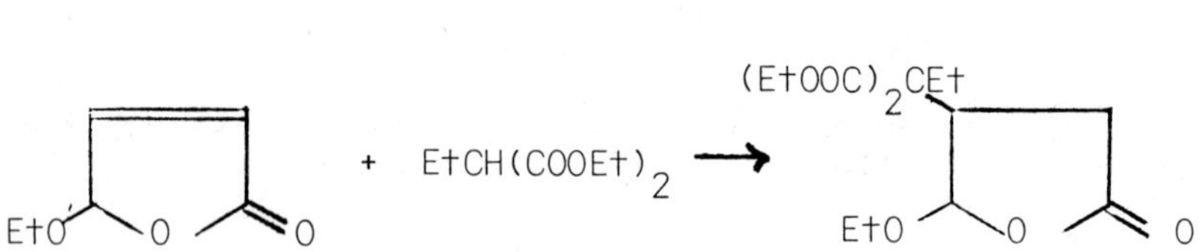
EtO
O
O
+ EtCH(COOEt)2
(EtOOC)2CEt
EtO
O
O

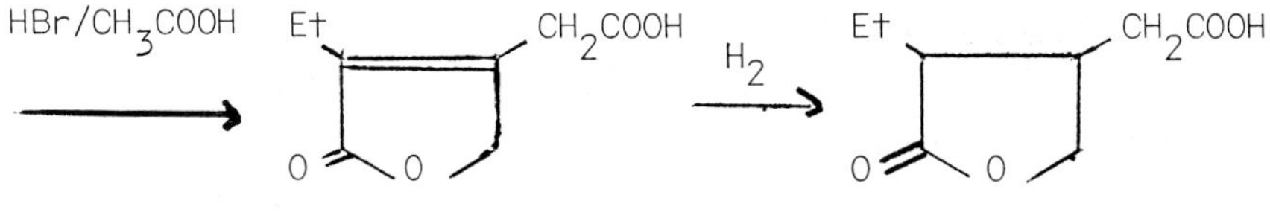

HBr/CH$_3$COOH

(±)-Homopilopic acid

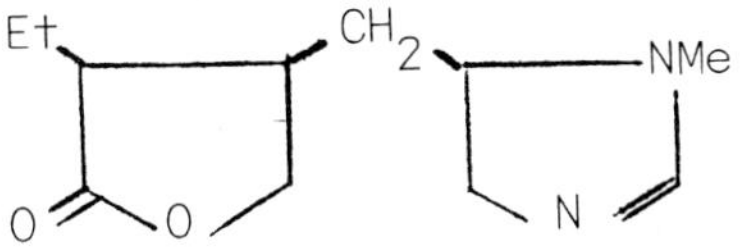

Pilocarpine

The reaction of **α**-phenylamino- Δ$^{\alpha\beta}$ -butenolide with diphenylketen gives 2,3,3a,4,6,6a-hexahydro-2,6-dioxo-1,3,3-triphenyl-1H-furo 3,4-b pyrrole (T.Minami *et al.*, J. org. Chem., 1974, <u>39</u>, 3236).

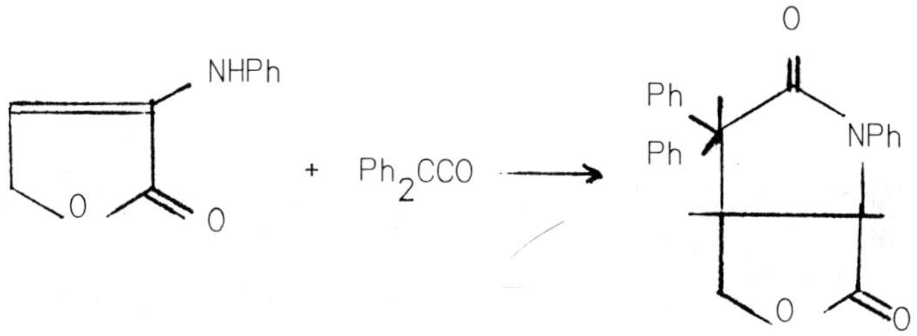

Tetronic acids react with arylamines to provide **β**-arylaminobutenolides (T.A.Favorskaya *et al.*, Zh. org. Khim., 1970, <u>6</u>, 2015).

When a carboxyl group is present in the lactone the
corresponding amide is obtained (A.A.Avatisyan *et al.*, Chem.
Abs., 1972, 78, 33845b).

Acetylbutenolides react with amines to provide
iminobutenolides (A.A.Avatisyan, Chem. Abs., 1974, 80,
3300v).

When carboxyethylbutenolides are heated with dimethylamine in water at 150°C they give the corresponding
carboxamide (A.A. Avetisyän, Chem. Abs., 1971, 74, 124878q).

Benzyltetronic acids react with thionyl chloride to yield α-benzylidene- β-hydroxy- Δ^βγ -butenolide (V.Olhoff *et al.*,Z. Chem., 1970, 10, 341).

The treatment of α-azido- γ -butyrolactones with catalytic amounts of sodium ethoxide leads to the elimination of nitrogen with rearrangement to yield aminobutenolides (A.Schmitz *et al.*, Ber., 1975, 103, 1010).

Under similar conditions α-4-nitrobenzylideneamino-γ-butyrolactone can be converted into a butenolide derivative (B.Reichert and H.Schafer, Arch. Pharm., 1958, 291, 134).

Diketones such as biacetyl react with carbon suboxide in the presence of trialkyl phosphites to yield γ-acetyl-γ-methyl-β-methoxy-α-phosphoryl butenolides (F.Ramirez and G.V.Loewengert, J. Amer. chem. Soc., 1969, 91, 2293).

Also, when carbon suboxide is added to an adduct of biacetyl and methyl diphenyl phosphinite, a butenolide is produced.

The reaction of thymoquinone with sodium azide in trichloroacetic acid gives α-methyl- β-amino- γ-(1-cyano-2-methylpropylidene)- $\Delta^{\alpha\beta}$ -butenolide (H.W.Moore and H.R.Sheldon, J. org. Chem., 1968, 33; A.H.Rees, Chem. and Ind., 1964, 931; 1965, 1298).

On the other hand, 3-amino-6-azido-2-methyl-5-isopropyl-1,4-benzoquinone and 2-amino-5-azido-3-isopropyl-6-methyl-1,4-benzoquinone conversion (I.A.Kuzovnikova *et al.*, Chem. Abs., 1974, 81, 151,301r).

A similar stereoselective transformation of 2-azido-3-methyl-6-isopropyl-1,4-benzoquinone gives α -isopropyl-γ -(1-cyanoethylidene)- $\Delta^{\alpha\beta}$ -butenolide in which the cyano group on the exocyclic double bond is *trans* to the lactone oxygen (H.W.Moore and A.R.Sheldon, J.Amer.chem. Soc., 1970, 92, 1675).

The mechanism of these transformations has been suggested to be as follows.

This rearrangement can be used in the synthesis of vulpinic acid (H.W.Moore and W.Weyler, J. Amer. chem. Soc., 1971, 93, 2812; H.W.Moore *et al.*, *ibid.*,1970, 92, 1675). The pyrolysis of 2-azido-3,6-diphenyl-5-hydroxy-1,4-benzoquinone leads to the formation of α-phenyl-γ-(1-cyanobenzylidene)- $\Delta^{\alpha\beta}$ - butenolide (H.W. Moore *et al.*, Tetrahedron Letters, 1969,3947).

2-Aminoethyl- α -tetronic acid can be made from γ-butyrolactone as follows (E.E.Galantay *et al.*, J. org. Chem., 1970, 35, 4277).

The reaction of β-triphenylmethylthiomethyl-α-hydroxy-Δ$^{\alpha\beta}$-butenolide similarly provides the corresponding α-amino derivative (J.E.Dolfini *et al.*, J. org. Chem., 1968, 34, 1588; D.N.Reinhoudt *et al.*, Rec. trav. Chim., 1966, 85, 347).

Pulvinic acid derivatives

The lichens, about 15,000 species of which are known, are symbiotic forms of life composed of a fungal part (mycobiont) and an algal part (phycobiont). Symbiosis generally represents a close and enduring association of unlike organisms. Lichens are the most advanced form of this symbiotic process in which both organisms demonstrateably gain from the association.

Lichen substances are generally low-molecular weight organic compounds. the important aromatic lichen substances i.e. depsides, depsidones, depsones, dibenzofurans and pulvinic acid derivatives almost all occur solely in lichens as shown by the following biosynthetic pathways (K.Mosbach, Acta. Chem. Scand., 1967, 21, 2331).

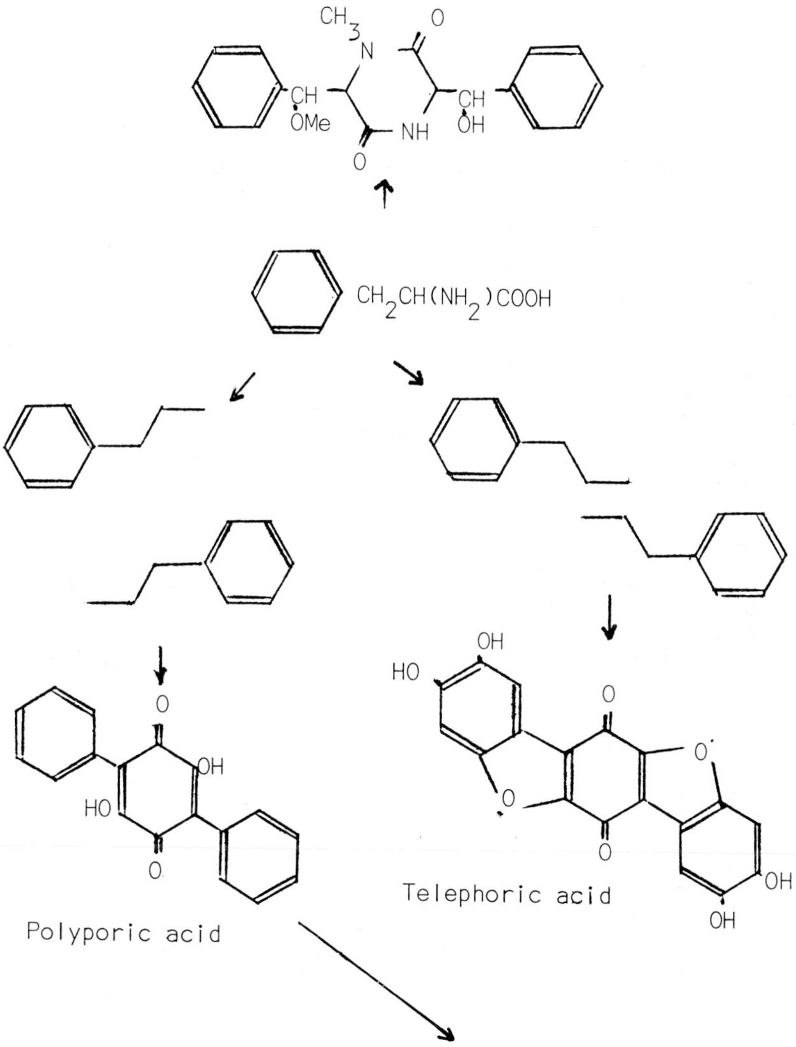

Polyporic acid

Telephoric acid

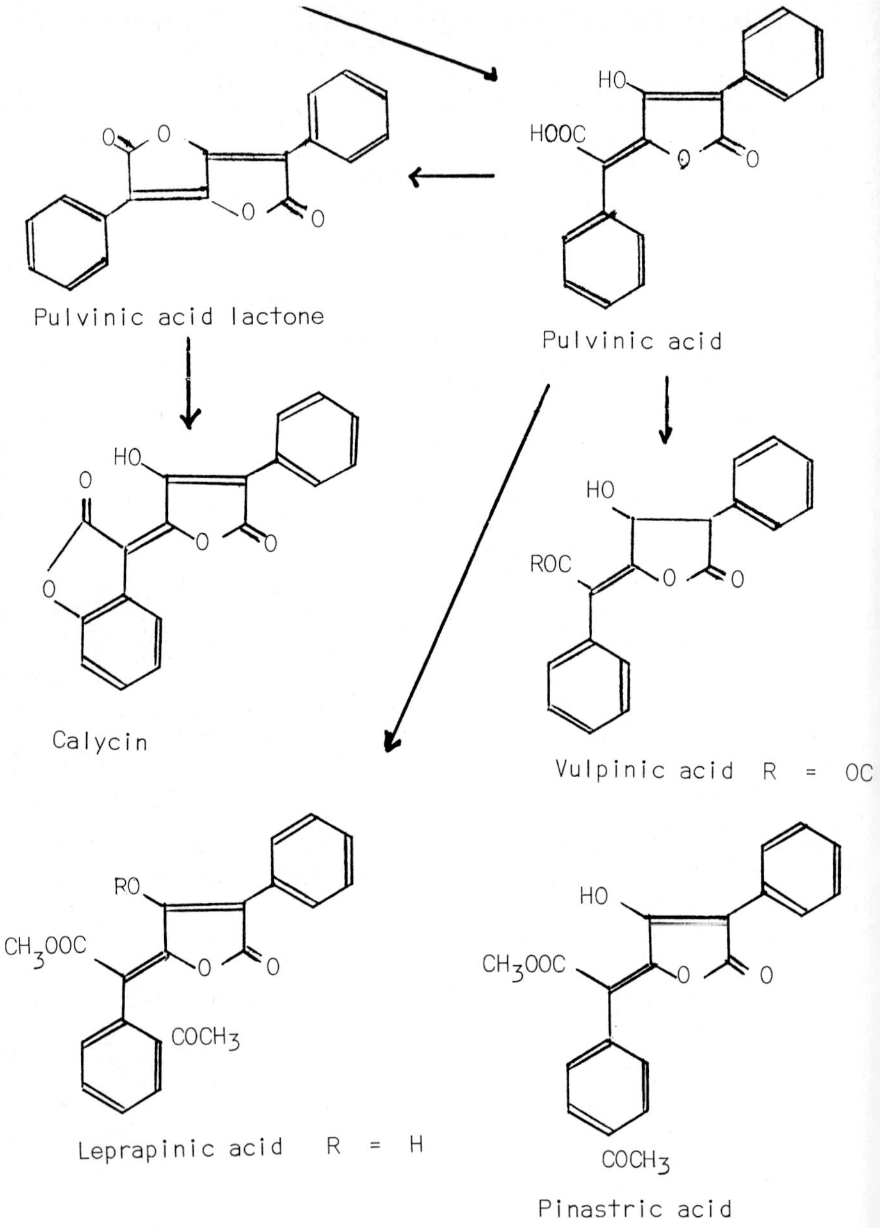

Pulvinic acid lactone

Pulvinic acid

Calycin

Vulpinic acid   R = OC

Leprapinic acid   R = H

Pinastric acid

Biosynthesis of Lichen compounds

The higher fatty acids occurring in lichens and their lactones can be divided into three main groups.

a) Lactone acid derivatives (lichesterinic acid)

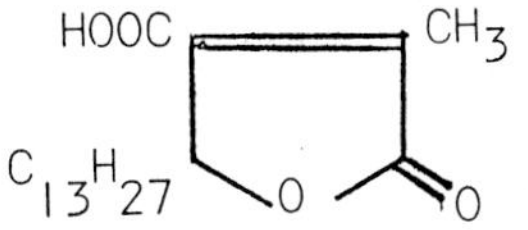

b) Dibasic fatty acids, e.g. rocellic acid,

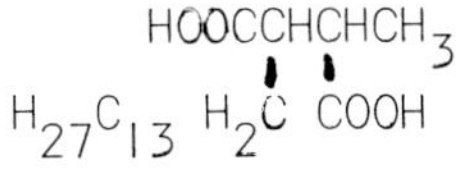

c) Tribasic fatty acids, eg. rangformic acid

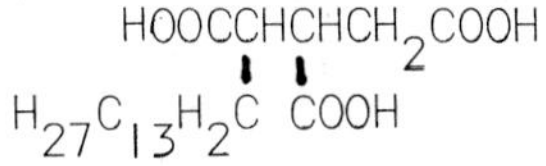

All aliphatic lichen acids are optically active.

The formation of lichen acids is considered to proceed by an aldol condensation of long chain alkanoyl-SCoA units with keto acids which, in turn, are formed by condensation of "starter" acetyl SCoA with molecules of malonyl-SCoA. The lichen substances shown in the following diagram are probably formed by the condensation of myristoyl or palmitoyl SCoA with oxaloacetic acid or pyruvic acid. In the lactone acids the corresponding hydroxy fatty acids are poss possible condensation units. The protolichesterinic acid occurring in lichens is formed by condensation of palmitic acid with pyruvic acid (J.L.Bloomer, W.R.Eder and W.F.Hoffmann, Chem. Comm., 1968, 354).

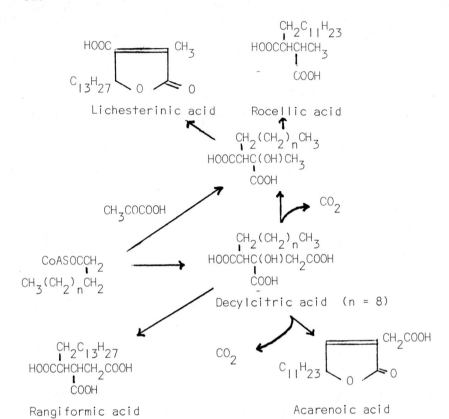

Proposed biosynthesis of lichesterinic acid, rocellic acid
rangiformic acid, acarenoic acid and decylcitric acid

Methods of Synthesis of Pulvinic acids

a) By the condensation of diethyl oxalate with phenyl
acetonitrile (J.Volhard, Ann. Chem., 1894, 282; H.D.Stachel
Ann. Chem., 1965, 689, 119; Arch. Pharm., 1963, 296, 479).
This method can be modified so that phenyl acetonitrile
can be condensed with diethyl oxalate to give ethyl
cyanophenyl pyruvate. This may then be condensed with any
aryl acetonitrile under Claisen conditions to provide
unsymmetrical ketipic acid derivatives (E.C.Agarwal and
T.R.Seshadri, Indian J. Chem., 1964, 2, 17; P.C.Beaumont
et al., J. chem. Soc. (C), 1968, 2968; F.Bohlman, Ber.,
1966, 99, 3544; P.K.Grover and T.R.Seshadri, Tetrahedron,
1958, 4, 105).

$$2 \text{ PhCH}_2\text{CN} \quad + \quad \overset{\displaystyle \text{COOEt}}{\underset{\displaystyle \text{COOEt}}{|}} \quad \xrightarrow{\text{EtONa}} \quad \text{PhCH(CN)COCOCH(CN)Ph}$$

$$\xrightarrow{\text{H}_2\text{SO}_4} \quad \text{PhCH(COOH)COCOCH(COOH)Ph} \longrightarrow$$

b) By the condensation of phenylacetyl chloride with oxalyl chloride in the presence of triethylamine and pyridine to give pulvinic dilactone (T.Sakan et al., Tetrahedron Letters, 1967, 1623; D.G.Farnum et al., J. Amer. chem. Soc., 1966, 88, 3075).

$$2 \text{ PhCH}_2\text{COCl} \quad + \quad \overset{\displaystyle \text{COCl}}{\underset{\displaystyle \text{COCl}}{|}} \quad \xrightarrow{\text{C}_5\text{H}_5\text{N}}$$

c) Phenylhydroxycyclopropenone is converted into pulvinic dilactone by treatment with thionyl chloride in benzene at 40°C (E.E.Galantay et al., J. org. Chem., 1970, 35, 4277).

oxidation

d) From polporic acid by oxidation with lead tetraacetate in glacial acetic acid to provide pulvinic dilactone (K.Mossbach, Biochem. Biophys. Res. Comm., 1964, 17, 363; B.F.Cain, J.chem. Soc., 1961, 936; 1963, 356; R.L.Frank et al., J. Amer. chem. Soc., 1950, 72, 1823). Polyporic acid and its analogues have been used as sources for the synthesis of lichen tetronic acids. Biogenetic pathways for these substances involve polyporic acid. Analogues of pulvinic dilactone such as 4,4'-dimethoxypulvinic dilactone can also be made by this method.

The oxidation of 2,5-di-p-hydroxyphenyl-3,6-dihydroxy-1,4-benzoquinone may be carried out in acetic acid using hydrogen peroxide.

e) By heating polyporic acid, 2,5-dihydroxy-3,6-diphenyl-1,4-benzoquinone, with dimethyl sulphoxide/ acetic anhydride to give pulvinic dilactone (R.J.Wikholm and H.W.Moore, J. Amer. chem. Soc., 1972, 94, 6152)

Reactions of Pulvinic Acid

Pulvinic acid can be decarboxylated to give pulvinone, α-phenyl- β -hydroxy- γ -benzylidene- $\Delta^{\alpha\beta}$ -butenolide, which, on further heating at 270°C, is converted into a cyclopentanetrione derivative. This latter reaction is reversible (A.Schoneberg and A.Sina, J. chem. Soc., 1946, 601).

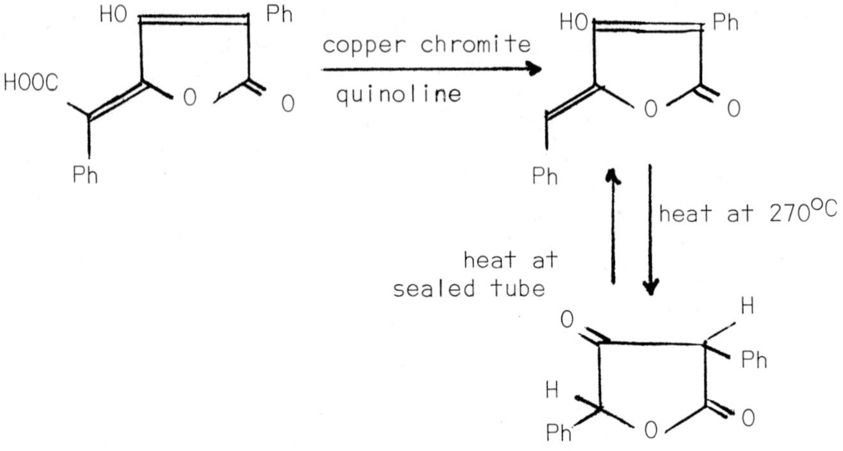

Pulvinone derivatives can be isolated from a culture of *Asperigillus terreus*. The structures of these pulvinone derivatives have been established as shown below (N.Ojima, S.Takenaka and S.Seto, Phytochem., 1973, 12, 2527; 1975, 14, 573).

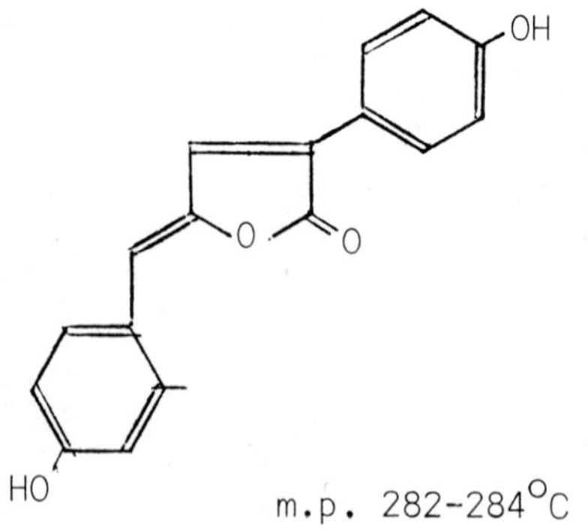

m.p. 282-284°C

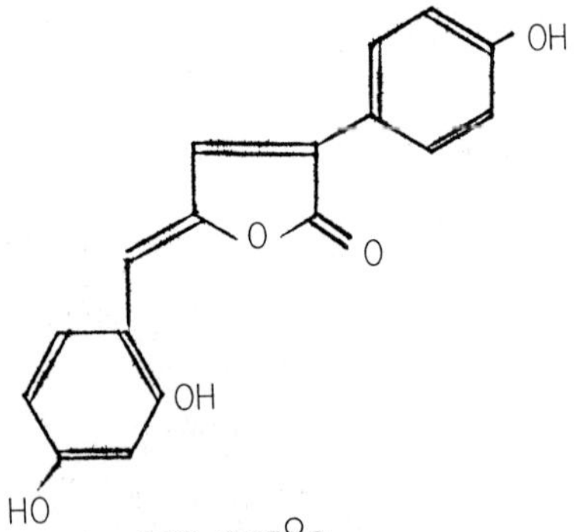

m.p. 262-263°C

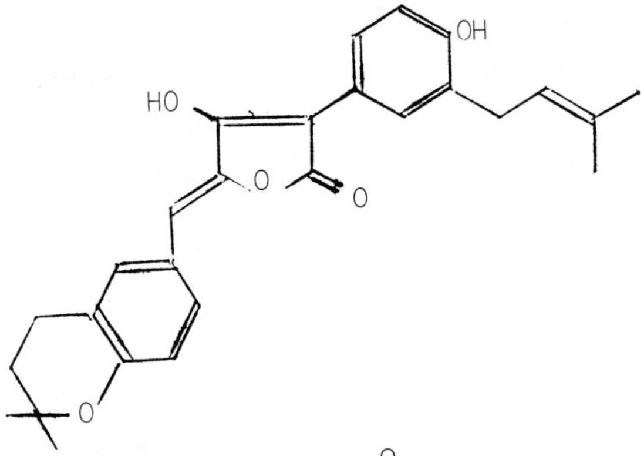

m.p. 187-189°C

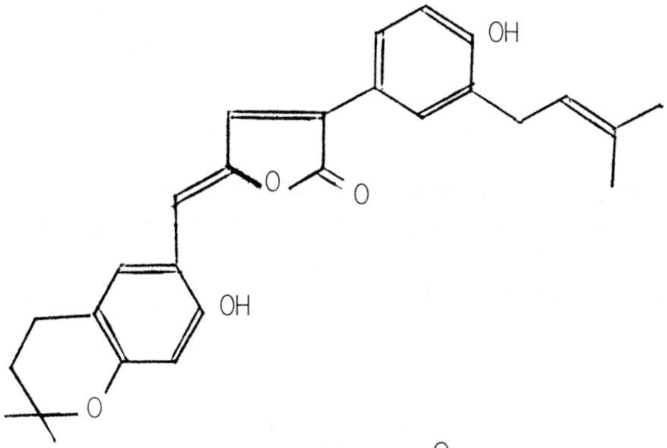

m.p. 257-259°C

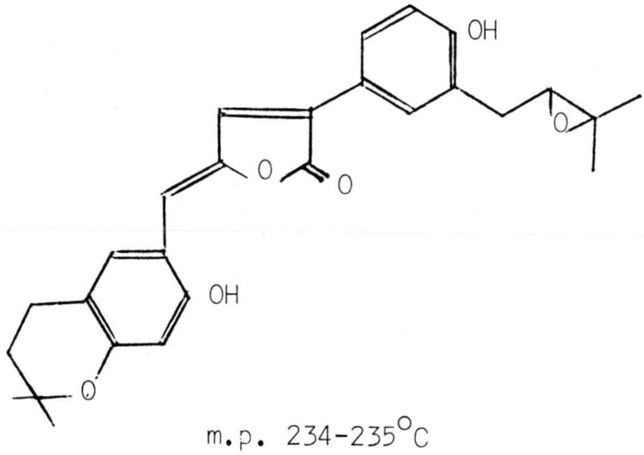

m.p. 234-235°C

3,4,4'-Trimethoxypulvinic acid lactone (1)

This compound, orange needles, m.p. 230-231°C, can be made made by the reaction of 2-(3,4-dimethoxyphenyl)-5-(4-methoxyphenyl)-3,4-dioxoadiponitrile with acetic acid and sulphuric acid (R.L.Edward and M.Gill, J. chem. Soc. Perkin I, 1973, 1529). A similar hydrolysis of the dinitrile yields 4,4'dimethoxypulvinic acid (II), 3,4,4'-trimethoxypulvinic acid (III), 3',4',4-trimethoxypulvinic acid (IV) and 3,3',4,4'-tetramethoxypulvinic acid (V) (M.Asano and Y.Kameda, Ber., 1934, 67b, 1522).

Also, this lactone (I) can be converted into methyl 3,4,4'-trimethoxypulvinate (VI), orange needles, m.p. 177°C, by its reaction with methanolic potassium hydroxide. Methyl 3',4',4-trimethoxypulvinate (VII) can be also isolated as lemon platelets, m.p. 168°C.

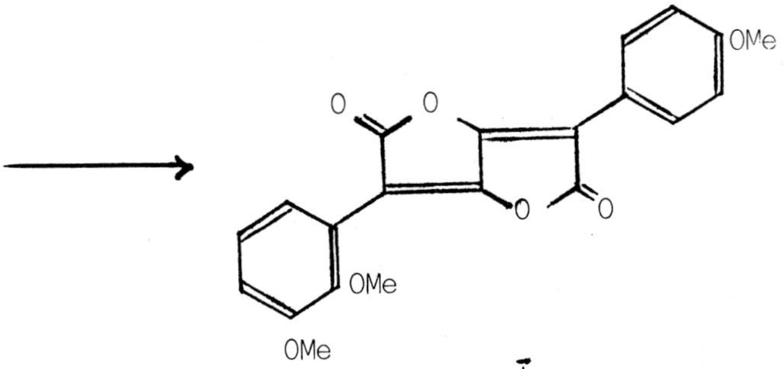
O
O
OMe
O
O
OMe
OMe

I

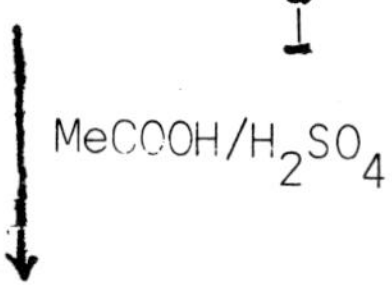
MeCOOH/H2SO4

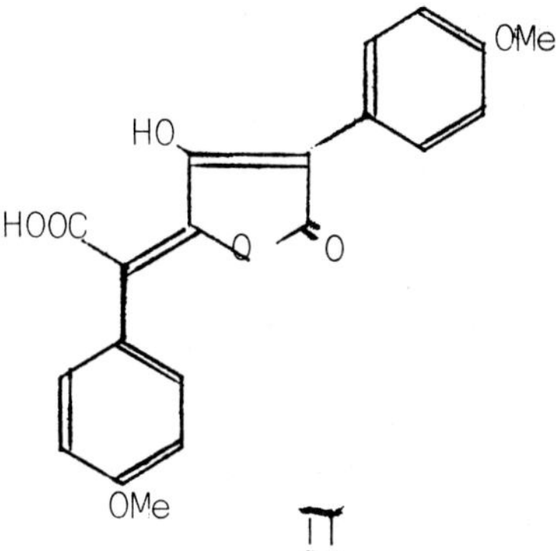
OMe
HO
HOOC
O
O
OMe

II

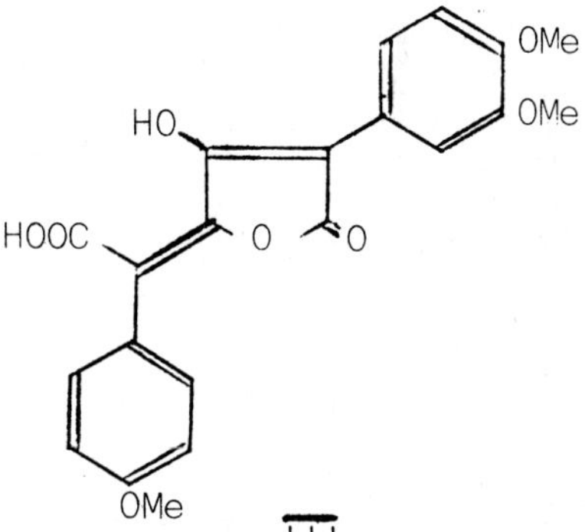
OMe
OMe
HO
HOOC
O
O
OMe

III

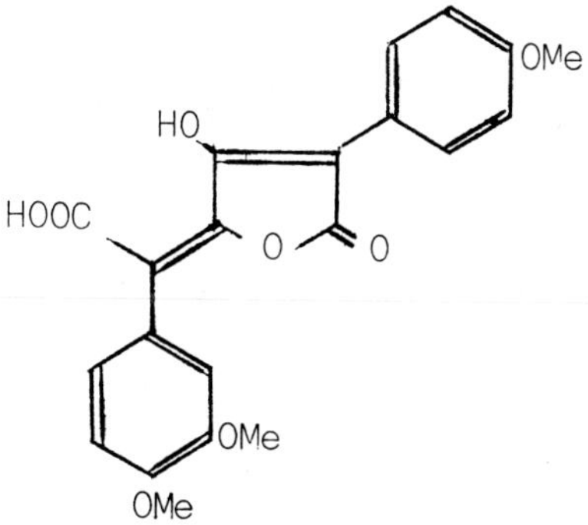
OMe
HO
HOOC
O
O
OMe
OMe

IV

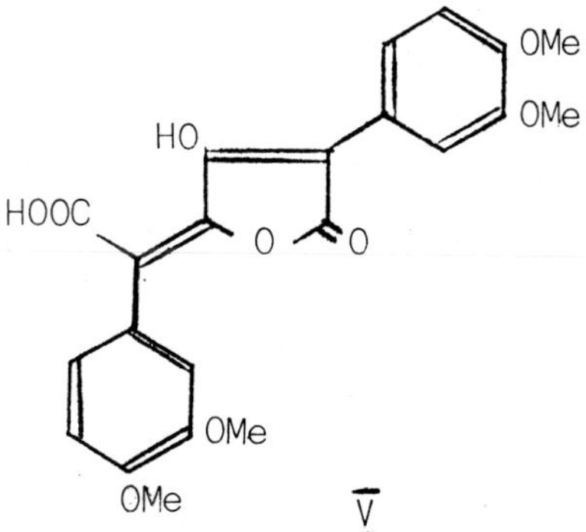
OMe
OMe
HO
HOOC
O
O
OMe
OMe

V

Methyl 3,4,4'trimethoxypulvinate (VI) reacts with diazomethane to give methyl O-methyl-3,4,4'-trimethoxypulvinate (VII) as cadmium yellow needles, m.p. 146-148°C. Methyl O-methyl-3',4',4-trimethoxypulvinate (IX), yellow rods, m.p. 150-151°C, can be made similarly.

$$\xleftarrow{\quad CH_2N_2 \quad}$$

IX   VIII

3,4,4'-Trimethoxypulvinic lactone (I) reacts with
aqueous potassium hydroxide to give 3',4',4-trimethoxy-
pulvinic acid (IV), glistening yellow plates, m.p. 249-253°C,
and 3,4,4'-trimethoxypulvinic acid (III), orange needles,
m.p. 200-205°C.   When a solution of trimethoxypulvinic
acid reacts with copper(II) acetate it gives a copper
chelate, m.p. 210°C.   This copper chelate, on heating in
refluxing quinoline, gives a mixture of 3,4,4'- (X) and
3',4',4-trimethoxypulvinone (XI) and 2-(3,4-dimethoxyphenyl)
-5-(4-methoxyphenyl)cyclopentane-1,3,4-trione (XII) as
lemon needles, m.p. 209-210°C.
    A mixture of 3,4,4'-trimethoxy- (X): and 3',4',4-
trimethoxypulvinone (XI) reacts with diazomethane to give
3,4,4'-trimethoxy-O-methylpulvinone (XIII) tan coloured
needles, m.p. 179-181°C,   and 3',4',4-trimethoxy-O-methyl-
pulvinone (XIV)', tan coloured leaflets, m.p. 153-155°C.

III → (copper acetate / quinoline) → X

XIII (CH₂N₂)

IV → (copper acetate / quinoline) → XI

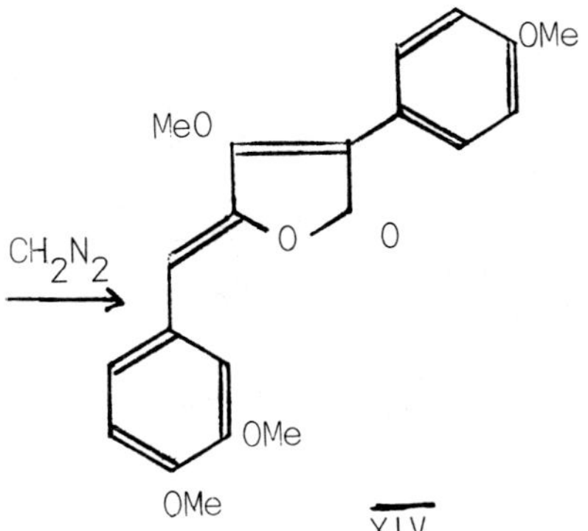

$\underline{XIV}$

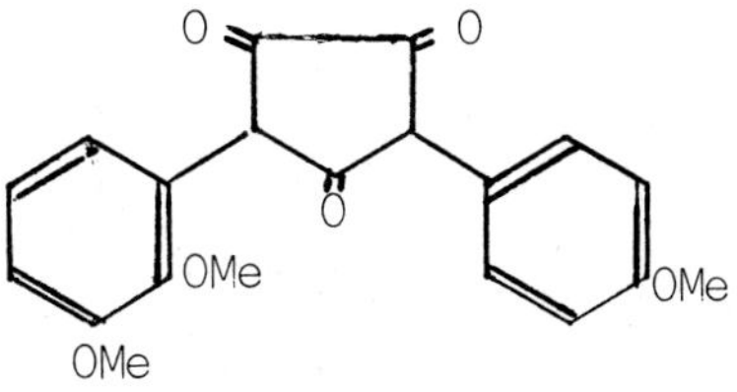

$\underline{XII}$

A mixture of trimethoxypulvinone (X and XI) can be acetylated with acetic anhydride to give a mixture of O-acetyl-3,4,4'-trimethoxypulvinone (XV) and O-acetyl-3',4', 4-trimethoxypulvinone (XVI) (R.L.Edward and M.Gill, J. chem. Soc. Perkin I, 1973, 1529).

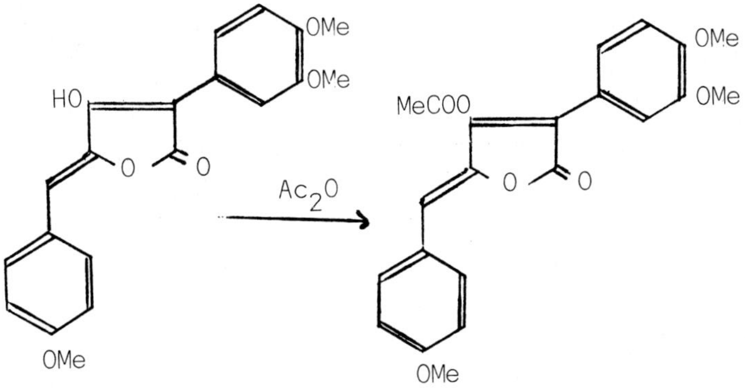

$\underline{X}$

$\underline{XV}$

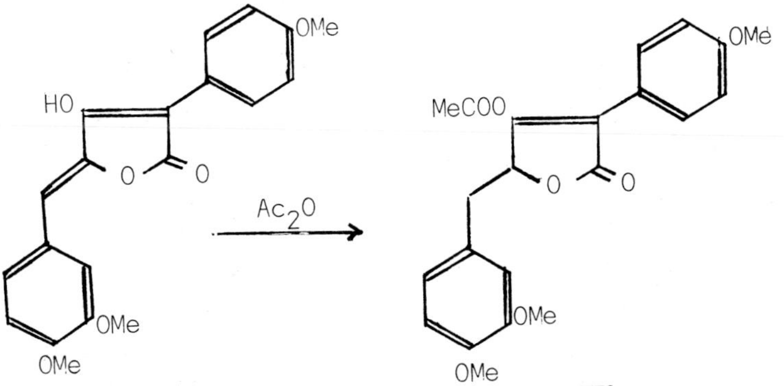

$\underline{XI}$

$\underline{XVI}$

Trihydroxypulvinone (XVII), isolated from the fungus, *Suillus grevillei*, is methylated with diazomethane to give mainly 3',4',4-trimethoxy-*O*-methylpulvinone (XIV)), tan leaflets, m.p. 153-155°C, and is acetylated with acetic anhydride to yield mainly 3',4',4-triacetoxy-*O*-acetylpulvinone (XVIII) pale yellow needles, m.p. 200-203°C. This compound can be oxidised with chromium trioxide in acetic acid to give 3,4-dihydroxybenzoic acid and 4-hydroxybenzoic acid.

Methyl *O*-methyl-3,4,4'-trimethoxypulvinate (VIII)yields
5-(3,4-dimethoxyphenyl)-4-methoxy-2-(4-methoxyphenyl)-cyclo-
pent-4-ene-1,3-dione (XIX) , yellow leaflets, m.p. 151-153ºC,
by reaction with methanolic potassium hydroxide. Methyl
O-methyl-3',4',4-trimethoxypulvinate (IX) similarly reacts
to give the potassium salt of dimethyl 4-(3,4-dimethoxyphenyl)
-3-hydroxy-2-methoxy-1-(4-methoxyphenyl)buta-1,3-diene-1,4-
dicarboxylate (XXI)), yellow needles, m. p. 230ºC, together
with the corresponding dione (XX) , pale yellow needles,
m.p. 163-165ºC (R.L. Edward and M. Gill, J. chem Soc. Perkin I,
1973,1529).

$$CH_3OOCC=C(OH)C(OCH_3)=CCOOCH_3$$

3,4,4'-Trimethoxypulvinic lactone (I) can be converted into 3,4,4'-trihydroxypulvinic acid (xerocomic acid) (XXII) red needles, m.p. 295°C, and 3',4',4-trihydroxypulvinic acid (isoxerocomic acid)(XXIII) orange-red needles, m.p. 300-305°C, by reaction with a mixture of acetic acid and hydriodic acid (R.L. Edward and M. Gill, J. chem Soc. Perkin I, 1973,1529).

3⁴,4⁴,4- Trihydroxypulvinic acid(XXIII)can be oxidised with hydrogen peroxide to yield xerocomorubin red-brown needles, m.p. 285°C, which , with diazomethane, gives gives the trimethyl ether (XXIV), red needles, m.p. 213-216°C, and with acetic anhydride gives the triacetate (XXV) as yellow yellow needles, m.p. 232-235°C (R.L. Edward and M. Gill, J. chem Soc. Perkin I, 1973,1529).

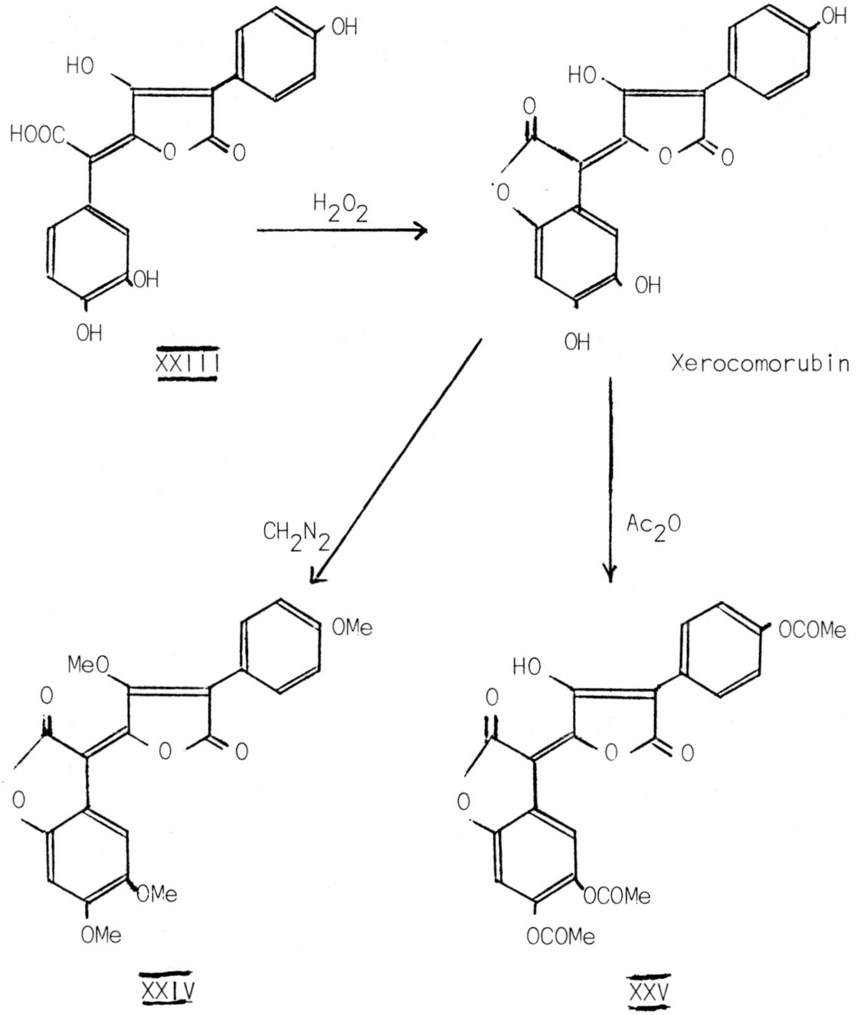
OH
HO
HOOC
O
O
OH
OH
XXIII
H_2O_2
OH
HO
O
O
O
O
OH
OH
Xerocomorubin
CH_2N_2
Ac_2O
OMe
MeO
O
O
O
O
OMe
OMe
XXIV
OCOMe
HO
O
O
O
O
OCOMe
OCOMe
XXV

When a mixture of 3',4',4-trihydroxy- (XXIII) and
3,4,4'-trihydroxypulvinic acid (XXII) and chloromethyl
methyl ether are heated with anhydrous potassium carbonate,
3,4,4'tris(methoxymethoxy)pulvinic acid lactone (XXVI)
yellow needles, m.p. 142-145°C, is formed. This lactone
reacts with methanolic potassium hydroxide to give methyl
3',4',4-tris(methoxymethoxy)pulvinate (XXVII), yellow
needles, m.p. 138-140°C, and methyl 3-hydroxy-4,4'-
bis(methoxymethoxy)pulvinate(XXVIII), an orange oil.

Methyl 3-hydroxy-4,4'-bis(methoxymethoxy)pulvinate
(XXVIII)can be converted, by refluxing with acetic acid
and sulphuric acid, into methyl xerocomate (XXIX) , red
rhombs, m.p. 258-261°C. Similar hydrolysis of methyl
3',4',4-tris(methoxymethoxy)pulvinate (XXVII) gives methyl
isoxerocomate (XXX) as lustrous orange leaflets, m.p. 219-
222°C.

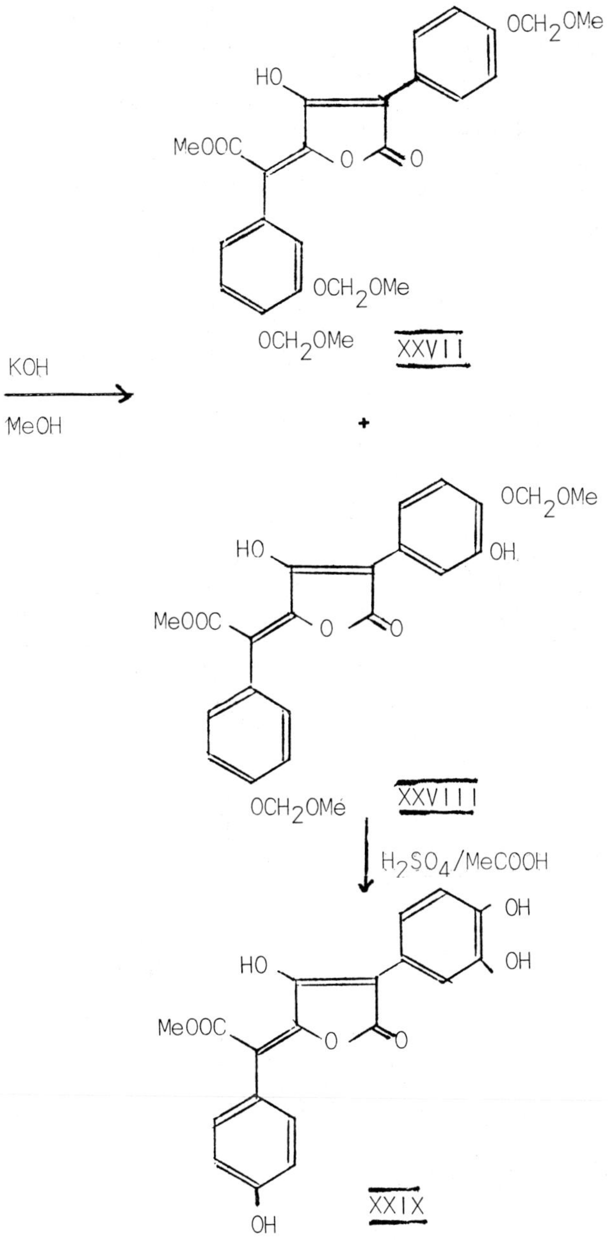

OCH2OMe
HO
MeOOC
O
O
OCH2OMe
OCH2OMe
XXVII
KOH
MeOH
+
OCH2OMe
HO
OH
MeOOC
O
O
OCH2OMe
XXVIII
H2SO4/MeCOOH
OH
HO
OH
MeOOC
O
O
OH
XXIX

Methyl 3-hydroxy-4,4'-bis(methoxymethoxy)pulvinate
(XXVIII)reacts with chloromethyl methyl ether to give methyl
O-(methoxymethoxy)-3,4,4'-tris(methoxymethoxy)pulvinate
(XXXI), a pale yellow gum, and methyl O-methyl-3,4,4'-
tris(methoxymethoxy)pulvinate (XXXII), pale yellow rosettes,
m.p.106-109°C, after reaction with diazomethane. Similar
methylation of methyl 3',4',4-tris(methoxymethoxy)pulvinate
(XXVII)gives methyl O-methyl-3',4',4-tris(methoxymethoxy)
pulvinate (XXXIII), pale yellow needles, m.p. 91-93°C.

Methyl O-methyl-3,4,4'-tris(methoxymethoxy)pulvinate
(XXXII) undergoes rearrangement by reaction with methanolic
potassium hydroxide to yield 5- 3,4-bis(methoxymethoxy)
phenyl -4-methoxy-2-(4-methoxymethoxyphenyl)cyclopent-4-ene-
1,3-dione (XXXIV), yellow needles, m.p. 95°C. A similar
rearrangement of methyl O-methyl-3',4',4-tris(methoxymethoxy)
pulvinate (XXXIII) gives 2- 3,4-bis(methoxymethoxy)phenyl
-4-methoxy-5-(4-methoxymethoxyphenyl)cyclopent-4-ene-1,3-
dione (XXXV), pale yellow needles, m.p. 62-64°C

2- 3,4-Bis(methoxymethoxyphenyl) -4-methoxy-5-
(4-methoxymethoxyphenyl)cyclopent-4-ene-1,3-dione (XXXV)
reacts with sodium borohydride to give (±)- *cis*-5-
(3,4-dihydroxyphenyl)-3,4-dihydroxy-2-(4-hydroxyphenyl)
cyclopent-2-enone (involutin) which is a metabolite
extracted from the fungus,*Paxillus involutus*.

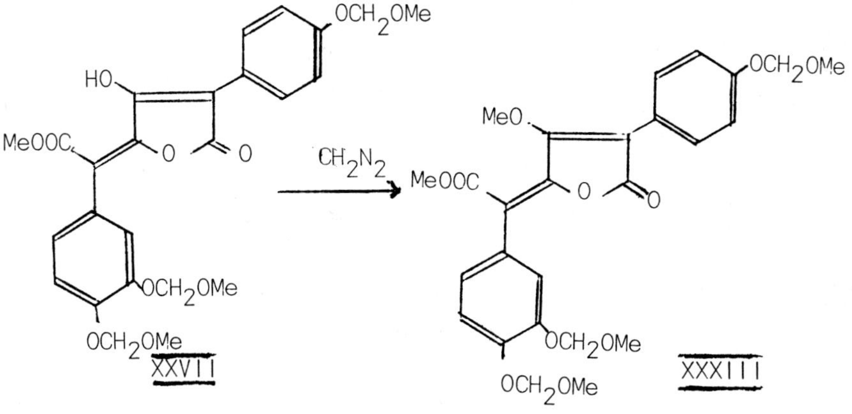
OCH2OMe
HO
MeOOC
O
O
OCH2OMe
OCH2OMe
XXVII
CH2N2
OCH2OMe
MeO
MeOOC
O
O
OCH2OMe
OCH2OMe
XXXIII

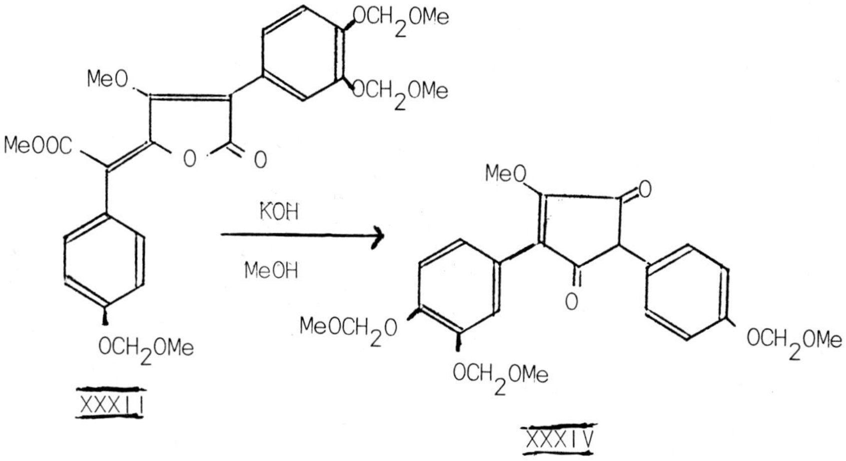
OCH2OMe
OCH2OMe
MeO
MeOOC
O
O
OCH2OMe
XXXII
KOH
MeOH
MeO
O
O
MeOCH2O
OCH2OMe
OCH2OMe
XXXIV

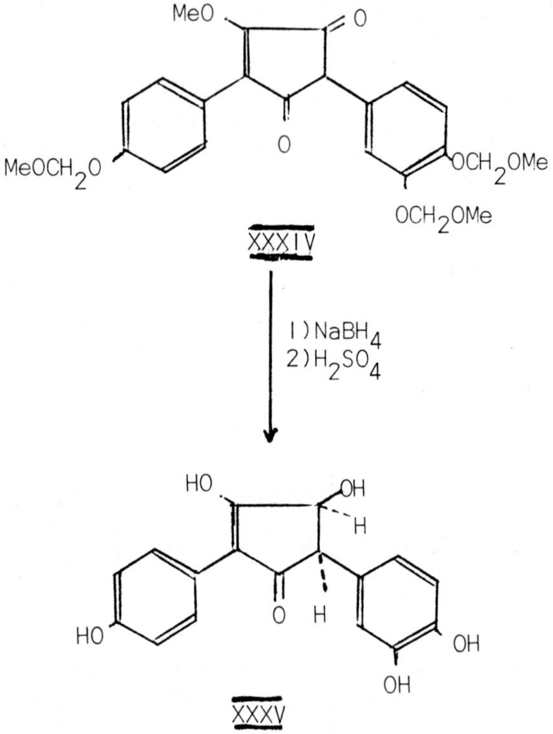
MeO
O
O
MeOCH2O
OCH2OMe
OCH2OMe
XXXIV
1) NaBH4
2) H2SO4
HO
OH
H
O
H
HO
OH
OH
XXXV

Halogen substituted Furanones

Mucochloric and mucobromic acids exist as hydroxy lactones lactones (E.I.Vinogradova and M.M.Shemyakin, Zh. Obshch. Khim., 1946, 16, 709).

X = Y = Br mucobromic acid
X = Y = Cl mucochloric acid

Methods of Preparation

a) Chlorination of furfural with hydrochloric acid and manganese dioxide provides β-chloro- γ-hydroxy- $\Delta^{\alpha\beta}$ - butenolide and thence mucochloric acid (Y.Hachihama *et al.*, J. org. Chem., 1964, 29, 1371).

b) Dichloromaleic anhydride with sulphur tetrafluoride gives αβ-dichloro- γγ-difluoro- $\Delta^{\alpha\beta}$ -butenolide (W.J.Feast *et al.*, J. chem. Soc., C, 1970, 2428)

Maleic anhydride reacts with sulphur tetrafluoride to give γγ-difluoro- $\Delta^{\alpha\beta}$ -butenolide.

c) Treatment of *cis*-perchloro-hexa-1,3-diene-6-carboxylic acid with silver ions (A.Roedig *et al.*, Ann., 1970, 733, 105).

d) By the treatment of tetrachloro-2,2-dialkyl-2H-pyrans with fuming nitric acid (A.Roedig and T.Neukam, Ber., 1974, 107, 3463).

$$R = Me, Et$$

e) By the reaction of chlorofumaroyl chloride with 1,1-dichloroethylene in the presence of aluminium chloride (M.L (M.Levas, Compt. rend., 1970, 270C, 1524)

f) From aroylacrylic acids, made by the condensation of glyoxylic acid with ketones in dioxane in the presence of sulphuric acid (W.Engel *et al.*, Chem. Abs., 1972, 77, 19,386s).

g) By reaction of halogen substituted lactones. Mucobromic acid reacts with substituted acetophenones to give αβ-dihalo-γ-phenacyl- Δ^αβ -butenolides (B.Kakac *et al.*, Coll. Czech. chem. Comm., 1968, 33, 1256; V.Zikan *et al.*, *ibid.*,1970, 35, 3475).

h) From pyruvic acid by condensation with aromatic aldehydes and then treatment with bromine (I.M.Roushidi *et al.*, Pharmazie, 1972, 27, 731).

MeCOCOOH + ArCHO ⟶ ArCH=CHCOCOOH

Br₂
⟶ ArCHBrCHBrCOCOOH −HBr ⟶

Reactions of Mucohalic Acids

Mucochloric or mucobromic acid react with alkyl naphthyl ethers in the presence of phosphoric acid and phosphorus pentoxide to give β-naphthyl derivatives (M.Semonsky *et al.*, Coll. Czech. chem. Comm., 1970, 35, 96).

Alkoxybenzenes react with mucohalic acids in the presence of phosphorus pentoxide to give γ-4-alkoxyphenyl butenolides (M.M.Rao *et al.*, Phytochem., 1976, 14, 1071).

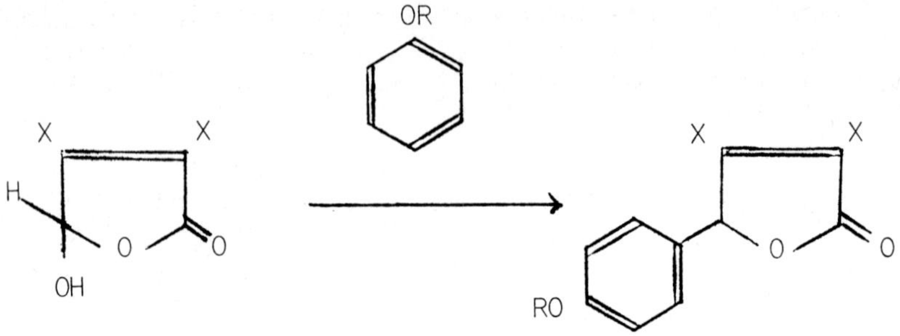

Mucochloric acid reacts with anilines to give γ-anilino derivatives (G.L.Pachmann, Chem. Abs., 1971, 75, 48,891s).

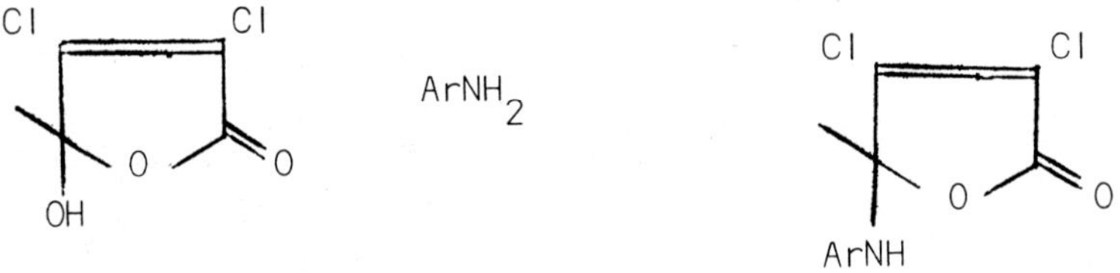

β-Bromo-γ-4-methoxyphenyl-γ-acetoxy-$\Delta^{\alpha\beta}$-butenolide reacts with amino acids to give N-(β-4-methoxy-benzoyl-β-bromoacryloyl)amino acids (N.Kucharezyk *et al.*, Coll. Czech. chem. Comm., 1969, 34, 3637).

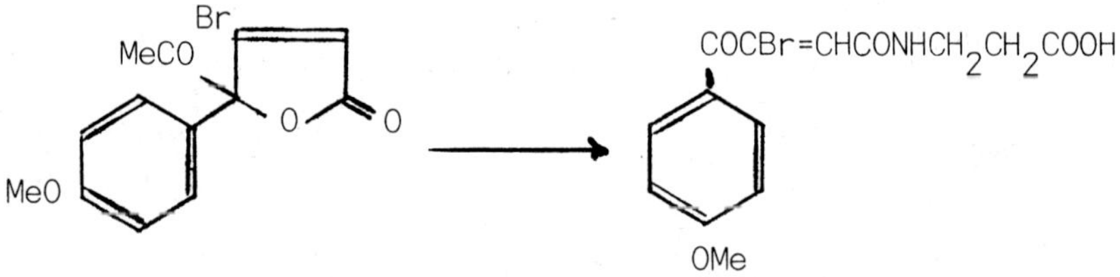

The reaction of γ-alkoxy-α,β-dihalo-$\Delta^{\alpha\beta}$-butenolides with trialkyl phosphites is shown below (K.W.Ratts and W.C.Philips, J. org. Chem., 1974, 39, 3300).

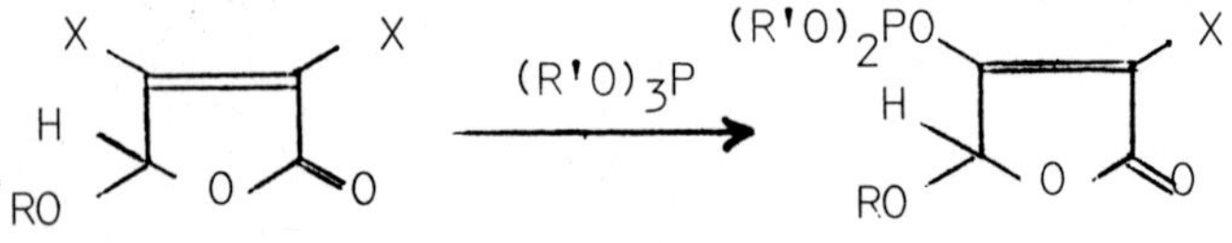

In the presence of sodium hydrogen carbonate mucobromic acid condenses with malonic, acetoacetic esters, nitromethane and pentane-2,4-dione (E.Gudriniece *et al.*, Chem. Abs., 1971, 71, 124,208j).

γ-Bromomethylene-$\alpha\beta$-dibromo-$\Delta^{\alpha\beta}$-butenolide reacts with alcohols (I.Kalnina *et al.*, Chem., Abs., 1971, 77, 20,061x).

Thiovulpinic Acids

Pulvinic acid lactone reacts with the sodium salt of methyl mercaptan to yield the thio-ester, orange needles, m.p. 156-158°C, together with a yellow product, m.p. 166.5-167°C (J.Weinstock, J.E.Blank and B.M.Sutton, J. org. Chem., 1974, 39, 2454).

The reaction of pulvinic acid lactone with hydrogen sulphide in the presence of sodium methoxide gives a base soluble product, which , on attempted recrystallisation, yields pulvinic acid lactone as the product. This suggests that pulvinic acid lactone is ring-opened to give a thiol acid which behaves as a mixed anhydride to reform thermally to pulvinic acid lactone.

The treatment of pulvinic acid lactone with thioacetic acid and pyridine in chloroform provides the monothiolactone, yellow needles, m.p. 175-176,5°C (*idem.*, *ibid.*, *loc. cit..*).

With excess thioacetic acid the dithiolactone is obtained as an orange solid, m.p. 218-221.5°C.

In this reaction the thioacetate attacks pulvinic acid lactone.

TThe intermediate (B) is able to revert readily to pulvinic acid lactone by loss of the thioacetate group, whereas the other intermediate (A) is unable to regenerate the starting material. The carboxylate anion can be acylated either by the adjacent thioacyl group or by thioacetic acid. The mixed anhydride (C) can now form the monothiolactone by a sterically favoured cyclisation. When the monothiolactone reacts with methanolic sodium methoxide it gives thiovulpinic acid, m.p. 115-116°C.

Chapter 20

TRIHYDRIC ALCOHOLS AND THEIR OXIDATION PRODUCTS
(continued)

B.J.Coffin

1 *Hydroxy- and amino-dicarboxylic acids*

Aminomalonic acid, $NH_2CH(COOH)_2$, reacts as its dimethyl
ester hydrochloride with phenyl isocyanate to yield
initially the corresponding hydantoin. Further reaction with
phenyl isocyanate gives the imidazoimidoazole derivative
which exists mainly in its keto form (L.Capuano *et al.*,
Ber., 1974, <u>107</u>, 3237).

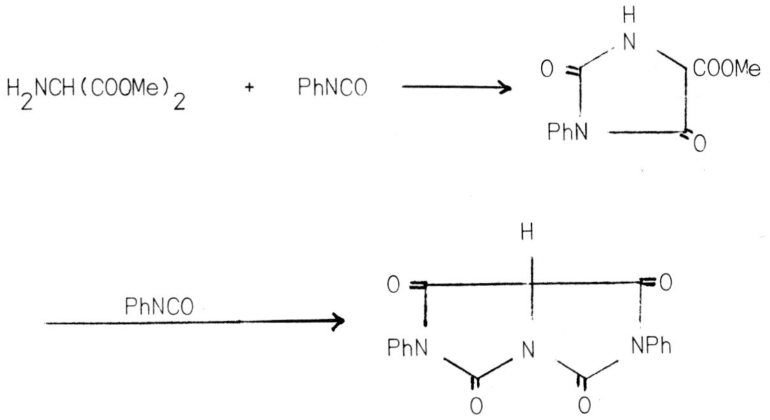

$H_2NCH(COOMe)_2$ + PhNCO $\longrightarrow$

$\xrightarrow{\quad PhNCO \quad}$

Benzylidene amines react with diethyl aminomalonate to
yield benzodiazepine derivatives (T.Perlotto, Canad. Pat.,
948,627)

Diethyl pyrrolquinazoline dicarboxylate can be synthesised from diethyl aminomalonate (G. De Martino *et al.*, Farmaco Ed. Sci., 1974, *29*, 579).

Aminomalonic acid diamide can be converted enzymatically with leucine aminopeptidase into aminomalonic acid monamide (N.Nishino *et al.*, Mein. Fac. Sci., Kyushu Univ., Ser C, 1975, *9*, 311).

Diethyl formamidomalonate, $HCONHCH(COOEt)_2$ undergoes a Michael addition with atropaldehydes to give a pyrrolidine as a single diastereoisomer with a *cis*-configuration (U.Hengartner *et al.*, J. org. Chem., 1979, 44, 3748).

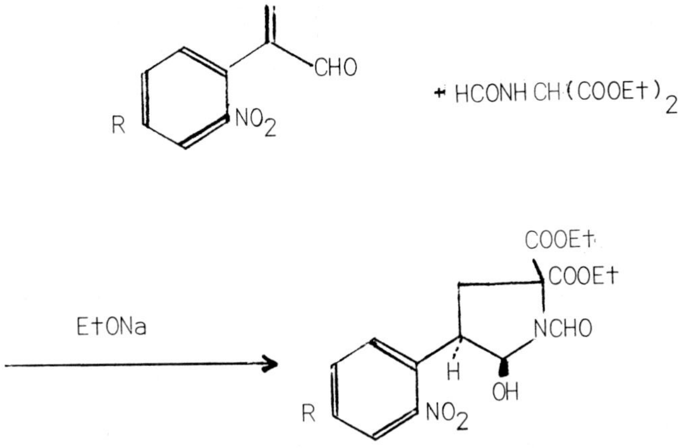

+ $HCONH\ CH(COOEt)_2$

$EtONa$ →

R = H, Me or Cl

The atropaldehyde (R = H), 2-(2-nitrophenyl)acrolein can be made from 2-nitrotoluene and dimethyl formamide dimethylacetal, giving the corresponding enamine. This, on reaction with aqueous formaldehyde and dimethylamine provides 2-(2-nitrophenyl)acrolein, m.p. 54.5-56°C. 2-(4-methyl-2-nitrophenyl)acrolein, m.p. 59-61°C, and 2-(4-chloro-2-nitrophenyl)acrolein, m.p. 88.5-90°C can be prepared similarly.

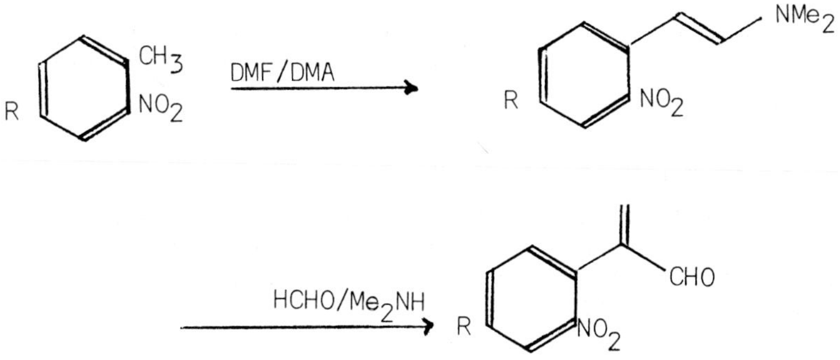

DMF/DMA →

$HCHO/Me_2NH$ →

Diethyl *cis*-1-formyl-5-hydroxy-4-(2-nitrophenyl)
pyrrolidine-2,2-dicarboxylate, m.p. 113-115°C, can be
made by the reaction of 2-(2-nitrophenyl)acrolein and
diethyl formamidomalonate. Hydrogenation of this compound
yields ethyl 2-formamido-3-(3-indolyl)-2-carbethoxy-
propionate as white crystals, m.p. 179-180°C, which, on
hydrolysis with aqueous sodium hydroxide, yields tryptophan
(R = H) as white crystals, m.p. 292°C.

6-Methyltryptophan (R = Me), m.p. 298°C, and
6-chlorotryptophan (R = Cl) can be prepared similarly,
m.p. 278°C.

The sodio-derivative of diethyl formamidomalonate
can be used as a precursor for tryptophans (H-Li Sum *et al.*,
Yao Hsuch Hsuch Pao, 1979, 14, 253).

1) EtONa
2) hydrolysis

$$O_2N\text{-indole-}CH_2CH(NH_2)COOH$$

Diethyl acetamidomalonate can be used in the preparation of substituted diaminocarboxylic acids by its initial condensation with 1-nitro-2-phenylethane, followed by hydrogenation and hydrolysis (I.N.Kiseleva *et al.*, Metody Sint., Str. Khim. Prevrashch, Nitrosoedin, Gertsenorskie Cheniya 31st, 1978, 24).

$$MeCONHCH(COOEt)_2 \quad + \quad PhCH=CHNO_2$$

$$\xrightarrow{MeONa} O_2NCH=CPhC(COOEt)_2NHCOMe$$

$$\xrightarrow{H_2/Ni}$$

Ph—[pyrrolidinone ring]—COOEt, NHCOMe, N-H, =O

$$\xrightarrow{hydrolysis} H_2NCH_2CH_2CHPhCH(NH_2)COOH$$

Amidine hydrochlorides in a methanolic solution of diethyl acetamidomalonate at room temperature for 24 hours yield pyrimidone derivatives (W.Dymek and R.Zimon, Acta Pol. Pharm., 1972, 29, 243).

Diethyl acetamidomalonate reacts with substituted benzylidene malonic acid esters to produce cyclic derivatives (P.Pachaly and H.P.Westfeld, Arch. Pharm., 1976, 309, 385).

$$\text{MeCONHCH(COOEt)}_2 \quad + \quad 4\text{-R } C_6H_4CH=C(COOEt)_2$$

R = H, Me, MeO or Cl

With cyclohexen-3-one diethyl acetamidomalonate undergoes a Michael condensation (I.Iko *et al.*, Nagoya Shiritsu Daigaka Yakugakuba Kenkyu Nempo, 1975, 23, 38).

The sodio-derivative of diethyl acetamidomalonate can be used in the synthesis of a hydrolysis product of the antibiotic, SF 1293 (Y.Ogawa *et al.*,Meiji Seiha Kenkyu Nempo, 1973, No. 13, 49).

$$MeCONHCH(COOEt)_2 \xrightarrow{\text{Na}} MeCONHCNa(COOEt)_2$$

$$\xrightarrow{Br(CH_2)_2P(O)MeOEt} MeCONHC(COOEt)_2CH_2CH_2P(O)MeOEt$$

$$\xrightarrow{\text{hydrolysis}} DL-HOP(O)MeCH_2CH_2CH(NH_2)COOH$$

Pyrroline derivatives can be synthesised by the reaction of diethyl aminomalonate with unsaturated ketones (J.Koch*et al.*,Compt. rend., 1978, **286**, 95).

$$PhCH=CHCOPh \quad + \quad H_2NCH(COOEt)_2 \longrightarrow$$

Tryptophan-5,7-$^3H_2$ and -4,6-$^3H_2$ can be obtained from aniline-2,4,6-$^3H_3$ and -3,5-$^3H_2$ respectively by conversion to the labelled substituted acetamidomalonate, $PhNHN=CH(CH_2)_2C(NHAc)(COOEt)_2$, followed by sequential cyclisation with sulphuric acid, alkaline hydrolysis, decarboxylation and finally deacetylation (G.P.Gardini and G.Palla, J. Labelled Compounds Radiopharm., 1977, **13**, 339).

A diethyl 2-acyl (or 2-benzyloxycarbonyl) amino-2-alkyl (or 2-aralkyl)malonate may be hydrolysed to the corresponding DL-monoester which, on heating, is decarboxylated to the corresponding amino-carboxylic acid derivative (A.Berger *et al.*, J. org. Chem., 1973, **38**, 457).

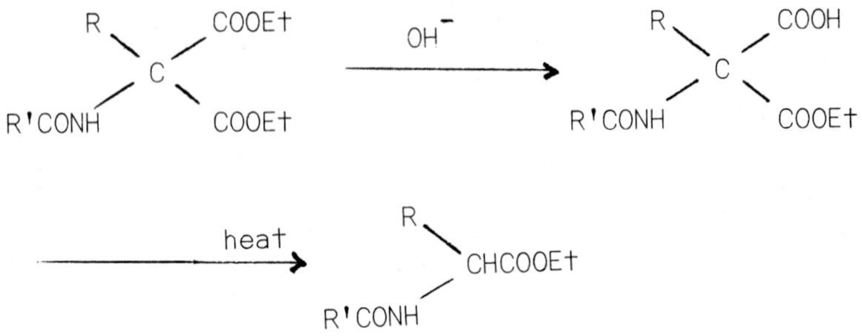

Aspartic acid, $HOOCCH_2CH(NH_2)COOH$.

    Erythro- β -hydroxy-L-aspartic acid, $HOOCCH(OH)CH(NH_2)-$
$COOH$, occurs in the unripe seeds of the legume, *Astragalus
sinicus* (H.Inatomi *et al.*, Chem. pharm. Bull., 1971, **19**,
216). β -Methylaspartic acid, $HOOCCH(CH_3)CH(NH_2)COOH$, occurs
in the bound form in the antibiotic, amphomycin, and
possesses the L-*threo*-configuration (M.Bodanszky and
G.G.Marconi, J. Antibiotics, 1970, **23**, 238).

    The asymmetric synthesis of pure L-aspartic acid can
be achieved by the following scheme, starting with one
of the optical isomers of 1,2-diphenylethanolamine
(J.P.Vigneron *et al.*, Tetrahedron Letters, 1968, 5681).

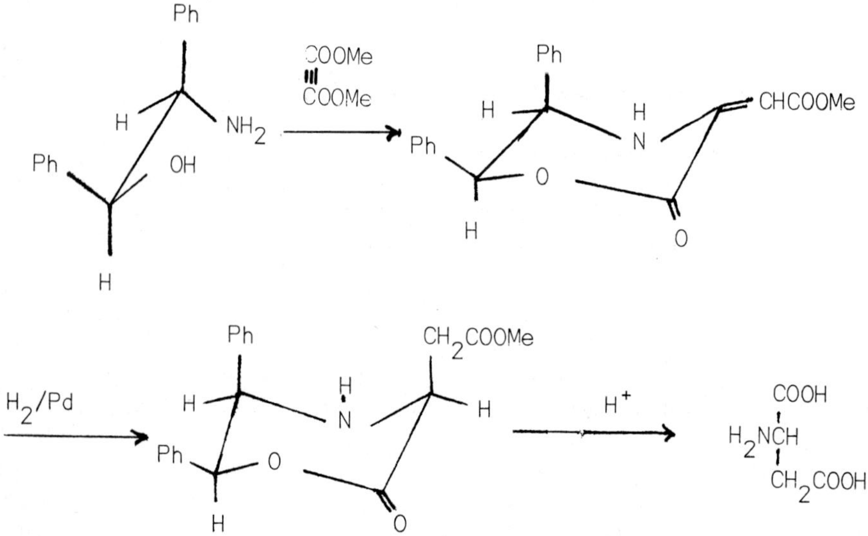

(R)- and (S)-Aspartic acids can be prepared by the following scheme (K.Haradon et $al.$, Chem. Letters, 1978, <u>10</u>, 1171).

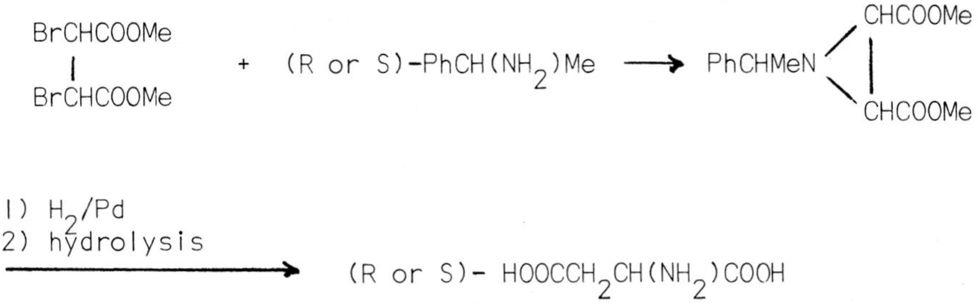

BrCHCOOMe
|
BrCHCOOMe
  + (R or S)-PhCH(NH$_2$)Me ⟶ PhCHMeN
$\diagup$ CHCOOMe
$\diagdown$ |
CHCOOMe

1) H$_2$/Pd
2) hydrolysis
⟶ (R or S)- HOOCCH$_2$CH(NH$_2$)COOH

The salicylideneglycinato complexes of copper or cobalt can be converted into DL-aspartic acid (O.S. Vostrikova and U.M.Dzhemilev, Khim. Vysokomol. Soedin-Neftekhim., 1973, 38).

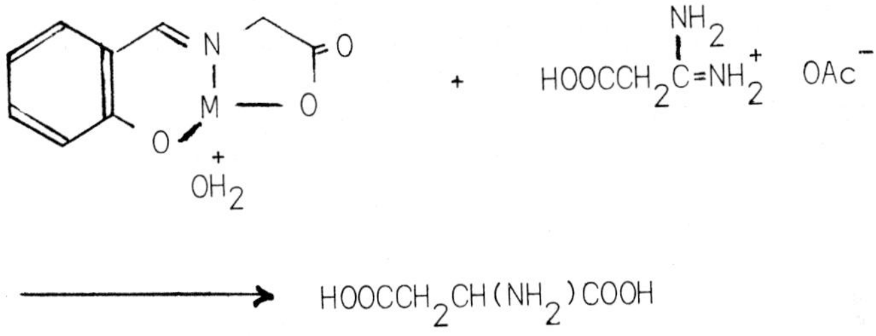

  + HOOCCH$_2$C=NH$_2^+$ OAc$^-$ (NH$_2$)

⟶ HOOCCH$_2$CH(NH$_2$)COOH

The intramolecular cyclisation of N-chloracetyl-2-phenyl glycine ester, using a base, yields the corresponding azetidinone derivative which can then be ring-opened to 2-phenyl aspartic acid derivatives.

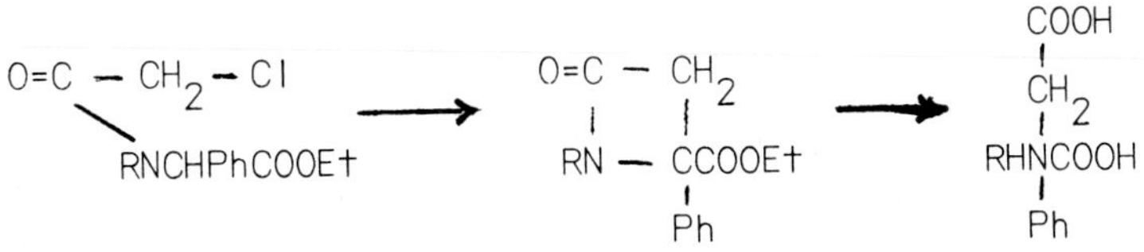

O=C — CH$_2$ — Cl
$\diagdown$
RNCHPhCOOEt
  ⟶
O=C — CH$_2$
| |
RN — CCOOEt
     |
     Ph
  ⟶
COOH
|
CH$_2$
|
RHNCOOH
|
Ph

N,N-Disubstituted aspartic acids can be prepared by a Stevens rearrangement of the relevant quaternary ammonium salt (A.T.Babayan *et al.*, Armyan. Khim. Zhur., 1976, <u>29</u>, 456).

$$\begin{array}{c} R' \\ \diagdown \\ \diagup \\ R \end{array} \overset{+}{N}(CH_2COOMe)_2X^- \longrightarrow MeOOCCH_2NRR'COOMe$$

β,β-Disubstituted aspartic acids may be made by the alkylation of ethyl isonitriloacetate with α-bromo-esters, followed by hydrolysis of the product (B.Weistein *et al.*, J. org. Chem., 1976, <u>44</u>, 3634).

$$BrCRR'COOEt \quad + \quad EtOOCCH_2NC \longrightarrow \begin{array}{c} EtOOCCHNC \\ | \\ CRR' \\ | \\ COOEt \end{array}$$

$$\xrightarrow{\text{hydrolysis}} HOOCCH(NH_2)CRR'COOH$$

β-Hydroxy aspartic acid can be synthesised by the ammonolysis of *cis*-epoxysuccinic acid to yield *threo*-hydroxy-DL-aspartic acid. The *trans*-epoxysuccinic acid provides a 2:1 mixture of the *erythro*- and *threo*-isomers (H.Okai *et al.*, Bull. chem. Soc. Japan, 1967, <u>40</u>, 2154).

$$O \diagup \diagdown \begin{array}{c} CHCOOH \\ | \\ CHCOOH \end{array} \xrightarrow{NH_3} \begin{array}{c} COOH \\ | \\ CHNH_2 \\ | \\ CHOH \\ | \\ COOH \end{array}$$

This hydroxy acid can also be synthesised by treating
β-furylserine, made from glycine and furfural, with
phosgene to give the cyclic urethane. This, on oxidation
with potassium permanganate and hydrolysis, gives either
the *threo-* or the *erythro-*β-hydroxy-DL-aspartic acid
(T.Inui *et al.,* Bull. chem. Soc. Japan, 1968, 41, 2148).

$$\text{(furfural)} \quad + \quad H_2NCH_2COOH$$

$$\xrightarrow{\quad\quad} \quad
\begin{array}{c} COOH \\ | \\ CHNH_2 \\ | \\ CHOH \end{array}
\quad \xrightarrow{\ COCl_2\ } \quad
\begin{array}{c} COOH \\ | \\ CHNH \\ | \\ CHO \end{array}\!\!>\!=\!O$$

$$\xrightarrow{\ KMnO_4\ } \quad
\begin{array}{c} COOH \\ | \\ CHNH \\ | \\ CHO \\ | \\ COOH \end{array}\!\!>\!=\!O
\quad \xrightarrow{\ \text{hydrolysis}\ } \quad
\begin{array}{c} COOH \\ | \\ CHNH_2 \\ | \\ CHOH \\ | \\ COOH \end{array}$$

The resolution of *erythro-*β-hydroxy aspartic acid can be
achieved by preparing its *N*-benzyl derivative (Y.Liwschitz
et al., Israel J. Chem., 1968, 6, 647).
    The Strecker synthesis can be used to prepare DL-
glutamic acid-1-^{14}C, DL-ornithine-1-^{14}C and DL-arginine-1-
^{14}C, using 3-cyanopropanal (I.Mezo *et al.,* Acta. Chem. Acad.
Sci. Hung., 1969, 60, 399).

$$NCCH_2CH_2CHO \xrightarrow{\text{HCN}} NCCH_2CH_2CH(OH)CN$$

$$\xrightarrow{\text{NH}_4Cl} NCCH_2CH_2CH(NH_2)CN$$

$$\xrightarrow{\text{hydrolysis}} HOOCCH_2CH_2CH(NH_2)COOH$$

Aspartic acid-2-^{2}H can be made from L-asparto-bis-(ethylenediamine)cobalt(III) perchlorate by equilibration with ^2H_2O (W.E.Keyes and J.I.Legg, J. Amer. chem. Soc., 1973, 95, 3431).

Aspartic acid, labeled with ^{15}N, can be made by aspartase catalysed addition of $^{15}NH_3$ to fumaric acid. This labelled acid can then be used with a transaminase to prepare other ^{15}N labelled amino-acids (J.A.Zintel et al., Canad. J. Chem., 1969, 47, 4117).

Aspartic acid undergoes thermal degradation at 350-600°C to provide a mixture of succinic acid and dimethylmaleic anhydride (A.W.Fort et al., J. org. Chem., 1976, 41, 3697).

N-Bromosuccinimide oxidises aspartic acid to the nor-aldehyde which can then undergo α-bromination and decarboxylation to bromoform (W.L.Parker et al., Experientia, 1970, 26, 242).

By anodic oxidation, β -alkyl aspartic acids are converted into 3-acetoxy-2-acylaminoalkanoic acids (T.Iwawski *et al.*, J. org. Chem., 1972, 42, 2419).

$$HOOCCHRCH(NH_2)COOH \xrightarrow{HOAc} HOOCCH(COMe)CH(NHCOMe)COOH$$

L-aspartic acid, dissolved in concentrated sulphuric acid and ethyl acetate, followed by treatment with an alcohol, gives a product which can be cyclised with acetic anhyride to give L-aspartic acid anhydride. This can be isolated as its alkyl hydrogen sulphate salt (Y.Arijoshi *et al.*, Bull. chem. Soc. Japan, 1973, 46, 2611).

Stereospecific complex formation forms the basis of a procedure for the resolution of DL-aspartic or DL-glutamic acids. These acids can be combined with (L-arginato)copper perchlorate, followed by crystallisation and destruction of the complex with hydrogen sulphide. This method provides amino-acids of moderate optical purity (T.Sakurai *et al.*, J. chem. Soc. Chem. Commun., 1976, 553; O.Yamauchi *et al.*, Bull. chem. Soc. Japan, 1977, 50, 1776).

Glutamic and aspartic acids can be *N*-acetylated and esterified with monoglycerides of palmitic or stearic acids to provide *N*-acetylamino acid esters of the monoglycerides (K.Horikawa *et al.*, Kagaku To Kogyo (Osaka), 1977, 51, 281).

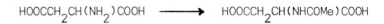

Asparagine, $H_2NCOCH_2CH(NH_2)COOH$, can be made as its N-ethyl-L-form by a reliable procedure (R.W.Dineen and D.O.Gray, Org. Prep. Proceed. Internat., 1977, <u>9</u>, 39).

Thioasparagine, $H_2NCSCH_2CH(NH_2)COOH$, can be made from β -cyanoalanine derivatives by a conventional thioamide synthesis. Thioamidation of N-p-methoxybenzyloxycarbonyl-L- β -cyanoalanine with hydrogen sulphide and ammonia provides N-p-methoxybenzyloxy-L-thioasparagine.

$$NCCH_2CH(NH)COOH \quad \xrightarrow{H_2S/NH_3} \quad H_2NCSCH_2CH(NH)COOH$$
$$CO \qquad\qquad\qquad\qquad\qquad CO$$
$$CH_2 \qquad\qquad\qquad\qquad\qquad CH_2$$
$$Ar \qquad\qquad\qquad\qquad\qquad Ar$$

t-Butyloxycarbonyl-L-thioasparagine and benzyloxycarbonyl-L-thioasparagine can be made from the corresponding β -cyanoalanine derivatives by similar treatment, leading to thioasparagine. During the deprotection of N-p-methoxy-benzyloxycarbonyl-L-thioasparagine in trifluoroacetic acid, major side reactions give aspartic-p-methoxybenzyl- β - imidothiolic ester, and from this β -S-p-methoxybenzyl

aspartic thioester (due to anisyl transfer). This transfer
reaction can also be used to make δ -p-methoxybenzylcysteine.

$$
\begin{array}{c}
\text{COOH} \\
| \\
\text{CHNHR} \\
| \\
\text{CH}_2 \\
| \\
\text{CN}
\end{array}
\xrightarrow{\text{H}_2\text{S/NH}_3}
\begin{array}{c}
\text{COOH} \\
| \\
\text{CHNHR} \\
| \\
\text{CH}_2 \\
| \\
\text{C=S} \\
| \\
\text{NH}_2
\end{array}
$$

$$
\longrightarrow
\begin{array}{c}
\text{COOH} \\
| \\
\text{CHNH}_2 \\
| \\
\text{CH}_2 \\
| \\
\text{C=S} \\
| \\
\text{NH}_2
\end{array}
+
\begin{array}{c}
\text{COOH} \\
| \\
\text{CHNH}_2 \\
| \\
\text{CH}_2 \\
| \\
\text{C=NH} \\
| \\
\text{SCH}_2\text{C}_6\text{H}_4\text{OMe-p}
\end{array}
\xrightarrow{\text{H}^+}
\begin{array}{c}
\text{COOH} \\
| \\
\text{CHNH}_2 \\
| \\
\text{CO} \\
| \\
\text{S} \\
| \\
\text{CH}_2 \\
| \\
\text{C}_6\text{H}_4\text{OMe-p}
\end{array}
$$

$$
\begin{array}{c}
\text{COOH} \\
| \\
\text{CHNH}_2 \\
| \\
\text{C=S} \\
| \\
\text{NH}_2
\end{array}
+ \quad \text{p-MeOC}_6\text{H}_4\text{CH}_2\text{OCOCHMeCOOH}
\qquad \xrightarrow{\text{TFA}}
$$

Asparagine can be labelled with tritium by use of
tritioammonia (K.Bloss, J. labelled Compounds, 1969, 5
555).

N-o-Nitrophenylsulphonyl asparagine reacts with
dicyclohexylcarbodiimide (DCC) to provide the corresponding
protected 3-cyanoamino acid (A.Chimiak and J.J.Pastuszak,
Chem. Ind. Internat., 1971, 427).

L-asparagine can be converted into S-isoserine by reaction with alkali nitrite in acid to provide L-β - malamidic acid which is then reacted with sodium hypochlorite (T.Miyazawa, Jap. Pat., 1975, 37,723).

$$H_2NCOCH_2CH(NH_2)COOH \xrightarrow{\text{KNO}_2} H_2NCOCH_2CH(OH)COOH$$

$$\xrightarrow{\text{NaOCl}} H_2NCH_2CH(OH)COOH$$

Glutamic acid, $HOOCCH_2CH_2CH(NH_2)COOH$, in the L-form can be made by the alkylation of the cobalt(III) complex of N-salicylidene glycine (Y.N.Belokon *et al.*, Izvest. Akad. Nauk. SSSR, Ser. Khim., 1977, 428).

The synthesis of (±)-glutamic acid can be achieved by the reduction and hydrolysis of the product from the reaction of methyl nitroacetate and methyl 3-bromopropanoate (E.Kaji and S.Zen, Bull. chem. Soc. Japan, 1973, 337).

$$O_2NCH_2COOMe \quad + \quad BrCH_2CH_2COOMe$$

$$\longrightarrow MeOOCCH_2CH_2CH(NO_2)COOMe$$

$$\longrightarrow HOOCCH_2CH_2CH(NH_2)COOH$$

Acrylonitrile can be converted industrially into L-glutamic acid (T.Yoshida, Chem. -Ing. -Tech., 1970, <u>42</u>,

641; T.Watanabe *et al.*, Kogyo Kagaku Zasshi, 1969, <u>77</u>, 1080).

$$CH_2=CHCN \xrightarrow{\text{CO/H}_2} OHCCH_2CH_2CN$$

$$\xrightarrow{\text{NH}_3\text{/KCN}} H_2NCH(CN)CH_2CH_2CN$$

$$\xrightarrow{\text{H}^+} H_2NCH(COOH)CH_2CH_2COOH$$

The *N*-acyl methyl glutamate can be synthesised by the reaction of the appropriate aldehyde or amide with carbon monoxide in the presence of cobalt carbonyl (H.Wakamatsu *et al.*, Ger. Pat., 1969, 2,115,985).

D-glutamic acid can be prepared by using the hydrogenolytic asymmetric transamination approach. This involves the reaction between the resolved D-α -phenyl-glycine with 3-oxoglutaric acid to produce D-glutamic acid (K.Harada *et al.*,Bull. chem. Soc. Japan, 1973, <u>46</u>, 1901).

$[\alpha\beta\beta -^2H_3]$ -L-glutamic acid may be formed by incubation of L-glutamic acid in D_2O in the presence of pig-heart glutamate-oxaloacetate aminotransferase and catalytic amounts of pyridoxal-5'-phosphate and oxaloacetic acid (D.J.Whelan and G.J.Long, Austral. J. Chem., 1969, <u>22</u>, 1729).

L-glutamic acid is converted into L-(+)-2,4-diamino-butanoic acid by reaction with hydrazoic acid. This, in turn, can be converted into L-(-)-4-amino-2-hydroxybutanoic acid (T.Naito and S.Nakagawa, Jap. Pat., 74 24,914).

$$HOOCCH_2CH_2CH(NH_2)COOH \xrightarrow{\text{N}_3\text{H}} H_2NCH_2CH_2CH(NH_2)COOH$$

$$\xrightarrow{\hspace{2cm}} H_2NCH_2CH_2CH(OH)COOH$$

It is possible to convert D-glutamic acid into D-ornithine according to the following scheme (V.Gut and K.Poduska, Coll. Czech. chem. Comm., 1971, <u>36</u>, 3470).

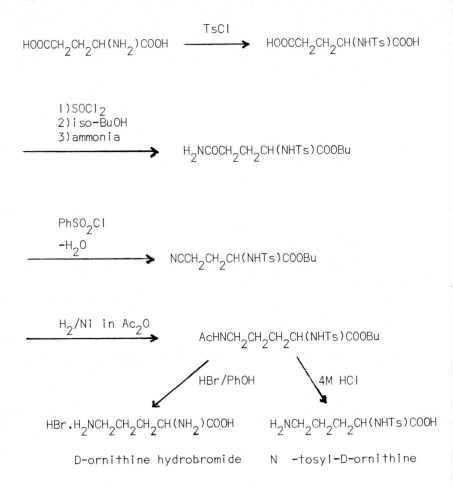

$$HOOCCH_2CH_2CH(NH_2)COOH \xrightarrow{\text{TsCl}} HOOCCH_2CH_2CH(NHTs)COOH$$

1) $SOCl_2$
2) iso-BuOH
3) ammonia

$$\longrightarrow H_2NCOCH_2CH_2CH(NHTs)COOBu$$

$PhSO_2Cl$

$-H_2O$

$$\longrightarrow NCCH_2CH_2CH(NHTs)COOBu$$

H_2/Ni in Ac_2O

$$\longrightarrow AcHNCH_2CH_2CH_2CH(NHTs)COOBu$$

HBr/PhOH   4M HCl

$HBr \cdot H_2NCH_2CH_2CH_2CH(NH_2)COOH$ $H_2NCH_2CH_2CH_2CH(NHTs)COOH$

D-ornithine hydrobromide   N -tosyl-D-ornithine

Glutamic acid can be condensed with two molecules of diketene to produce 3-acetyl-6-methyl(IH,3H)-pyridine-3,4-dione-l-acetic acid (T.Kato and Y.Kubota, Yakugaku Zasshi, 1967, 87, 1219).

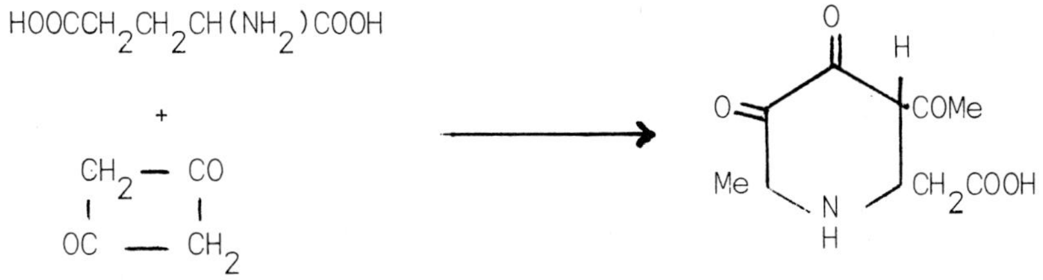

HOOCCH$_2$CH$_2$CH(NH$_2$)COOH

+

CH$_2$ — CO
| |
OC — CH$_2$

$\longrightarrow$

The hydrolysis of *N*-benzoylglutamic acid in dilute acid solution occurs with the intermediate formation of 2-pyrrolidone-5-carboxylic acid, whereas the hydrolysis of *N*-benzoylaspartic acid proceeds via the β-benzoylaspartic anhydride intermediate rather than the corresponding azetidinone derivative (J.B.Capindale and H.S.Fan, Canad. J. Chem., 1967, 45, 192).

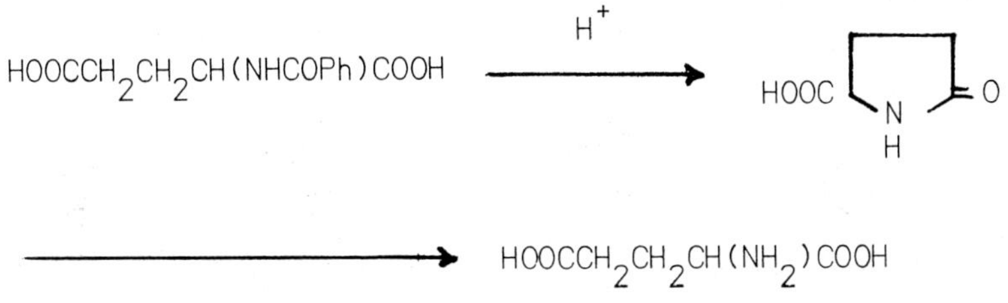

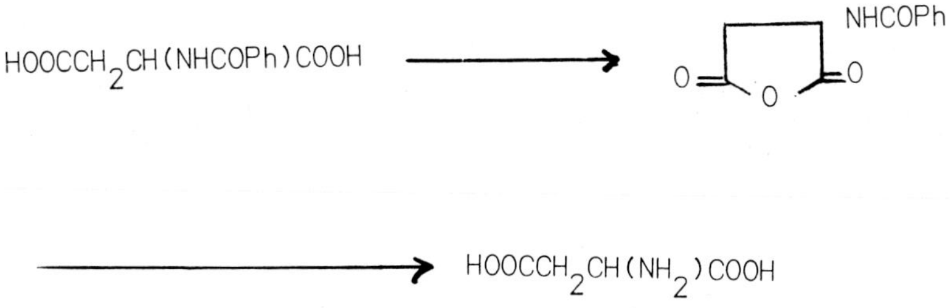

HOOCCH$_2$CH$_2$CH(NHCOPh)COOH $\xrightarrow{H^+}$

HOOC

$\longrightarrow$ HOOCCH$_2$CH$_2$CH(NH$_2$)COOH

HOOCCH$_2$CH(NHCOPh)COOH $\longrightarrow$

$\longrightarrow$ HOOCCH$_2$CH(NH$_2$)COOH

Oxidation of *N*-acylprolines with permanganic acid leads to the formation of pyroglutamates, whilst the oxidation of side-chain protected ornithines yields glutamates (I.Murematsu *et al.*, Chem. Letters, 1977, 1253).

$$H_2NCH_2CH_2CH_2CH(NHCOR)COOH \longrightarrow HOOCCH_2CH_2CH(NHCOR)COOH$$

The photo-oxidation of methyl DL-pyroglutamate in benzene yields the dipyrrolidone derivative (*ibid, loc. cit.*).

Vitamin K dependant blood clotting proteins such as prothrombin and factor X contain 𝛄-carboxyglutamic acid. This can be made as follows (B.Weinstein *et al.*, J. org. Chem., 1976, 41, 3634).

$$PhCH_2OCONHCH(CH_2OH)COOCH_2Ph$$

$$\xrightarrow{SOCl_2}$$

$$PhCH_2OCONHCH(CH_2Cl)COOCH_2Ph$$

$$\xrightarrow{NaOH}$$

$$PhCH_2OCONHC(COOCH_2Ph)=CH_2$$

$$\xrightarrow{NaCH(COOCH_2Ph)_2}$$

$$PhCH_2OCONHCH(COOCH_2Ph)CH_2CH(COOCH_2Ph)_2 \quad ,$$

$$\xrightarrow{H_2/Pt}$$

$$H_2NCH(COOH)CH_2CH(COOH)_2$$

A Dakin West reaction of anhydrides of the form (I) (R = Ph, Me or CF_3) with acetic anhydride or trifluoroacetic anhydride yield furaoxazoles (R' = H or COOMe), furanones or oxazinones (R = CF_3, R'' = $CHCF_3OOCCF_3$, depending on the conditions of the reaction) (L.Lapschy *et al.*, Ann. Chem., 1975, 11, 1753).

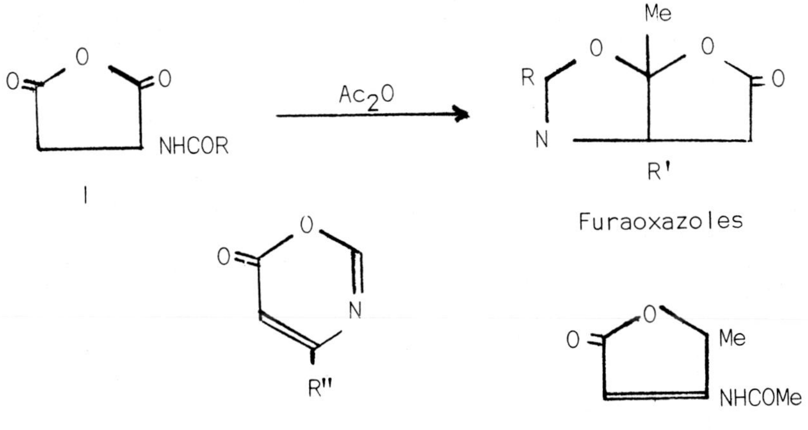

Furaoxazoles

Oxazinone

Furanone

The reaction of glutamic acid with acetic anhydride yields the pyrrolidinone whilst the reaction of N-(thiobenzoyl)aspartic acid with acetic anhydride in the presence of 3-picoline gives the corresponding thioxazinones (*ibid. loc. cit.*).

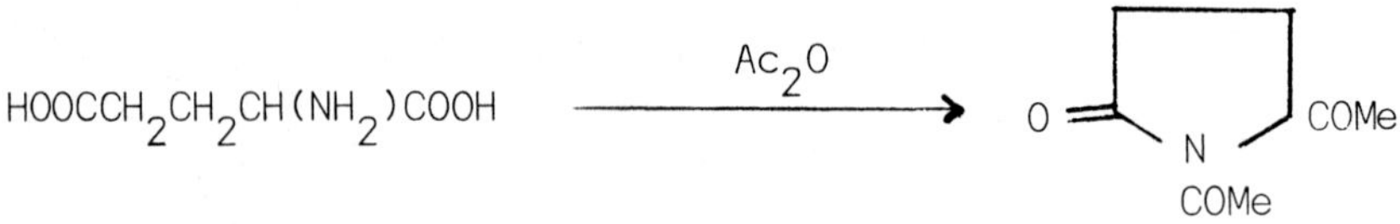

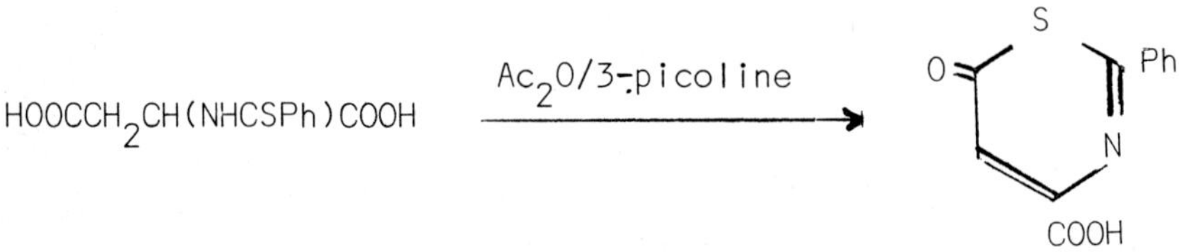

$$HOOCCH_2CH_2CH(NH_2)COOH \xrightarrow{Ac_2O}$$

$$HOOCCH_2CH(NHCSPh)COOH \xrightarrow{Ac_2O/3\text{-picoline}}$$

2 *Keto dicarboxylic acids*

Formylsuccinic acid, $OHCCH(COOH)CH_2COOH$, can be made as its diethyl ester by the reaction of sodium ethanolate with a mixture of diethyl succinate and ethyl formate (C.G. Wermuth, J. org. Chem., 1979, **44**, 2406; E.Carriere, Ann. Chem., 1922, <u>17</u>, 43). From its ir and nmr spectra, it exists as a mixture of the aldehyde and enolic forms.

When diethyl formylsuccinate is heated with ethyl orthoformate in the presence of toluene-p-sulphonic acid diethyl formylsuccinate diethyl acetal, b.p. 87-88°C/0.01mm., is formed. Alkaline hydrolysis of the diethyl acetal yields formylsuccinic acid diethyl acetal (E.E.Blaise and E.Carriere, Compt. Rend., 1913, <u>156</u>, 239) which, on heating under vacuum, yields by decarboxylation and loss of ethanol γ-ethoxybutyrolactone, b.p. 102-103°C/14mm.. This reaction is considered to proceed by the mechanism shown below.

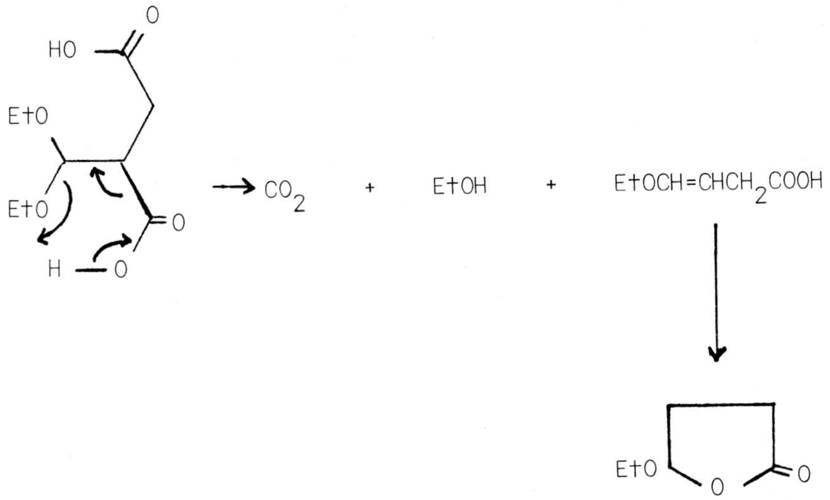

When steam is passed through an aqueous solution of
γ-ethoxybutyrolactone, succinic semialdehyde, b.p. 91-92°C/
0.05mm., is formed. The ir spectrum of this compound shows,
in addition to acidic and aldehydic bands at 1720 and 1750
cm^{-1}, a lactonic carbonyl band at 1780 cm^{-1} which suggests
that it exists as the equilibrium mixture shown below.

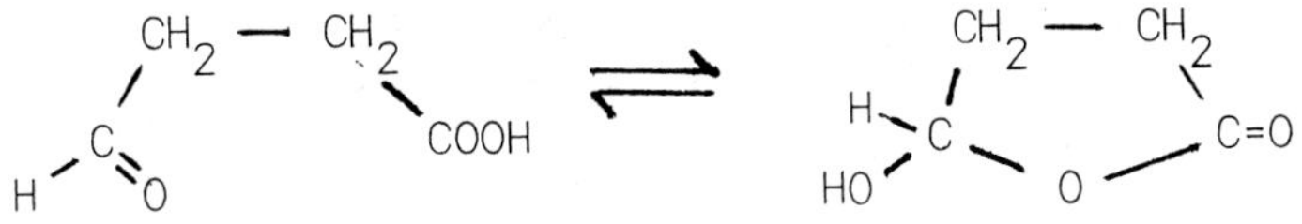

Succinic semialdehyde is produced by transamination of
γ-aminobutanoic acid (GABA) in a neural metabolic route
known as the GABA shunt, and is of considerable interest in
neurochemistry and central pharmacology (E.Roberts and
R.Hammerschlag, "Basic Neurochemistry", R.Wayne Albers,
G.J.Siegel, R.Katzman and B.W.Agranoff, Eds., Little
Brown and Co., Boston 1972, p. 131-135; N.Davidson,
Neurotransmitter Amino Acids", Academic Press, New York, 1976;

M.Maitre, L.Ciesielski, C.Cash and P.Mandel, Eur. J. Biochem., 1975, 52, 157; J.W.Kosh and G.D.Appelt, J. Pharm. Sci., 1972, 61, 1963; H.Laborit in "Progress in Neurobiology", Vol. I, Part 4, G.A.Kerkut and J.W.Phillis, Eds., Pergammon Press, Oxford, 1973, p. 255-274).

Diethyl formylsuccinate reacts readily with variously substituted 2-aminopyridines to yield substituted succinates (L.Vasvaric-Debreczy *et al.*, J. chem. Soc. Perkin I, 1978, 795).

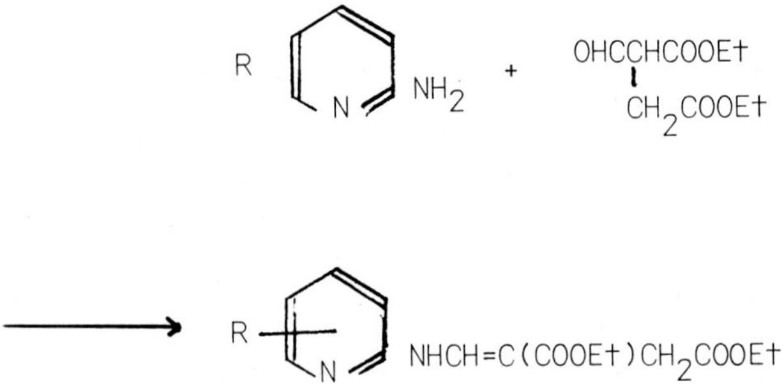

R ⟨pyridine⟩ NH₂ + OHCCHCOOEt / CH₂COOEt

$$R \text{—} \text{(pyridine)} \text{—} N\!H_2 \quad + \quad \begin{array}{c} OHCCHCOOEt \\ | \\ CH_2COOEt \end{array}$$

⟶ R ⟨pyridine⟩ NHCH=C(COOEt)CH₂COOEt

These condensation products can exist in three different tautomeric forms (i.e. Schiff's base, enamine or enimine) each of which exists as different geometric isomers.

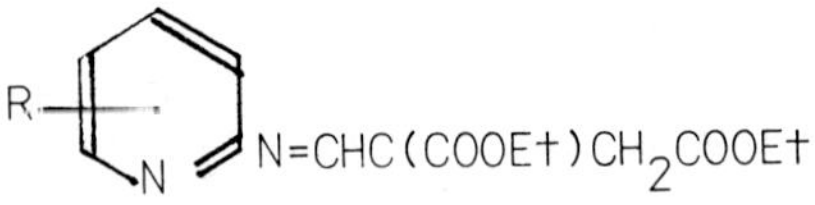

R ⟨pyridine⟩ N=CHC(COOEt)CH₂COOEt

Schiff's Base

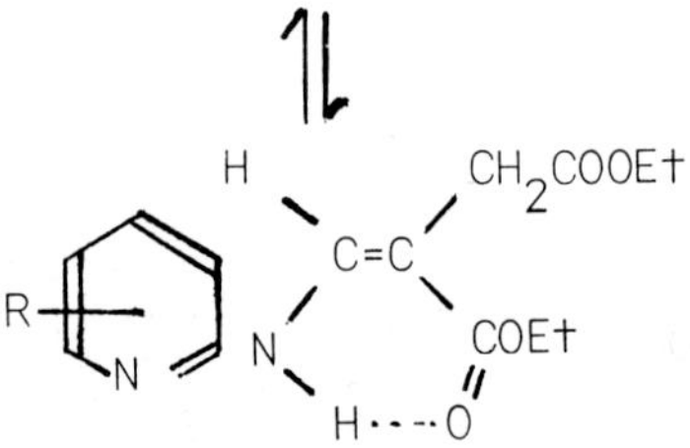

$\Updownarrow$

R ⟨pyridine⟩ structure with C=C, H, CH₂COOEt, COEt, H····O

Enamine
Z-form

Enimine

The ^{1}H nmr spectra of these condensation products provide
information on their tautomeric form and show that it is
the enamine form which predominates.

In the condensation reaction, mixtures of the
E and Z geometric isomers are formed. The geometric
isomers can be separated from their isomeric mixtures by
fractional crystallisation or by column chromatography. The
succinate geometric isomers can be distinguished and
identified by their ^{1}H nmr spectra. The NH proton of the
E-isomers resonates in the range δ = 7-9 while that of the
Z-isomers resonates over δ = 10, indicating a chelate
ring structure.

The condensation reaction is influenced by the choice
of solvent. The ratio of the amounts of the E and Z isomers
of these succinates is influenced considerably by the
polarity of the solvent used.

Isomerisation does occur in the presence of either
acid or alkali, and in equilibrium mixtures the Z-isomer
predominates. The predominance of the thermodynamically
less stable E-isomer in the reaction mixtures indicates
that generally steric approach control is present in the
first step of the addition-elimination reaction.

Ring closure of these substituted succinates, using phosphoryl chloride-polyphosphoric acid, yields the cyclic products pyrido [1,2-a] pyrimidines and pyridylpyrrolinones (*ibid.*, *loc. cit.*).

In some cases, a pyridylpyrrole may be formed from the pyridylpyrrolinone by way of OH ⟶ Cl exchange.

The ratio of the yields of pyridopyrimidines and pyridylpyrroles is independant of the geometry of the starting succinate, but is dependent on the substituent of the pyridine ring and on its position. Substituents in the 6-position hinder the formation of pyridopyrimidines while those in the 3-position inhibit the formation of pyridylpyrrolinones.

Diethyl 2-formylglutarate, OHCCH(COOEt)CH$_2$CH$_2$COOEt, can be made by the method described by J.Biggs and P.Sykes (J. chem. Soc., 1959,1849). Cyclisation of substituted glutarates,made by reaction with substituted 2-amino-pyridines, with phosphoryl-polyphosphoric acid mixture gives only one product, the pyrido [1,2-a] pyrimidine (*ibid., loc. cit.*).

R = H or 6-Me

Diethyl acetylsuccinate, CH$_3$COCH(COOEt)CH$_2$COOEt, reacts with thiosemicarbazide to yield 5-hydroxy-3-alkyl-1-1-(thiocarbamoyl)pyrazole-4-alkanoic acid derivatives (A.A.Santilli and D.H.Kim, U.S.Pat., 3,704,242).

Veratraldehyde can be condensed readily with diethyl acetylsuccinate (P.A.Ganeshpure, Curr. Sci., 1976, _45_, 494).

MeO—[ring]—CHO + MeCOCHCOOEt | CH$_2$COOEt (MeO substituents on ring)

$\longrightarrow$ MeO—[ring]—CH=C(COOEt)CH(COMe)COOEt (MeO substituents on ring)

In the presence of a base, diethyl acetylsuccinate can be condensed with the unsaturated ketone, Me$_2$CHCH$_2$COCH=CH$_2$, to provide a product which, after hydrolysis, can be cyclised to a cyclohexenone derivative which exhibits anti-inflammatory activity (J.H.Vipond _et al._, Brit. Pat., 1,265,800).

Me$_2$CHCH$_2$COCH=CH$_2$ + MeCOCHCOOEt | CH$_2$COOEt

$\longrightarrow$ Me$_2$CHCH$_2$C(=CH=CH$_2$)CCOCH(COOEt)CH$_2$COOEt

$\longrightarrow$ Me$_2$CHCH$_2$—[cyclohexenone ring with CH$_2$COOEt and O]

Oxaloacetic acid, 2-oxobutanedioic acid, $HOOCCOCH_2COOH$, undergoes self-condensation followed by decarboxylation to yield 4-carboxy-4-hydroxy-2-keto-hexane-1,6-dioic acid (citroylformic acid) (R.H.Wiley and K.S.Kim, J. org. Chem., 1973, 38, 3582).

$$HOOCCOCH_2COOH \longrightarrow HOOCCOCH_2C(OH)(COOH)CH_2COOH$$

Diethyl oxaloacetate can also undergo condensation with aldehydes to yield products which can be cyclised with a mixture of ammonium acetate and acetic acid to give substituted pyridines after aromatisation (R.Balicki *et al.,* Pol. J. Chem., 1979, 53, 893).

$$RCHO \quad + \quad 2\ EtOOCCOCH_2COOEt \longrightarrow RCH\left[CH(COOEt)COCOOEt\right]_2$$

Diethyl oxaloacetate reacts with ethyl aminoformate in the presence of phosphoryl chloride to provide ethyl oxaorate (S.S.Washburne and K.K.Park, Tetrahedron Letters, 1976, 4, 243).

Pyrrolidine derivatives can be made from diethyl oxaloacetate with isocyanates in the presence of triethylamine (L.Caperano *et al.,* Ber., 1973, 106, 3677).

EtOOCCOCH$_2$COOEt   +   RNCO $\xrightarrow{\text{Et}_3\text{N}}$ EtOOCCOCH(CONHR)COOEt.

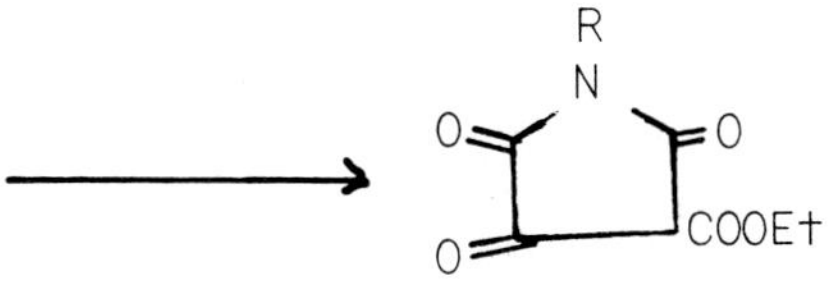

$\xrightarrow{\hspace{3cm}}$

Diethyl oxaloacetate undergoes condensation with phenols in concentrated sulphuric acid to provide a mixture of the coumarin and spirolactone (J.L.Belletire *et al.*, J. hetercyclic. Chem., 1979, 16, 1680).

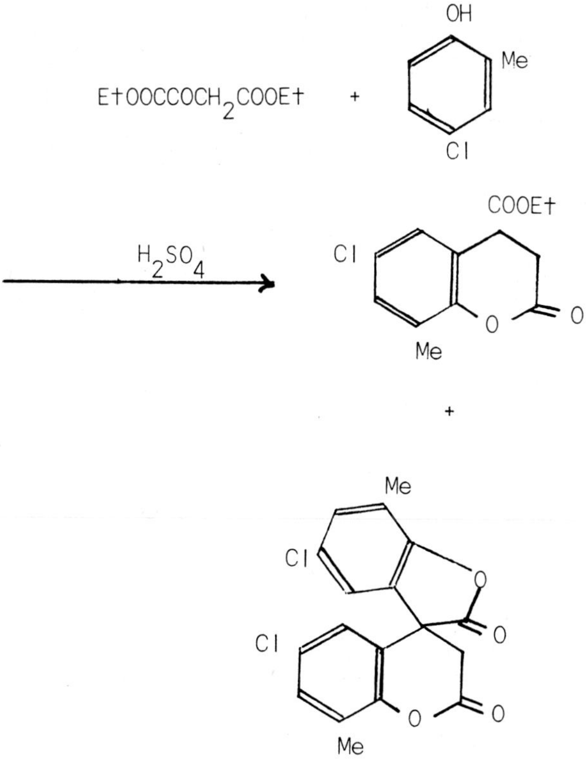

EtOOCCOCH$_2$COOEt   +

$\xrightarrow{\text{H}_2\text{SO}_4}$

+

3-Deoxy-D-*erythro*- and 3-deoxy-D-*threo*-hex-2-ulosonic acid 5-(dihydrogen phosphates)

can be made by the condensation of D-glyceraldehyde 2-phosphate with oxaloacetic acid at pH 7 (F.Trigalo *et al.*, J. chem. Soc. Perkin Trans. I, 1979, **3**, 649).

3-Oxoglutaric acid, 3-oxopentanedioic acid,
$HOOCCH_2COCH_2COOH$, can be made as its diethyl ester by the reaction of phosgene with diketene at $0°C$, followed by reaction with ethanol (Lonza Ltd., Fr. Pat., 1970, 2,107,696).

Diethyl 3-oxoglutarate reacts with aniline at $100°C$ to provide 4-anilino-1-phenyl-2,6-(1H,3H)-pyridine dione. This can be hydrolysed with concentrated hydrochloric acid to 1-phenyl-piperidine-2,4,6-trione which can tautomerise to the enol form (J.D.Mee, J. org. Chem., 1975, **40**, 2135).

The reaction of diethyl 3-oxoglutarate with benzene diazonium chloride, followed by hydroxylamine, gives rise to a number of different products (V.M.Belikov *et al.*, Izv. Akad. Nauk. SSSR, Ser. Khim., 1973, **9**, 2060).

When diethyl 3-oxoglutarate is treated with an alkyl
isocyanate, followed by phosphorus oxychloride, pyridone
derivatives are obtained (Metallidus, Chem. Chron.,
1972, 1, 151).

Diethyl 3-oxoglutarate is capable of undergoing a
double aldol condensation and the product then reacting
further by a double Michael reaction to give tetraethyl
7,10-dihydroxy-8,9-dihydro-9,5-propano-5H-benzocycloheptene-
6,8,10,12-tetracarboxylate which can be hydrolysed and
decarboxylated to 5,6,8,9-tetrahydro-5,9-propano-7H-
benzocycloheptene-7,11-dione. This dione in turn can undergo
nucleophilic transannular ring closure with phosphorus
pentachloride to provide oxabenzoadamantene (B.Foehlisch
et al., Ann. Chem., 1973, 11, 1839).

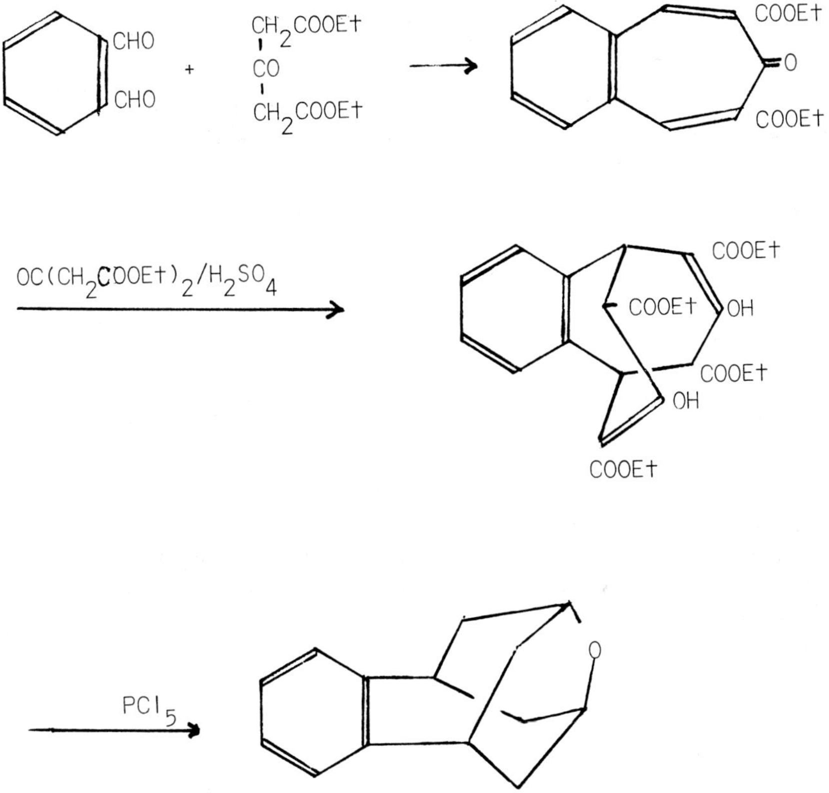

It appears that the above reaction in general for alicyclic 1,2-dicarbonyl compounds proceeds with 1,2-stoichiometry to give tetramethylpropellanedione tetracarboxylate derivatives from methyl 3-oxoglutarate in good yield.

Tetramethyl [10.3.3] propellane-14,17-dione-13,15,16, 18-tetracarboxylate, m.p. 156.5-158°C, can be made by the reaction of cyclododecane-1,2-dione with dimethyl 3-oxo-glutarate. Hydrolysis and decarboxylation with dilute hydrochloric acid gives [10.3.3] propellane-14,17-

dione, m.p. 53-55°C. The reduction of this dione with
hydrazine and potassium hydroxide in diethylene glycol
yields [10.3.3] propellane, m.p. 33-34°C (S.Yang and
J.M.Cook, J. org. Chem., 1976, 41, 1903).

Tetramethyl [6.3.3] propellane-10,13-dione-9,11,
12,14-tetracarboxylate can be prepared from dimethyl
3-oxoglutarate and cyclooctane-1,2-dione, m.p. 160-163°C.
Hydrolysis and decarboxylation gives [6.3.3] propellane-
10,13-dione, m.p. 80-82°C.

Tetracyclo [10.3.3.0.0^{14,17}] octadecan-14-ol,
m.p. 142-143°C, can be made from [10.3.3] propellane-
14,17-dione by reduction with zinc amalgam. Similarly,
tetracyclo [6.3.3.0.0^{10,13}] tetradecan-10-ol can be made
by reduction of [6.3.3] propellane-10,13-dione and has
m.p. 87-88°C.

The preferential 1,2-stoichiometry observed in the
reaction of 1,2-diketones and dimethyl 3-oxoglutarate
can be explained through a sequence of reactions shown in
the following scheme (ibid. loc. cit.).

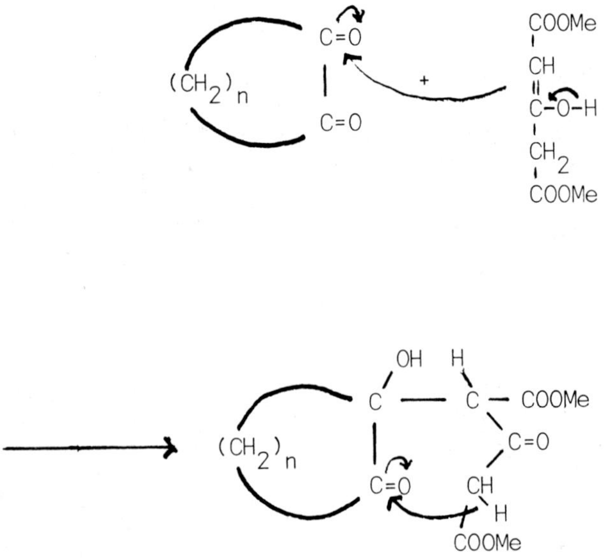

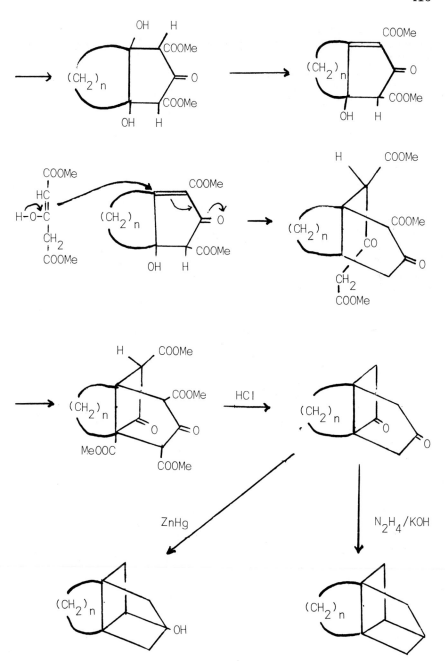

Attempts to isolate the 1:1 intermediates in this sequence have been unsuccessful. However, a 1:1 adduct, 2,5-dicarbomethoxy-4-hydroxy-3,4-diphenylcyclopent-2-enone, m.p. 136-140°C, can be obtained from the reaction of benzil with dimethyl 3-oxoglutarate.

Diethyl 3-oxoglutarate condenses with diethyl oxalate in the presence of sodium ethoxide to give a cyclic trione (B.C.Chowdhri *et al.*, Labder, Part A, 1972, 10, 35).

Formaldehyde and dimethylamine react with diethyl 3-oxoglutarate to give diazabicyclononanedicarboxylates (H.Henig and N.Besch, Arch. Pharm., 1974, <u>307</u>, 569).

The total synthesis of (RS)-zearalone can be achieved by using dimethyl 3-oxoglutarate as the starting material (R.N.Hurd and D.H.Shah, J. med. Chem., 1973, <u>16</u>, 543).

CH$_2$COOMe
|
CO
|
CH$_2$COOMe

→

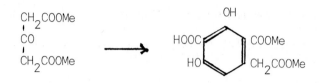

1) −CO$_2$
2) benzylate
3) 4−benzyloxy−
   butyraldehyde/
   NaH
——————————→

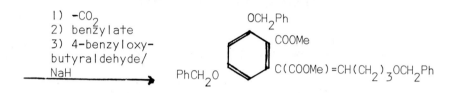

1) hydrolysis
2) −CO$_2$
3) 5−chloropentan−2−ol
——————————→

1) H$_2$/catalyst
2) hydrolysis
3) Convert to dichloro
   compound
4) Convert to dicyano
   compound
Internal Dieckmann reaction
——————————————————→

3 Alkane and alkene tricarboxylic acids

Aconitic acid, 2-carboxyglutaconic acid, propene-1,2,3-tricarboxylic acid, $HOOCCH_2C(COOH)=CHCOOH$.

A 1:1 mixture of *cis*- and *trans*-aconitic acid can be made from a malonic acid derivative according to the following scheme (N.Gutierrez and V.Lamberti, U.S. Pat., 4,123,458).

$$MeOOCCH_2CH(COOH)CH(COOMe)_2$$

$$\xrightarrow{NaOCl} MeOOCCH_2CH(COOH)CCl(COOMe)_2$$

$$\xrightarrow{-HCl} MeOOCCH_2C(COOH)=C(COOMe)_2$$

$$\xrightarrow{Saponification} HOOCCH_2C(COOH)=C(COOH)_2$$

$$\xrightarrow{H^+} HOOCCH_2C(COOH)=CHCOOH$$

Trans-aconitic acid can be converted into citric acid by its reaction with calcium hydroxide, followed by hydrogenation (N.Gutierrez and V.Lamberti, Neth. Pat., 77 04,006).

By reaction with hydrogen peroxide in the presence of tungstic acid, *trans*-aconitic acid is epoxidised to the (±)-*threo*-form of the following tricarboxylic acid (R.W.Guthrie *et al.*, Ger. Pat., 2,258,955).

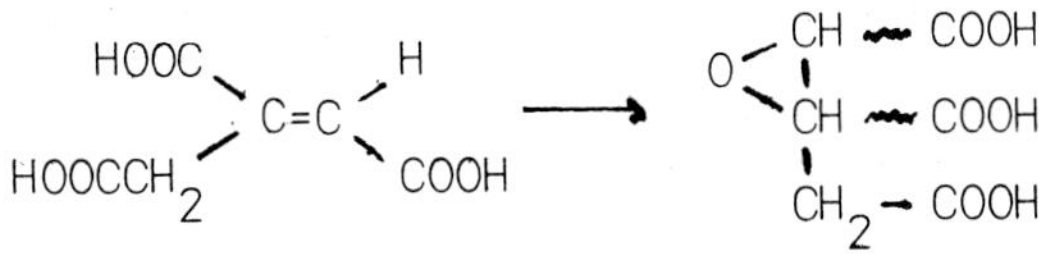

Propane-1,2,3-tricarboxylic acid, $HOOCCH_2CH(COOH)CH_2COOH$, can be converted into its trimethyl ester by reaction with an orthoester. Similarly, dihydroxymalonic acid can be esterified (H.Cohen and J.D.Mier, Chem. and Ind., 1965, 349).

CH₂COOH → Me₂C(OMe)₂/HCl → CH₂COOMe (structures)

$$CH_2COOH \quad CHCOOH \quad CH_2COOH \xrightarrow{Me_2C(OMe)_2/HCl} CH_2COOMe \quad CHCOOMe \quad CH_2COOMe$$

$$(HO)_2C(COOH)_2 \xrightarrow{HC(OMe)_3} (MeO)_2C(COOMe)_2$$

The base-catalysed condensation of trimethyl propane-1,2,3-tricarboxylate with dimethyl dimethoxy-malonate using sodium hydride yields trimethyl 4,6-dihydroxy-5,5-dimethoxycyclohexa-1(6),3-diene-1,2,3-tricarboxylate which can then undergo hydrolysis and decarboxylation to gallic acid by using 48% hydrobromic acid (M.T.Shipchandler and C.A.Peters, J. chem. Soc. Perkin I, 1975, 1400).

$$CH_2COOMe \quad CHCOOMe \quad CH_2COOMe + (MeO)_2C(COOMe)_2 \xrightarrow{NaH} $$

(cyclohexadiene structure with COOMe, MeOOC, COOMe, HO, OH, MeO, OMe)

$$\xrightarrow{H^+} \xrightarrow{HBr} \text{gallic acid}$$

(structures of intermediate and gallic acid with COOH, HO, OH, OH)

The formation of the cyclic ester in the scheme can be thought to involve a dianion or equivalent monoanions in two steps.

$$MeOOCCH_2CH(COOMe)CH_2COOMe$$

$$MeOOC\bar{C}HCH(COOMe)\bar{C}HCOOMe \quad + \quad MeOOCCH_2\bar{C}(COOMe)\bar{C}HCOOMe$$

Tricarballylic acid can be used as a precursor for the total synthesis of the mold product, avenaciolide

which can be isolated from *Aspergillus avenaceus* H. Smith (D.Brookes, B.K.Tidd and W.B.Turner, J. chem. Soc., 1963, 5385) and from cultures of *Aspergillus kischeri* var. glaber (J.J.Ellis, F.H.Stodola, R.F.Vesonder and C.A.Glass, Nature, 1964, 203, 1382).

Acylative decarboxylation of tricarballylic acid with nonanoic anhydride provides the dilactone of 3-(1,1-dihydroxynonyl)glutaric acid. Reduction of this dilactone by means of alkaline borohydride leads to the formation of *trans*-tetrahydro-2-octyl-5-oxo-3-furanacetic acid which can be converted *via* its acid chloride into a pyrrolidine amide, *trans*-β -(1-hydroxynonyl)-δ -oxo-1-pyrrolidine valeric acid-γ -lactone. Carbo-methoxylation of this compound gives the amidolactonic ester, methyl [2-hydroxy-1-(1-pyrrolidinylcarbonyl)methyl] decylmalonic acid γ -lactone. Treatment of this material with sodium hypochlorite solution followed by both the neutral and acidic products with aqueous hydrobromic acid leads to the formation of the dilactone 2-hydroxy-3-(1-hydroxynonyl)glutaric acid di-γ -lactone. Carboxylation of this dilactone with methyl methoxy-magnesium carbonate yields the dilactonic acid, 3-hydroxy-2-(1-hydroxynonyl)-1,1,3-propane tricarboxylic acid 1,3-di-γ -lactone, which, on tratment with formaldehyde and diethylamine in buffered acetic acid, yields avenaciolide (W.L.Parker and F.Johnson, J. org. Chem.,1973, 38, 2489).

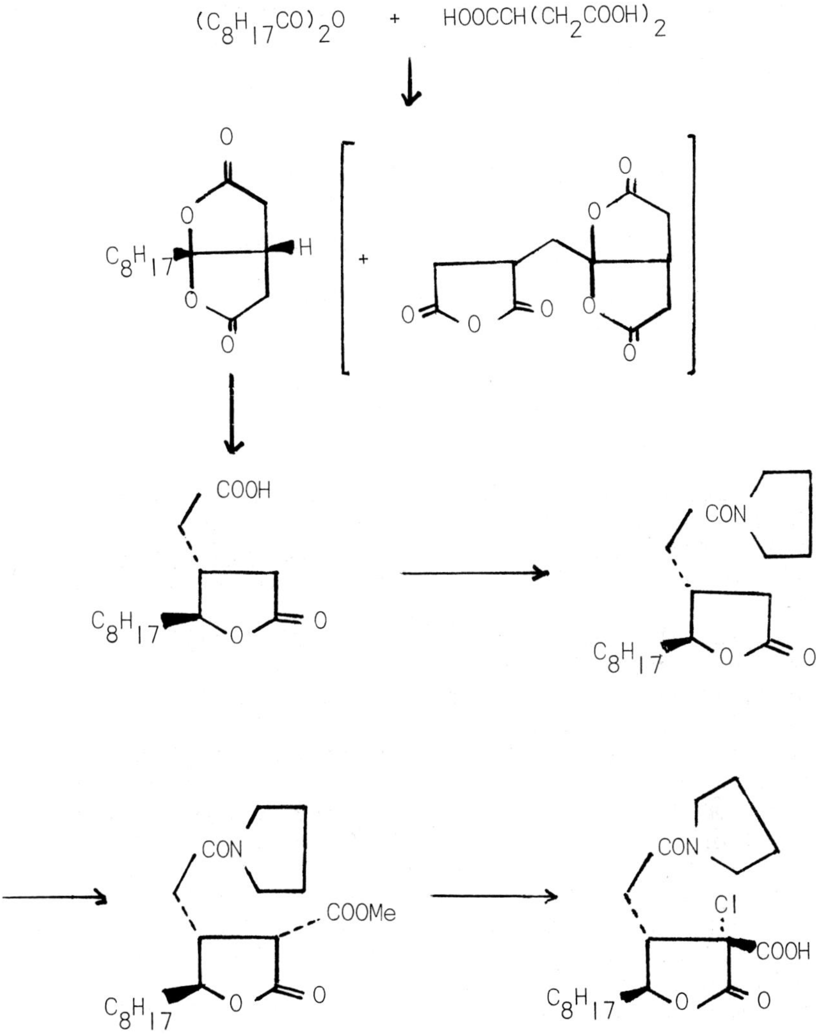
(C8H17CO)2O
+
HOOCCH(CH2COOH)2
C8H17
H
COOH
CON
COOMe
Cl
COOH

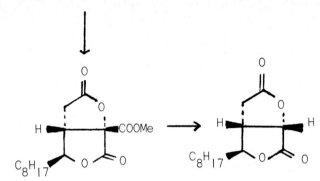

Chapter 21

PHOSPHOLIPIDS AND GLYCOLIPIDS

R.H. GIGG

Introduction

The order of the chapter follows that of the original
article which should be consulted for the formulae of the
lipids since these have been omitted if they were present
in the original article.   Most Russian journals are fully
translated into English and the titles of the translated
journals are given here in parentheses: Zhur. org. Khim.
(J. org. Chem. USSR);   Zhur. obshchei Khim. (J. gen. Chem.
USSR);   Bioorg. Khim. (Soviet J. bioorg. Chem.);   Uspekhi
Khim. (Russ. chem. Rev.);   Khim. prirod Soed. (Chem. nat.
Products);   Doklad. Akad. Nauk SSSR (Proc. Acad. Sci. USSR –
which in the translated edition is in various parts e.g.
Biochem.).   In the references to Russian journals the page
numbers of the translated editions are given (when available)
in parentheses after the page number of the original Russian
journal and for Doklad Akad. Nauk SSSR the part of the translated journal is also given in parentheses.

General references

(a) *Lipids*:   ed. R.M. Burton and F.C. Guerra, 'Fundamentals
of Lipid Chemistry', BI–Science Publications, Webster Groves,
1974;   M.I. Gurr and A.T. James, 'Lipid Biochemistry: an
Introduction', 2nd. ed., Chapman and Hall, London 1976;   ed.
L.A. Witting, 'Glycolipid Methodology', Amer. oil Chem. Soc.,
1976;   ed. G.D. Fasman, C.R.C. Handbook of Biochem. and Mol.
Biol., 3rd. ed., 'Lipids, Carbohydrates and Steroids', 1975;
ed. N.G. Bazan, R.R. Brenner and N.M. Giusto, 'Function and

Biosynthesis of Lipids', Plenum Press, New York, 1977; J.D.
Weete, 'Fungal Lipid Biochemistry', Plenum Press, New York,
1974; ed. T. Galliard and E.I. Mercer, 'Recent Advances in
the Chem. and Biochem. of Plant Lipids', Academic Press, New
York, 1975; ed. F. Snyder, 'Lipid Metabolism in Mammals',
Vols. 1 and 2, Plenum Press, New York, 1977; ed. P.K. Stumpf,
'The Biochemistry of Plants', Vol. 4, 'Lipids: structure and
function', Academic Press, 1980; J.D. Weete and D.J. Weber,
'Lipid Biochemistry of Fungi and other Organisms', Plenum
Press, 1980; ed. L.D. Bergelson, 'Lipid Biochemical Prepar-
ations', Elsevier-N.Holland, 1980; ed. C.C. Sweeley, 'Cell
Surface Glycolipids', ACS Symposium Series, No. 128, 1980;
J.B.M. Rattray, A. Schibeci and D.K. Kidby, 'Lipids of
yeasts', Bacteriol. Rev., 1975, 39, 197; A.F. Rosenthal,
'Chem. synthesis of phospholipids containing carbon-
phosphorus bonds', Methods in Enzymology, 1975, 35, 429;
R.G. Jensen and R.E. Pitas, 'Synthesis of some acylglycerols
and phosphoglycerides', Adv. Lipid Res., 1976, 14, 213; M.
Kates, 'Synthesis of stereoisomeric phospholipids for use in
membrane studies', Methods in Membrane Biol., 1977, 8, 219;
R.A. Klein and P. Kemp, 'Recent methods for elucidation of
lipid structures', *ibid.*, 1977, 8, 51; S. Rottem, 'Membrane
lipids in mycoplasma', Biochim. biophys. Acta, 1980, 604, 65;
R.M. Bell and R.A. Coleman, 'Enzymes of glycerolipid synthes-
is in eukaryotes', Ann. Rev. Biochem., 1980, 49, 459; ed.
G.V. Marinetti', 'Lipid Chromatographic Analysis', 2nd. ed.
Vols. 1-3, Decker, New York, 1976; IUPAC-IUB recommendations
for lipid nomenclature, Lipids, 1977, 12, 455; K. Sandhoff,
'The biochemistry of the lipid storage diseases', Angew.
Chem. internat. Edn., 1977, 16, 273; ed. R.H. Glew and S.R.
Peters, 'Practical Enzymology of the Sphingolipidoses', A.R.
Liss Inc., New York, 1977; R.O. Brady, 'Sphingolipidoses',
Ann. Rev. Biochem., 1978, 47, 687; P.G. Pentchev and J.A.
Barranger, 'Sphingolipidoses, molecular manifestations and
biochem. strategies', J. lipid Res., 1978, 19, 401; D.M.
Marcus and G.A. Schwarting, 'Immunochemical properties of
glycolipids and phospholipids', Adv. Immunol., 1976, 23,
203; R. Gigg 'Synthesis of glycolipids', Chem. Phys.
Lipids, 1980, 26, 287; H. Eibl, 'Synthesis of glycerophospho-
lipids, *ibid.*, 1980, 26, 405; Articles on the mass-
spectrometry of lipids in Chem. Phys. Lipids, 1978, 21, 289-
416.
(b) *Membranes*: H. Tien, 'Bilayer lipid membranes, theory
and practice', Decker, New York, 1974; ed. S. Fleischer

and L. Packer, 'Biomembranes' in Methods in Enzymology, Vol.
32, 1974; ed. G. Weissman and R. Claiborne, 'Cell membranes.
Biochem., cell biology and pathology', Hospital Practice
Publishing, New York, 1975; D. Chapman, 'Phase transitions
and fluidity characteristics of lipids and cell membranes',
Quart. Rev. Biophysics, 1975, $\underline{8}$, 185; J. Seelig and A.
Seelig, 'Lipid conformation in model membranes and biological membranes', *ibid.*, 1980, $\underline{13}$, 19; M.S. Bretscher
and M.C. Raff, 'Mammalian plasma membranes', Nature, 1975,
$\underline{258}$, 43; ed. D. Chapman and D.F.H. Wallach, 'Biological
membranes', Vol. 3, Academic Press, New York, 1976; ed.
A.H. Maddy, 'Biochemical analysis of membranes', Chapman
and Hall, London 1976; J.E. Rothman and J. Lenard,
'Membrane Asymmetry', Science, 1977, $\underline{195}$, 743; L.D.
Bergelson and L.I. Barnskov, 'Topological asymmetry of
phospholipids in membranes', *ibid.*, 1977, $\underline{197}$, 224; R.B.
Gennis and A. Jonas, 'Protein-lipid interactions', Ann.
Rev. Biophys. and Bioengineering, 1977, $\underline{6}$, 195; H.
Brockerhoff, 'Molecular design of membrane lipids', in ed.
E. van Tamelen, 'Bioorganic Chemistry, Vol. 3', Academic
Press, New York 1977, p.1; S.C. Kinsky and R.A. Nicolotti,
'Immunological properties of model membranes', Ann. Rev.
Biochem., 1977, $\underline{46}$, 49; H.C. Anderson, 'Probes of membrane
structure', *ibid.*, 1978, $\underline{47}$, 359; C.R. Alving, 'Immune
reactions of lipids and lipid model membranes', in ed. M.
Sela, 'The Antigens', Vol. 4, Academic Press, New York, 1977,
p.1; C. Tanford, 'The hydrophobic effect: formation of
micelles and biological membranes', 2nd. ed. Wiley-
Interscience, New York, 1980; C. Tanford, 'The hydrophobic
effect and the organisation of living matter', Science, 1978,
$\underline{200}$, 1012; ed. D. Papahadjopoulos, 'Liposomes and their
use in biology and medicine', Ann. New York Acad. Sci., 1978,
Vol. 308; ed. G. Gregoriadis and A.C. Allison, 'Liposomes
in biological systems', Wiley, 1980; B.E. Ryman and D.A.
Tyrrell, 'Liposomes - bags of potential', Essays in Biochem.,
1980, $\underline{16}$, 49; P.L. Yeagle 'Phospholipid headgroup behaviour
in biological assemblies', Acc. chem. Res., 1978, $\underline{11}$, 321;
G. Büldt and R. Wohlgemuth, 'The headgroup conformation of
phospholipids in membranes', J. membrane Biol., 1981, $\underline{58}$,
81; D. Chapman, 'Liquid crystals and cell membranes', Pure
applied Chem., 1978, $\underline{50}$, 627; P.J. Quinn, 'The fluidity of
cell membranes and its regulation', Progr. Biophys. Mol.
Biol., 1981, $\underline{38}$, 1; ed. J.B. Finean, R. Coleman and R.H.
Michell, 'Membranes and their cellular function', 2nd. ed.,

Blackwell, London, 1978; H.G. Khorana, 'Chemical studies of biological membranes', Bioorg. Chem., 1980, 9, 363; D.F.H. Wallach *et al.*, 'Application of laser Raman infra-red spectroscopy to the analysis of membrane structure', Biochim. biophys. Acta, 1979, 559, 153.

Phospholipids and glycolipids based on glycerol

Dihydroxyacetone phosphate is an intermediate in the biosynthesis of phospholipids and the 1-deoxy-1-fluoro-, 1-bromo-, and 1-chloro-analogues have been synthesised (T.P. Fondy *et al.*, Biochemistry, 1975, 14, 2252). Dihydroxyacetone phosphate is reduced by glycerol phosphate dehydrogenase (EC 1.1.1.8) to give *sn*-glycerol 3-phosphate, the precursor of phosphatidic acids and this biosynthetic step has been investigated in *E. coli* (J.R. Edgar and R.M. Bell, J. biol. Chem., 1978, 253, 6348; 1979, 254, 1016). Phosphonate and phosphate analogues of glycerol 3-phosphate have been synthesised (K.-C. Tang *et al.*, Tetrahedron, 1978, 34, 2873) and the analogue 3,4-dihydroxybutyl 1-phosphonate is incorporated into the phospholipids of *Bacillus subtilis* (B.E. Tropp *et al.*, J. Bacteriol., 1977, 129, 550). Enantiomerically pure *sn*-glycerol 1-phosphate is readily synthesised from glycerol, ATP and glycerol kinase (V.M. Rios-Mercadillo and G.M. Whitesides, J. Amer. chem. Soc., 1979, 101, 5828).

Phosphatidic Acids

The biosynthesis of phosphatidic acids in various tissues by acylation of *sn*-glycerol 3-phosphate or from 1-*O*-acyldihydroxyacetone 3-phosphate by enzymic reduction (EC 1.1.1. 101) and transacylation has been reviewed (A.K. Hajra, Biochem. Soc. Trans., 1977, 5, 34; J. Neurochem., 1978, 31, 125). The specificity of the enzymic incorporation of different fatty acids into *sn*-glycerol 3-phosphate has also been studied (K.J. Kako *et al.*, Canad. J. Biochem., 1977, 55, 308; K.S. Bjerve *et al.*, Biochem. J., 1976, 158, 249; M. Kito *et al.*, J. Biochem., 1979, 85, 1527; H.O. Kuyama *et al.*, J. biol. Chem., 1977, 252, 6682; M. Morikawa and S. Yamashita, Europ. J. Biochem., 1978, 84, 61). Unsaturated mixed-acid phosphatidic acids have been synthesised using bis (2,2,2-trichloroethyl) phosphorochloridate for the phosphorylation of diglycerides {M.S. Sadovnikova *et al.*,

Zhur. obshchei Khim., 1976, 46, 1389 (1365) } and a spin-
labelled phosphatidic acid has been prepared by enzymic
incorporation of 12-doxyl-octadecanoic acid into sn-{2-^{3}H}-
glycerol 3-phosphate (N.Z. Stanacev et $al.$, Canad. J.
Biochem., 1974, 52, 884). Phosphonic acid analogues of
phosphatidic acids (4-acyloxy-3-hydroxy- and 3,4-diacyl-
oxybutyl 1-phosphonic acid) and acyl dihydroxyacetone
phosphate (4-palmitoxy-3-oxo-butyl 1-phosphonic acid)
have been synthesised (J.-C. Tang et $al.$, Chem. Phys.
Lipids, 1976, 17, 169; 1977, 19, 99). The 4-acyloxy-3-
hydroxybutyl 1-phosphonic acid (a 'lysophosphotidic acid')
was a substrate for the enzyme acyl CoA:lysophosphatidate
acyl-transferase which converts lysophosphatidic acids into
phosphatidic acids.

Various isomers of 1,2-diacyloxy-3-hydroxy- cyclopentane
phosphate have been synthesised as analogues of phosphatidic
acids (A.J. Hancock et $al.$, J. lipid Res., 1977, 18, 81).

A simple method for the isolation and analysis of phosphatidic
acids by t.l.c. has been described (E.B. Rodriguez de Turco
and N.G. Bazan, J. Chromatog., 1977, 137, 194) and partial
purification of the enzyme phosphatidate phosphatase (EC 3.
1.3.4) which converts phosphatidic acids into diglycerides,
has been achieved (H. Hosaka et $al.$, J. Biochem., 1975, 77,
501). The positional specificity of the fatty acids and
the incorporation of ^{32}P into the pyrophosphate of phosphatidic acid ('pyrophosphatidic acid') present in $Cryptococcus$
$neoformans$ has been studied (T. Itoh and H. Kaneko, J.
Biochem., 1975, 77, 777; Lipids, 1977, 12, 809).

Chemically induced phase separations in model membranes
('liposomes') containing dipalmitoyl phosphatidic acids by
calcium ions and polylysine have been investigated by optical
and e.s.r. methods (H.-J. Galla and E. Sackmann, J. Amer.
chem. Soc., 1975, 97, 4114; Biochim. biophys. Acta, 1975,
401, 509). This separation is due to the formation of
calcium ion bound patches of phosphatidic acids which are
in rigid (or crystalline) structure.

Phosphatidic acids accumulate in the membranes of $E.$ $coli$
mutants defective in cytidine diphosphate diglyceride
synthetase (B.R. Ganong et $al.$, J. biol. Chem., 1980, 255,
1623).

Lysophosphatidic Acids

Lysophosphatidic acids injected intravenously affect the blood pressure of animals, the effect depending upon the chain length and unsaturation of the fatty acyl groups (A. Tokumura *et al.*, Lipids, 1978, 13, 572). They also cause the aggregation of human blood platelets which is thought to be due to binding with calcium ions (J.M. Gerrard *et al.*, Am. J. Pathol., 1979, 96, 423). The related long-chain cyclic acetals of glycerol 3-phosphate (known as 'Darmstoff') cause hypotension in rats. The analogues shown have been synthesised and exhibit similar cardiovascular activity to 'Darmstoff'. (A.N. Milbert and R.A. Wiley, J. med. Chem., 1978, 21, 245; see also Y. Wedmid *et al.*, J. org. Chem., 1980, 45, 1582 for the synthesis of the cyclic acetals of glycerol 2- and 3-phosphate.)

$$RHC \overset{O-CH_2}{\underset{O-C-H}{}} \quad CH_2O \cdot P \cdot (OH)_2 \quad \overset{O}{}$$

"Darmstoff"

$$R = -C_{15}H_{31}$$

Cytidine Diphosphate Diglyceride (CDPDG)

A biologically active spin-labelled CDPDG has been synthesised from the spin-labelled phosphatidic acid described above, by reaction with cytidine monophosphomorpholidate (L. Stuhne-Sekalec and N.Z. Stanacev, Canad. J. Biochem., 1976, 54, 553). The β-adrenergic receptor blocking agent DL-propranolol causes an accumulation of CDPDG and a rapid biosynthetic method for the preparation of labelled CDPDG has been developed using rat pineal glands and ^{32}P - orthophosphate in the presence of DL-propranolol (J. Eichberg and G. Hauser, J. lipid Res., 1978, 19, 778). In the absence of inositol, inositol-requiring strains of *Saccharomyces cerevisiae* accumulate phosphatidic acid and CDPDG, in quantities sufficient for chemical studies, before loss of cell viability ('inositol-less death') (G.W. Becker and R.L. Lester, J. biol. Chem., 1977, 252, 8684). Because of the small amounts of CDPDG normally present in tissues, the characterisation has been incomplete. A large scale isolation from beef liver has confirmed the structure

reported previously (W. Thompson and G. Macdonald, *ibid.*,
1975, 250, 6779).   The fatty acid composition of beef liver,
bovine brain and rat liver CDPDG has also been studied (W.
Thompson and G. Macdonald, Europ. J. Biochem., 1976, 65, 107;
Canad. J. Biochem., 1977, 55, 1153).   Hydrolysis of CDPDG
by phospholipase C gives diglycerides and the fatty acid
distribution in these has been established by hydrolysis
with pancreatic lipase.   Stearic acid is found mainly on
the primary hydroxy-group and arachidonic acid mainly on
the 2-hydroxy-group, which is the same distribution as
found in phosphatidyl inositol.   Since phosphatidic acid
(the biosynthetic precursor of CDPDG) has a different fatty
acid composition (see H.H. Bishop and K.P. Strickland,
Canad. J. Biochem., 1976, 54, 249) it is possible that
transacylation of CDPDG may occur in tissues (see W.
Thompson and G. Macdonald, J. biol. Chem., 1978, 253, 2712;
W. Thompson in 'Cyclitols and Phosphoinositides', ed. W.W.
Wells and F. Eisenberg, Academic Press, New York, 1978,
p.215).

Cephalins, Phosphatidylethanolamines (PE)

PE   {1,2-di-*O*-acyl-*sn*-glycerol 3-(2-aminoethyl phosphate)}
and lyso-PE {mono-acyl-*sn*-glycerol 3-(2-aminoethyl phosphate)}
have been detected at the 100nmol level by HPLC after conversion into their biphenylcarbonyl derivatives (F.B.
Jungalwala *et al.*, Biochem. J., 1975, 145, 517).   HPLC has
also been used for the large scale (*ca.* 10g) purification of
these two lipids from egg-yolk phospholipids with partial
resolution of the molecular species (R.S. Fager *et al.*, J.
lipid Res., 1977, 18, 704).   The molecular species of PE
have been separated by 'argentation t.l.c.' of their
trifluoroacetyl derivatives (giving separation on the basis
of unsaturation)(S.K.F. Yeung *et al.*, Lipids, 1977, 12, 529)
and the molecular species of PE biosynthesised in mitochondria and endoplasmic reticulum have also been investigated after phospholipase C hydrolysis to diglycerides by
acetylation followed by 'argentation chromatography' (J.G.
Parkes and W. Thompson, Canad. J. Biochem., 1975, 53, 698).
Photo-densitometric analyses of PE (as the *N*-2,4-dinitro-
phenyl derivative) have also been described (G. Heidemann
and D. Lekim, J. Chromatog., 1976, 128, 235). ^{13}C-Enriched
PE's have been prepared from *E. coli* grown in the presence
of {1-^{13}C} - or {2-^{13}C} -acetate and the ^{13}C-n.m.r. spectra
assigned (N.J.M. Birdsall *et al.*, Biochim. biophys. Acta,

1975, <u>380</u>, 344). Fluorescent derivatives of PE have been
prepared by its reaction with 4-chloro-7-nitrobenzofurazan
(J.A. Monti *et al.*, J. lipid Res., 1978, <u>19</u>, 222) and with
o-phthalaldehyde and β-mercaptoethanol (B.C. Chang and
L. Huang, Biochim. biophys. Acta, 1979, <u>556</u>, 52) and by
enzymic methods (Y.G. Molotovsky *et al.*, Bioorg. Khim.,
1980, <u>6</u>, 144). These are considered as probes for studying
membrane phospholipids. A spin-labelled derivative of PE
has been prepared by incorporating a spin-labelled fatty
acid (V.A. Sukhanov *et al.*, Chem. phys. Lipids, 1979, <u>23</u>,
155) and the immunological properties of various *N*-
substituted PE's (dinitrophenyl, dinitrophenyl-amino-caproyl,
fluoresceinthiocarbamoyl) have been investigated (R.A.
Nicolotti *et al.*, J. Immunol., 1976, <u>117</u>, 1898).

The direct amination of the bromide (shown below) with
ammonia and methylamine has been achieved in yields of
ca. 80% without attack on the ester groups (H. Eibl and
A. Nicksch, Chem. Phys. Lipids, 1978, <u>22</u>, 1).

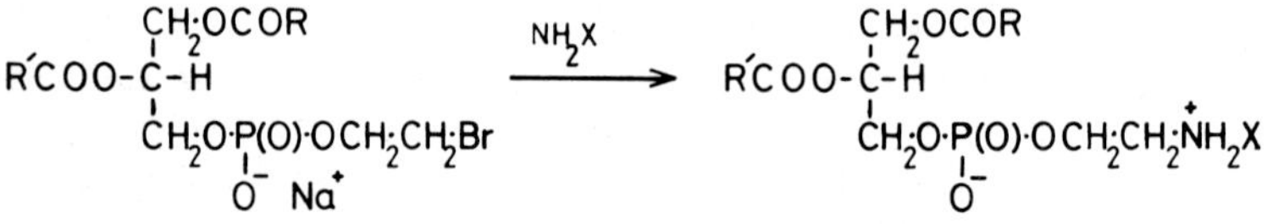

X = H or Me

This type of reaction had previously been achieved with
dimethyl- and trimethylamine. The partial synthesis of
PE by deacylation of tritylated natural PE and subsequent
reacylation with fatty acid anhydrides has been improved to
reduce phosphate migration which results in the formation of
glycerol 2-phosphate derivatives. In the improved method
ca. 4% of the 2-phosphate derivative was observed compared
with *ca.* 13% in the previous method {*cf.* V.A. Sukhanov *et al.*,
Zhur. obshch. Khim., 1977, <u>47</u>, 2130 (1943)}. The n.m.r.
and i.r. spectra of the 2- and the 3-phosphate have been
compared (J.G. Lammers *et al.*, Chem. Phys. Lipids, 1978, <u>22</u>,
293). PE has also been synthesised *via* phosphotriester
intermediates (J.G. Lammers and J.H. van Boom, Rec. trav.
Chim., 1977, <u>96</u>, 216) and *via* a 'cyclic enediolpyrophosphate'
intermediate (J.-L. Danan and L. Pichat, J. lab. compd. and
Radiopharm., 1980, <u>17</u>, 223) a method which was first introduced by Ramirez and his coworkers (Synthesis, 1975, 637;

J. Amer. chem. Soc., 1976, <u>98</u>, 5310).

A promising new method for the synthesis of PE *via* oxazaphos-
pholanes has been described (H. Eibl, Proc. nat. Acad. Sci.,
1978, <u>75</u>, 4074; Chem. Phys. Lipids, 1980, <u>26</u>, 405). The
oxazaphospholane, prepared by reaction of a phosphatidic
acid dichloride with 2-aminoethanol is hydrolysed by dilute
acid to give PE. The oxidation and subsequent hydrolysis
of oxazaphosphorinanes has also been used for the preparation of intermediates for the synthesis of homologues of
P.E. {M.K. Grachev *et al.*, Zhur. org. Khim., 1977, <u>13</u>, 1830
(1697); Bioorg. Khim., 1977, <u>3</u>, 68 (52)}.

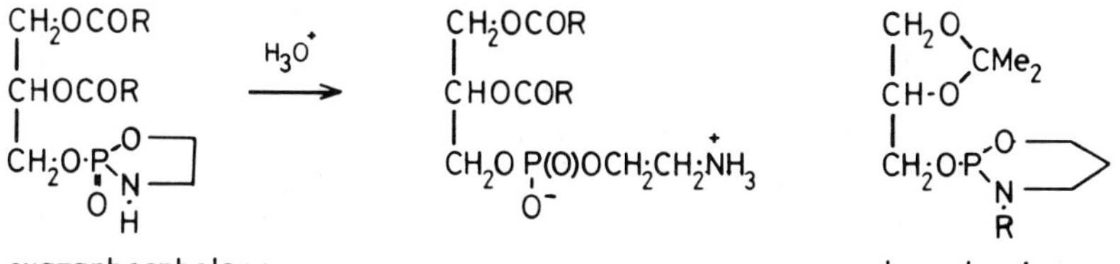

CH₂OCOR
|
CHOCOR
| H₃O⁺
CH₂O·P⟨O⟩
      N
     O H

→

CH₂OCOR
|
CHOCOR
|
CH₂O P(O)OCH₂CH₂NH₃
     O⁻

CH₂O⟩
| CMe₂
CH·O
|
CH₂O·P⟨O⟩
      N
      R

oxazaphospholane   oxazaphosphorinane

The isosteric phosphonate analogue ('phosphotidyl
ethanolamine' - shown below) of PE has been synthesised by
the condensation of 3.4-dipalmitoxybutyl-1-phosphonic acid
with *N*-trichloroethoxycarbonyl ethanolamine followed by
deprotection with zinc-acetic acid (D. Braksmayer *et al.*,Chem.
Phys. Lipids, 1977, <u>19</u>, 93).

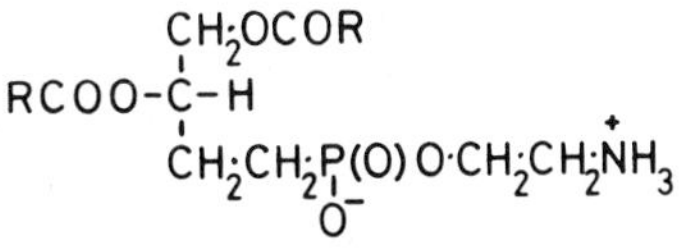

CH₂OCOR
|
RCOO-C-H
|
CH₂CH₂P(O)O·CH₂CH₂NH₃
         O⁻

The i.r., and Raman spectra (R. Mendelsohn *et al.*, Biochim.
biophys. Acta, 1975, <u>413</u>, 329; H. Akutsu and Y. Kyogoku,
Chem. Phys. Lipids, 1975, <u>14</u>, 113), X-ray crystallography
(M. Elder *et al.*, Proc. Roy. Soc., 1977, <u>354A</u>, 157), single
crystal electron diffraction (D.L. Dorset, Biochim. biophys.
Acta, 1976, <u>426</u>, 396), ³¹P n.m.r. (J. Seelig and H.-V. Gally,
 Biochemistry, 1976, <u>15</u>, 5199; P.R. Cullis and B. de Kruijff,
Biochim. biophys. Acta, 1978, <u>513</u>, 31) and ¹H n.m.r. (Y.
Lange *et al.*, Biochem. Biophys. res. Comm., 1975, <u>62</u>, 891)

of PE have been investigated.

N-Acyl derivatives of PE's have been isolated from mammalian
epidermis (G.M. Gray, Biochim. biophys. Acta, 1976,431,1)
and from BHK (baby hamster kidney cells)(P. Somerharju and
O. Renkonen, Biochim. biophys. Acta, 1979, 573, 83) and
synthesised by reaction of fatty acid anhydrides with PE
{V.I. Shvets *et al.*, Bioorg. Khim., 1975, 1, 758 (573)}.
N-Acyl derivatives of lyso-PE have been isolated from
cotton seeds {T.S. Kaplunova *et al.*, Khim. prirod. Soed.,
1978, 41 (14, 31)}.   An intermolecular transacylation of
PE in which two molecules of PE give one molecule of
N-acyl PE and one of lyso-PE occurs in *Butyrivibrio sp.*
(G.P. Hazelwood and R.M.C. Dawson, Biochem. J., 1975, 150,
521).

The different routes for the biosynthesis of PE have been
reviewed (B. Akesson, Biochim. biophys. Acta, 1979, 573,
481; P. Orlando *et al.*, Neurochemical Res., 1979, 4, 595).
2-Aminopropan-1-ol and 2-aminobutan-1-ol inhibit the biosynthesis of PE and are themselves incorporated into
analogues of PE (B. Akesson, Biochem. J., 1977, 168, 401)
and 2-aminoethyl phosphonic acid and 3-aminopropyl phosphonic
acid are competitive inhibitors of the conversion of ethanolamine phosphate into PE (M. Plantavid *et al.*, Biochimie,
1975, 57, 951).   The biosynthesis of PE in *E. coli* is
inhibited by alcohols (L.O. Ingram, Canad. J. Microbiol.,
1977, 23, 779).   A protein which transfers PE between
biolological membranes has been purified from rat liver
(B. Bloj and D.B. Zilversmit, J. biol. Chem., 1977, 252,
1613).

A series of deoxy analogues of lyso-PE (phosphoryl ethanolamine derivatives of the alcohols derived from linolenic
acid and arachidonic acid) have been prepared and shown to
be effective (like lyso-PE) as inhibitors of the activity
of renin (J.G. Turcotte *et al.*, J. med. Chem., 1975, 18,
1184).   Some commercial samples of synthetic lyso-PE have
been shown to be contaminated with phospholipase A_2 (F.C.
Lee *et al.*, Biochemistry, 1980, 19, 1934).

N-*Methyl*- *and* N,N-*Dimethylphosphatidylethanolamines*.

Improved conditions for the synthesis of these compounds
by direct interaction of methylamine and dimethylamine with

the bromoethyl ester of phosphatidic acid have been described
(H. Eibl and A. Nicksch, Chem. Phys. Lipids, 1978, 22, 1).
N-Isopropylethanolamine and N-methyl-N-isopropylethanolamine
are metabolised by rat liver to give the corresponding PE-
derivatives. (T.-C. Lee *et al.*, Biochim. biophys. Acta.,
1975, 409, 218; C. Moore *et al.*, Chem. Phys. Lipids, 1978,
21, 175). The enzymes responsible for the methylation of
PE to N-methyl-PE, N,N-dimethyl-PE and lecithin (using
S-adenosyl-L-methionine) have been investigated (W.J.
Schneider and D.E. Vance, J. biol. Chem., 1979, 254, 3886;
P. Hirata *et al.*, Proc. nat. Acad. Sci., 1978, 75, 1718;
B. Akesson, FEBS Letters, 1978, 92, 177; R. Mozzi and
G. Porcellati, *ibid*, 1979, 100, 363). PE is mainly
located in the inner layer of biological membranes whereas
lecithin is located in the outer layer; thus a trans-
location occurs during the methylation by these enzymes of
PE to lecithin (see F. Hirata and J. Axelrod, Nature, 1978,
275, 219). The kinetics of the conversion of N,N-dimethyl-
PE to lecithin in human liver microsomes has been investigated (O. Zierenberg and H. Betzing, Z. physiol. Chem., 1978,
359, 1239).

Lecithins, Phosphatidylcholine, PC

The analysis of PC, in the presence of sphingomyelin, by
HPLC (F.B. Jungalwala *et al.*, Biochem. J., 1976, 155, 55)
and the separation of PC species by HPLC (N.A. Porter *et al.*,
Lipids, 1979, 14, 20) have been studied. An improved
procedure for the preparation of egg lecithin by column
chromatography on alumina and silica gel has been described
(N.S. Radin, J. lipid Res., 1978, 19, 922) and the saturated
PC from lung tissue has been isolated after treating the
total lipids with osmium tetroxide (to complex with the
unsaturated lipids) and subsequent chromatography on neutral
alumina (R.J. Mason *et al.*, J. lipid Res., 1976, 17, 281).
The molecular species of PC in a chemically induced hepatoma
(G. Okano *et al.*, Tohoku J. exp. Med., 1977, 122, 21) and in
rat gastric mucosa (M.K. Wassef *et al.*, Biochim. biophys.
Acta, 1979, 573, 222) have been determined. The hepatoma
PC was analysed by t.l.c., g.l.c. and specific enzymic
hydrolysis. New methods for the determination of the
lecithin-sphingomyelin ratio in amniotic fluid (to assess
foetal lung development) have been described (D.S. Ng and
K.G. Blass, J. Chromatog., 1979, 163, 37; H.H. Kohl *et al.*,
Clin. Chem., 1978, 24, 174) and a method for the quantification

of dipalmitoyl lecithin in lung extracts, by adding a known
amount of ^{3}H-labelled dipalmitoyl PC before lipid extraction
(S. Mulay *et al.*, Anal. Biochem., 1977, 77, 350) has been
described.   Saturated PC was shown to be the major component
of the surfactant of the lung in other animal species (J.L.
Harwood *et al.*, Biochim. biophys. Acta, 1976, 424, 159).
Babies born with hyaline-membrane disease have a lower percentage of palmitic acid in the PC of their lung surfactant
(with increased C18 and C20 fatty acid content) and the level
returns to normal with postnatal maturation of the surfactant
system (S.A. Shelley *et al.*, New England J. Med., 1979, 300,
112).   The molecular species of PC in lung during the late
stages of foetal development have also been investigated
(G. Okano and T. Akino, Biochim. biophys. Acta, 1978, 528,
373).   The ether, phosphonate analogue of dipalmitoyl PC
(shown below)

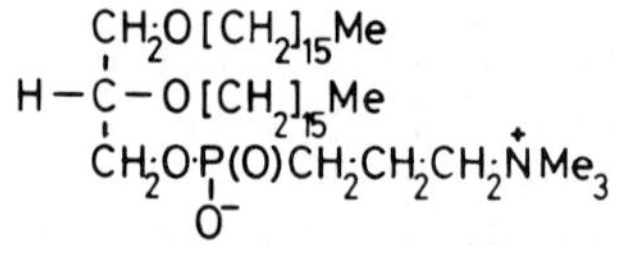

$$CH_2O[CH_2]_{15}Me$$
$$H-\overset{|}{\underset{|}{C}}-O[CH_2]_{15}Me$$
$$CH_2O\cdot\overset{}{\underset{\underset{O^-}{|}}{P}}(O)CH_2CH_2CH_2\overset{+}{N}Me_3$$

has been synthesised and its surface tension lowering and
respreading properties compared with dipalmitoyl PC with a
view to using it, in place of dipalmitoyl PC, as an aerosol
treatment for aiding the respiratory distress syndrome of
premature infants (J.G. Turcotte *et al.*, Biochim. biophys.
Acta, 1977, 488, 235;   see also C.J. Morley *et al.*, Nature,
1978, 271, 162 for discussions on the surface properties of
the lung surfactant).

The conditions for the direct acylation of glycerol phosphoryl choline for the synthesis of PC (see R.G. Jensen and
R.E. Pitas, Adv. lipid Res., 1976, 14, 213) have been
improved to avoid phosphate migration.   *N*-Acylimidazoles
have been used as acylating agents and the n.m.r. and i.r.
spectra of the PC from both glycerol 1- and 2- phosphate
have been investigated (J.G. Lammers *et al.*, Chem. Phys.
Lipids, 1978, 22, 293).   Acylations of glycerol phosphoryl choline with a fatty acid anhydride and 4-pyrollidino-
pyridine or 4-*N*,*N*-dimethylamino-pyridine have also improved
this method (K.M. Patel *et al.*, J. lipid Res., 1979, 20, 674;

C.M. Gupta *et al.*, Proc. nat. Acad. Sci., 1977, 74, 4315;
see also T.G. Warner and A.A. Benson, J. lipid Res., 1977,
18, 548; E.L. Pugh and M. Kates, *ibid.*, 1975, 16, 392 for
the direct acylation of glycerol phosphoryl choline).

PC analogues derived from glycerol 2-phosphate (A.J. Slotboom
et al., Chem. Phys. Lipids, 1976, 17, 128) and 4,5-dihydroxy-
pentan-1-ol phosphoryl choline{(V.V. Bezuglov *et al.*, Zhur.
org. Khim., 1974, 10, 1398 (1410)}have been synthesised.
The route to lecithins by the reaction of the bromoethyl
ester of phosphatidic acid with triethylamine has been reinvestigated (H. Eibl and A. Nicksch, Chem. Phys. Lipids,
1978, 22, 1; see also H. Eibl, *ibid.*, 1980, 26, 405 for a
critical review of methods for the synthesis of PC). An
improved route to PC from 1,2-di-*O*-acylglycerol, phosphorous
oxychloride and the toluene *p*-sulphonic acid salt of choline
has been described (H. Brockerhoff and N.K.N. Ayengar,
Lipids 1979, 14, 88) and a mild procedure for the methylation
of PE to PC using methyl iodide and 18-crown-6 with
potassium carbonate in benzene has been used for the isotopic
labelling of PC with ^{13}C -methyl iodide (K.M. Patel *et al.*,
Lipids, 1979, 14, 596). The method for the isotopic labelling of the choline residue by demethylation of PC with
sodium benzene thiolate and remethylation of the *N*,*N*-dimethyl-
PE has been described in detail (W. Stoffel, Methods in
Enzymology, 1975, 35B, 533). A completely deuteriated 1,2-
dimyristoyl-*sn*-glycerol 3-choline phosphate has been synthesised and its n.m.r. spectrum studied (P.B. Kingsley and G.W.
Feigenson, Chem. Phys. Lipids, 1979, 24, 135) and ^{32}P -
labelled PC has been isolated from soya beans germinated in
the presence of ^{32}P -orthophosphate (F.-H. Eubmann, Biochem.
J., 1979, 179, 713). Choline phosphate esters of octadecanol
and cholesterol have been prepared (as analogues of PC) by
means of cyclic enediol phosphates (F. Ramirez *et al.*, J.
org. Chem., 1978, 43, 2331) as shown below.

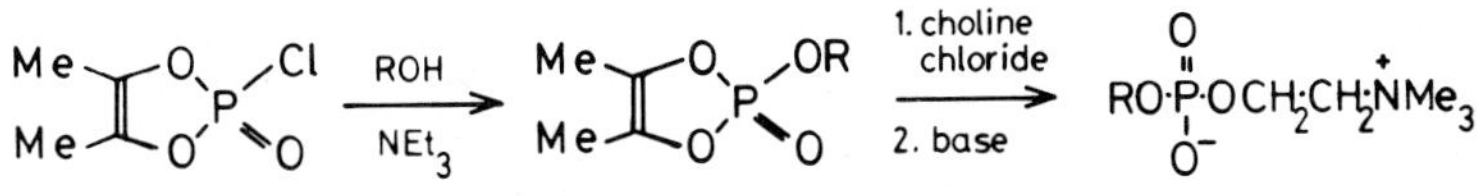

A thio-analogue of PC has been prepared from 1,2-di-*O*-acyl-3-deoxy-3-thioglycerol and was converted into the corresponding thio-analogue of PE by base exchange in the presence of phospholipase D (J.W. Cox *et al.*, Chem. Phys. Lipids, 1979, 25, 369).

Because of the intensive activity in the investigation of biolological membranes many analogues of PC with spin-labels {V.A. Sukhanov *et al.*, Bioorg. Khim., 1978, 4, 785 (570); P. Loiseau *et al.*, Biochimie, 1978, 60, 1201; L. Stuhne-Sekalec *et al.*, Canad. J. Biochem., 1979, 57, 408} or fluorescent labels { A.P. Kaplun *et al.*, Bioorg. Khim., 1978, 4, 1567 (1131); J.G. Molotkovsky *et al.*, *ibid.*, 1979, 5, 588 (437); H.J. Galla *et al.*, Chem. Phys. Lipids, 1979, 23, 239; J.A. Monti *et al.*, Life Sciences, 1977, 21, 345 } have been prepared. A PC containing an organotin analogue of palmitic acid (12, 12-dimethyl-12-stanna-hexadecanoic acid) has been synthesised as a probe for membranes (S.B. Andrews *et al.*, Biochim. biophys. Acta, 1978, 506, 1) and 1,2-di-(X,X-difluoromyristoyl)-PC's (where X = 4,8 or 12) have been prepared and their thermotropic behaviour examined. Some of the properties of the fluoro derivatives are equivalent to those exhibited by lipids containing *cis*-double bonds (J.M. Sturtevant *et al.*, Proc. nat. Acad. Sci., 1979, 76, 2239).

Racemic 2,3-diacyloxypropyl phosphonyl cholines ('A' below) (P.W. Deroo *et al.*, Chem. Phys. Lipids, 1976, 16, 60) and 'phosphotidyl cholines', 3,4-diacyloxybutyl phosphoryl cholines ('B' below) (D. Branksmayer *et al.*, *ibid.*, 1977, 19, 93) have been prepared.

$$CH_2OCOR$$
$$CH \cdot OCOR$$
$$CH_2 \cdot P(O) \cdot O \cdot CH_2CH_2 \overset{+}{N}Me_3$$
$$O^-$$

'A'

$$CH_2OCOR$$
$$CH \cdot OCOR$$
$$CH_2CH_2 P(O) \cdot O \cdot CH_2CH_2 \overset{+}{N}Me_3$$
$$O^-$$

'B'

The exchange protein which transfers PC between biological membranes (K.W.A. Wirtz and P. Moonen, Europ. J. Biochem., 1977, 77, 437; G. Schulze *et al.*, FEBS Letters, 1977, 74, 220) and factors influencing the rate of hemolysis of erythrocytes by dilauroyl- and didecanoyl-PC have been studied (T. Kitagawa *et al.*, Biochim. biophys. Acta, 1977, 467, 137). The effect of different molecular species of diglycerides on the biosynthesis of PC has been investigated (B.J. Holub, Canad. J. Biochem., 1977, 55, 700) and the glycol analogue of PC has been prepared biosynthetically from 2-O-acyloxyethanol (J.-T. Liu and W.J. Baumann, Biochem. Biophys. Res. Comm., 1976, 68, 211). N,N-Diethylethanolamine is incorporated into analogues of PC in biological systems (B. Akesson, Biochem. J., 1977, 168, 401).

The enzyme D-β-hydroxybutyrate dehydrogenase requires lecithin for activity. Various lecithin analogues have been tested with the enzyme and although the hydrophobic moiety can be changed, there is a high degree of specificity for the choline moiety (Y.A. Isaacson *et al.*, J. biol. Chem., 1979, 254, 117). The hydrolysis of PC by phospholipase A_2, immobilized on Celite columns (W.N. Marmer and K.A. Pietruszka, Lipids, 1978, 13, 840), the hydrolysis of PC derived from glycerol 2-phosphate ('β-lecithins') by phospholipase A_2 at water-lipid interfaces (A.J. Slotboom *et al.*, Chem. Phys. Lipids, 1976, 17, 128) and the hydrolysis of analogues of PC (e.g. 4,5-dipalmitoyloxy pentanol phosphoryl choline) by phospholipase A_2 {V.V. Bezuglov *et al.*, Khim., prirod Soed., 1975, 413 (11, 421)} have been studied.

Many physical studies on PC have been recorded e.g. Raman spectroscopy (B.P. Gaber *et al.*, Biophys. J., 1978, 21, 161; S. Sunder *et al.*, Chem. Phys. Lipids, 1978, 22, 279; D.A. Pink *et al.*, Biochemistry, 1980, 19, 349; R.C. Spiker and I.W. Levin, Biochim. biophys. Acta, 1975, 388, 361), Fourier transform i.r. spectroscopy (D.G. Cameron and H.H. Mantsch, Biochem. Biophys. Res. Comm., 1978, 83, 886), ^{31}P-n.m.r. (M.C. Uhing, Chem. Phys. Lipids, 1975, 14, 303; G. Klose, *ibid.*, 1975, 15, 9; R.G. Griffin *et al.*, Biochemistry, 1978, 17, 2718; A.C. McLaughlin *et al.*, J. Magnetic Res., 1975, 20, 146), ^{13}C-n.m.r. (P.G. Barton, Chem. Phys. Lipids, 1975, 14, 336; B. Sears, J. membrane Biol., 1975, 20, 59; C.F. Schmidt *et al.*, Biochemistry, 1977, 16, 3948), [1]H-n.m.r. (D. Lichtenberg *et al.*, *ibid.*, 1979, 18, 4169; R.E. London *et al.*, Chem. Phys. Lipids, 1979, 25, 7; R.D. Hershberg *et al.*, Biochim. biophys. Acta, 1976, 424, 73), [2]H-n.m.r. (R. Jacobs and E. Oldfield, Biochemistry, 1979, 18, 3280; J.H. Davis, Biophys. J., 1979, 27, 339), thermotropic behaviour (N. Albon and J.M. Sturtevant, Proc. nat. Acad. Sci., 1978, 75, 2258; J.M. Sturtevant *et al.*, *ibid.*, 1979, 76, 2239; J.R. Silvins

et al., Biochim. biophys. Acta, 1979, <u>555</u>, 175; T.Y. Song
and M.I. Kanehisa, Biochemistry, 1977, <u>16</u>, 2674; N.O.
Petersen *et al.*, Chem. Phys. Lipids, 1975, <u>14</u>, 343), field
desorption mass-spectrometry (G.W. Wood *et al.*, Biomed. Mass
Spectrom., 1976, <u>3</u>, 172), neutron diffraction (G. Buldt *et
al.*, J. Molecular Biol., 1979, <u>134</u>, 673, 693) and electron
diffraction on single crystals (I. Sakurai *et al.*, Chem.
Phys. Lipids, 1980, <u>26</u>, 41).

Monolayers of egg lecithin prevent mosquito larvae from sur-
facing with consequent death through oxygen starvation (P.
Reiter and A.I. McMullen, Ann. trop. Med. Parasitol., 1978,
<u>72</u>, 163, 169) and lecithin consumption by the rat leads to
an increased content of choline and acetylcholine in the
brain (M.J. Hirsch and R.J. Wurtman, Science, 1978, <u>202</u>,
223).

Lysolecithin (LPC)

2-*O*-t-Butyldimethylsilyl-*sn*-glycerol 3-phosphorylcholine
has been used as an intermediate for the synthesis of LPC
(K.Y. Yabusaki and M.A. Wells, Chem. Phys. Lipids, 1976, <u>17</u>,
120) and the enantiomer of natural LPC (3-*O*-acyl-*sn*-glycerol
1-phosphorylcholine) has been prepared from racemic PC.
Treatment of racemic PC with phospholipase A$_2$ gives the
enantiomer of natural PC (after separation from LPC).   The
1-*O*-acyl group of the enantiomer is removed by hydrolysis
and the 2-*O*-acyl group is then isomerised to the 1-position
by base (N.B. Smith and A. Kuksis, Canad. J. Biochem., 1978,
<u>56</u>, 1149).   A spin-labelled LPC has been synthesised (H.
Utsumi *et al.*, Chem. Phys. Lipids, 1977, <u>19</u>, 203) and [32]P-
labelled LPC has been prepared by germinating soya beans on
[32]P-orthophosphate (F.-H. Hubmann, Biochem. J., 1979, <u>179</u>,
713).   The thioanalogues, shown below, have been synthesised
and used in the assay of lysophospholipases (A.J. Aarsman *et
al.*, Bioorganic Chem., 1976, <u>5</u>, 241).

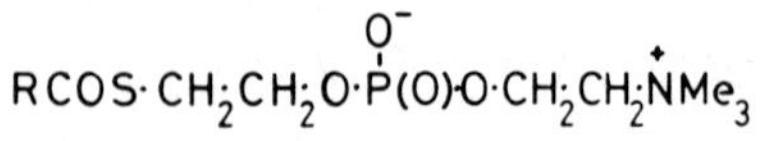

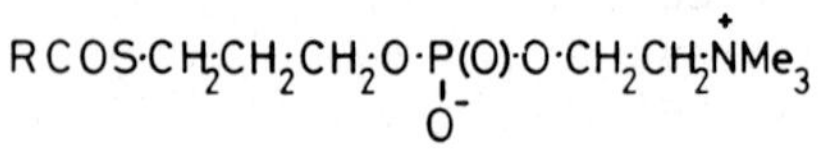

$$RCOS \cdot CH_2CH_2O \cdot \overset{O^-}{\underset{}{P}}(O) \cdot O \cdot CH_2CH_2 \overset{+}{N}Me_3$$

$$RCOS \cdot CH_2CH_2CH_2O \cdot \underset{\underset{O^-}{|}}{P}(O) \cdot O \cdot CH_2CH_2 \overset{+}{N}Me_3$$

The hemolytic activities of the synthetic ether analogues, shown below (H.V. Weltzien *et al.*, Biochim. biophys. Acta, 1977, 466, 411; 1978, 530, 47; Chem. Phys. Lipids, 1976, 16, 267) and the mechanism of the lysis of membranes by LPC have been investigated (T.J. Bierbaum *et al.*, Biochim. biophys. Acta, 1979, 555, 102; H.V. Weltzien, *ibid.*, 1979, 559, 259).

$$Me[CH_2]_n \cdot O \cdot CH_2CH_2CH_2O \cdot \overset{O^-}{\underset{}{P}}(O) \cdot O \cdot CH_2CH_2\overset{+}{N}Me_3$$

$$Me[CH_2]_{17} \cdot O \cdot CH_2CH(OCH_2Ph) \cdot CH_2O \cdot \underset{\underset{O^-}{|}}{P}(O) \cdot O \cdot CH_2CH_2\overset{+}{N}Me_3$$

The morphology of micelles containing LPC has been studied by electron microscopy (K. Inoue *et al.*, J. Biochem., 1977, 81, 1097). LPC and cholesterol form a stable bilayer in aqueous media containing equal molar proportions of each (R.P. Rand *et al.*, Canad. J. Biochem., 1975, 53, 189). The enzyme phosphatidyl choline: cholesterol acyl transferase has been purified from human plasma and characterised (J.J. Albers *et al.*, Biochemistry, 1976, 15, 1084) and the release of fatty acids from PC by the partially purified enzyme studied (V. Piran and T. Nishida, J. Biochem., 1976, 80, 887).

Sulphur, Phosphorus and Arsenic Analogues of Lecithin

The phosphonium analogue of PC (shown below) has been synthesised by an exchange reaction between PC and phosphonium choline chloride in the presence of phospholipase D (EC 3.1.4.4) and the ^{31}P-n.m.r. studied with the object of using the material as a probe for biological membranes (E. Sim *et al.*, Biochem. J., 1975, 151, 555)

$$RCOOCH_2CH(OCOR) \cdot CH_2O \cdot \underset{\underset{O^-}{|}}{P}(O) \cdot O \cdot CH_2CH_2\overset{+}{P}Me_3$$

Phosphonium choline chloride is also incorporated into the PC analogue by rats and by various cells in culture (E. Sim and C.A. Pasternak, *ibid.*, 1976, 154, 105).

The sulphonium analogue of PC (shown below) has been isolated from the diatom *Nitzschia alba* (where it replaces PC as the major phospholipid) and has also been synthesised (P.-A. Tremblay and M. Kates, Canad. J. Biochem., 1979, 57, 595). The biosynthesis of this lipid has been studied (R. Anderson *et al.*, Biochim. biophys. Acta, 1979, 573, 557).

$$RCOOCH_2CH(OCOR) \cdot CH_2O \cdot \overset{}{\underset{O^-}{P(O)}} \cdot O \cdot CH_2CH_2\overset{+}{S}Me_2$$

The arsonium phospholipid (shown below) has been isolated from algae (R.V. Cooney *et al.*, Proc. nat. Acad. Sci., 1978, 75, 4262).

$$RCOOCH_2CH(OCOR) \cdot CH_2O \cdot \overset{}{\underset{O^-}{P(O)}} \cdot O \cdot CH(CO_2^-) \cdot CH_2\overset{+}{A}sMe_3$$

Phosphatidyl Serine (PS)

PS has been analysed by HPLC (F.B. Jungalwala *et al.*, Biochem. J., 1975, 145, 517) and the molecular species (on the basis of unsaturation) of natural PS have been investigated by argentation t.l.c. of the PS (N. Salem *et al.*, Chem. Phys. Lipids, 1980, 27, 289 - who also investigated the various species by spectroscopic methods) or its trifluoroacetamido derivative (S.K.F. Yeung *et al.*, Lipids, 1977, 12, 529). PS has been prepared enzymically by transesterification of PC with phospholipase D in the presence of L-serine and purification by chromatography on carboxymethyl cellulose (P. Comfurius and R.F.A. Zwaal, Biochim. biophys. Acta, 1977, 488, 36).

The analogues of PS (shown below) containing ether linkages and phosphonate groups have been synthesised (I.L. Doer *et al.*, Chem. Phys. Lipids, 1977, 19, 185). Several deuteriated derivatives of PS (with ^2H in the serine or glycerol moieties) have been synthesised by coupling a phosphatidic acid with a protected serine derivative; the 360 MHz ^1H-n.m.r. spectra of these products have been investigated (J. Browning and J. Seelig, Chem. Phys. Lipids, 1979, 24, 103). PS (including the plasmalogen type) accumulates in anaerobic, lactate-fermenting bacteria (*Megasphaera elsdenii, veillonella parvula, Anaerovibrio lipolytica* and *Selenomonas ruminantium*)

(L.M.G. van Golde *et al.*, FEBS Letters, 1975, <u>53</u>, 57).

$$\begin{array}{ccc}
CH_2OR & CH_2OR & CH_2OCOR \\
CH\cdot OR & CH\cdot OR & CH\cdot OCOR \\
CH_2OP(O)OCH_2CH(CO_2H)\overset{+}{N}H_3 & (CH_2)_nP(O)OCH_2CH(CO_2H)\overset{+}{N}H_3 & (CH_2)_nP(O)OCH_2CH(CO_2H)\overset{+}{N}H_3 \\
\quad O^- & \quad O^- & \quad O^- \\
 & (n=1\ or\ 2) & (n=1\ or\ 2)
\end{array}$$

P.S. potentiates the release of histamine from mast cells exposed to various antigens but N-substituted derivatives of PS are not active (T.W. Martin and D. Lagunoff, Proc. nat. Acad. Sci., 1978, <u>75</u>, 4997; Science, 1979, <u>204</u>, 631). Various other pharmacological effects of PS in liposomes (G. Toffano *et al.*, Life Sciences, 1978, <u>23</u>, 1093) and the relationship between procoagulant activity (in blood clot formation) of PS and lipid phase transitions (G. Tans *et al.*, Europ. J. Biochem., 1979, <u>95</u>, 449) have been studied.

The enzyme PS decarboxylase, which converts PS to PE in *E. coli* (T.G. Warner and E.A. Dennis, J. biol. Chem., 1975, <u>250</u>, 8004; Arch. Biochem. Biophys., 1975, <u>167</u>, 761) and mutants of *E. coli* containing a thermolabile variety of this enzyme (which allows the accumulation of PS in the cell at 42°C)(E. Hawrot and E.P. Kennedy, Proc. nat. Acad. Sci., 1975, <u>72</u>, 1112) have been studied. Mutants of *E. coli* defective in the enzyme which forms PS from CDPDG and L-serine also have a low concentration of PE in their membranes (C.R.H. Raetz, J. biol. Chem., 1976, <u>251</u>, 3242). The formation of PS in mammalian systems by an exchange of L-serine with other phospholipids by means of an exchange enzyme (T. Taki and J.N. Kanfer, Biochim. biophys. Acta, 1978, <u>528</u>, 309; Z. Kiss, Europ. J. Biochem., 1976, <u>67</u>, 557) and the effect of morphine on this reaction (R. Natsuki *et al.*, Molecular Pharm., 1978, <u>14</u>, 448) have been studied. Proteins which transfer PS between biological membranes have also been studied (J. Baranska and Z. Grabarek, FEBS Letters 1979, <u>104</u>, 253). The binding of calcium and magnesium ions and anaesthetic amines with PS in 'liposomes' has been extensively studied by physical methods (R.C. MacDonald *et al.*, Biochemistry, 1976, <u>15</u>, 885; H. Hauser *et al.*, Biochim. biophys. Acta, 1975, <u>413</u>, 341; Europ. J. Biochem., 1976, <u>62</u>, 335; A. Portis *et al.*, Biochemistry, 1979, <u>18</u>, 780; J.S. Puskin and T. Martin, Molecular Pharm., 1978, <u>14</u>, 454;

444

R.J. Kurland *et al.*, Biochem. Biophys. Res. Comm., 1979, 88, 927; S.K. Hark and J.T. Ho, *ibid.*, 1979, 91, 665) and ^{2}H- and ^{31}P-n.m.r. spectra of bilayers of PS have been investigated (J.L. Browning and J. Seelig, Biochemistry, 1980, 19, 1262).

Phosphatidyl threonine is a major phospholipid of polyoma virus transformed cells (D. Mark-Malchoff *et al.*, Biochemistry, 1978, 17, 2684) and racemic phosphatidyl carnitine (shown below) has been synthesised (U. Hintze and G. Gerchen, Lipids, 1975, 10, 20).

$$RCOOCH_2CH(OCOR)\cdot CH_2O\cdot P(O)\cdot O\cdot CH(CH_2CO_2H)CH_2\overset{+}{N}Me_3$$
$$O^-$$

Phosphatidyl Inositol (PI)

Glyceryl phosphoryl inositol (GPI) accumulates in the culture medium of *Saccharomyces cerevisiae* and other yeasts (W.W. Angus and R.L. Lester, J. biol. Chem., 1975, 250, 22). An enzyme present in the outer cortex of rat kidney hydrolyses GPI to glycerol, inositol phosphate and inositol cyclic phosphate (R.M.C. Dawson and N. Hemington, Biochem. J., 1977, 162, 241) and a phosphodiesterase from rat tissue which hydrolyses GPI to glycerol phosphate and inositol has also been investigated (R.M.C. Dawson *et al.*, *ibid.*, 1979, 182, 39).

The *myo*-inositol-containing lipids (J.N. Hawthorne and D.A. White, Vitamins and Hormones, 1975, 33, 529) and various aspects of the biochemistry of PI - particularly the turnover of PI in stimulated cells - ('Cyclitols and Phosphoinositides', ed., W.W. Wells and F. Eisenberg, Academic Press, New York, 1978; J.W. Putney, Life Sciences, 1981, 29, 1183) have been reviewed. Russian workers have also reviewed their synthetic work on phosphoinositides (A.E. Stepanov and V.I. Shvets, Chem. Phys. Lipids, 1979, 25, 247).

HPTLC and fluorimetry have been used for the analysis of polyunsaturated PI in the picomole range (C.A. Harrington *et al.*, Anal. Biochem., 1980, 106, 307). New syntheses of PI have been reported {A.I. Lyutik *et al.*, Bioorg. Khim., 1975, 1, 684 (526); V.N. Krylova *et al.*, Zhur. org. Khim., 1979, 15, 2323; V.A. Sukhanov *et al.*, Zhur. obshchei Khim.,

1977, <u>47</u>, 2130 (1943))} and spin-labelled PI's have also been
prepared {V.A. Sukhanov *et al.*, Bioorg. Khim., 1977, <u>3</u>, 135
(103) ; V.I. Shvets *et al.*, Chem. Phys. Lipids, 1979, <u>23</u>,
163 }. Analogues of PI containing *muco*-inositol {V.P.
Schevchenko *et al.*, Bioorg. Khim., 1976, <u>2</u>, 923 (668)} ,
scyllo-inositol {V.P. Schevchenko *et al.*, *ibid.*, 1977, <u>3</u>,
252 (188)} and cyclopentanol derivatives (H.Z. Sable, *et al.*,
in 'Cyclitols and Phosphoinositides', p.3) instead of *myo*-
inositol have been synthesised.

The enzyme catalysing the reaction between CDPDG and inositol
to give PI has been purified from rat liver microsomes and
its properties studied (T. Takenawa and K. Egawa, J. biol.
Chem., 1977, <u>252</u>, 5419). An isosteric analogue of CDPDG
(shown below) has been synthesised (A.F. Rosenthal and L.A.
Vargas, J. chem. Soc. Chem. Commun., 1981, 976) to
investigate its properties in enzymic reactions of this kind.

The biosynthesis of PI in brain has been studied and the
results suggest that a transacylation of the initially
synthesised PI may account for the final fatty acid distribution in this phospholipid (G. MacDonald *et al.*, J.
Neurochem., 1975, <u>24</u>, 655). The acylation of a lyso-PI
(2-*O*-acylglycerol phosphoryl inositol) in rat liver microsomes shows a preference for stearic acid over palmitic acid
(B.J. Holub and J. Piekarski, Lipids, 1979, <u>14</u>, 529). A
phospholipase which specifically hydrolyses PI (to diglyceride
and inositol phosphate), but not other phospholipids, is
present in various bacteria (M.G. Low and J.B. Finean,
Biochim. biophys. Acta, 1978, <u>508</u>, 565; T. Ohyabu *et al.*,
Arch. Biochem. Biophys., 1978, <u>190</u>, 1; R. Sundler *et al.*,
J. biol. Chem., 1978, <u>253</u>, 4175). The hydrolysis of PI in
mammalian systems is predominantly *via* the diglyceride and
inositol phosphate route (see R.F. Irvine *et al.*, Biochem.
J., 1978, <u>176</u>, 475; 1979, <u>178</u>, 497; Biochem. Soc. Trans.,
1979, 7, 353). The exchange of PI between different biological membranes *via* a specific PI exchange protein has

been studied in detail (see A.D. MacNicoll *et al.*, Biochem.
J., 1978, 174, 421; P. Ruenwongsa *et al.*, J. biol. Chem.,
1979, 254, 9385).

Antibodies to PI have been prepared and used to detect PI in
brain membranes. The antibodies also react with cardiolipin
in mitochondrial membranes (M. Guarnieri, Lipids, 1975, 10,
294). Intensive studies are taking place on the stimulated
turnover of PI in membranes (see L.M. Jones *et al.*, Biochem.
J., 1979, 182, 669; J.N. Hawthorne *et al.*, Biochem. Soc.
Trans., 1977, 5, 514; 1979, 7, 348; Biochem. Pharmacol.,
1979, 28, 1143; M.E.M. Tolbert *et al.*, J. biol. Chem., 1980,
255, 1938; R.L. Bell and P.W. Majerus, *ibid.*, 1980, 255,
1790; R.H. Michell *et al.*, Biochem. Soc. Trans., 1979, 7,
861).

Polyphosphoinositides (Phosphatidyl Inositol Phosphates).

The mono- and diphosphate esters of PI have been purified by
chromatography of total lipid extracts on a column of the
amino-glycoside antibiotic, neomycin coupled to glass beads
(J. Schacht, J. lipid Res., 1978, 19, 1063). The chemical
synthesis of PI monophosphate {M.S. Sadovnikova *et al.*,
Bioorg. Khim., 1975, 1, 17 (9); Zhur. obshchei Khim., 1976,
46, 1389 (1365); V.P. Shevchenko *et al.*, Chem. Phys. Lipids,
1975, 15, 95} and its preparation by the enzymic hydrolysis
of PI diphosphate (F.B. St. C. Palmer, Preparative Biochem.,
1977, 7, 457) have been described and the enzymic synthesis
has been investigated in detail (Y.A. Lefebvre *et al.*, Canad.
J. Biochem., 1976, 54, 746; J.T. Buckley, *ibid.*, 1976, 54,
772; P.H. Cooper and J.N. Hawthorne, Biochem. J., 1976, 160,
97). A mixture of isomers of PI diphosphate has been synthesised (V.N. Krylova *et al.*, Zhur. obshchei Khim., 1981,
51, 210) and the biosynthesis studied (P.H. Cooper and J.N.
Hawthorne, Biochem. J., 1976, 160, 97; D.S. Deshmukh *et al.*,
J. Neurochem., 1978, 30, 1191). The very rapid enzymic
hydrolysis of PI phosphates in erythrocyte membranes and
methods for inhibiting the degradative enzymes (N.E. Garrett
et al., J. Cellular Physiol., 1976, 87, 63) and an enzyme
from the protozoan *Crithidia fasciculata* which hydrolyses
PI diphosphate but not the monophosphate (F.B. St. C. Palmer,
Biochim. biophys. Acta, 1976, 441, 477) have been studied.
The metabolism of PI phosphates is stimulated by cholinergic
drugs (J.F. Soukup *et al.*, Biochem. Pharmacol., 1978, 27,
1239; R.J. Hitzemann *et al.*, *ibid.*, 1978, 27, 2519; R.A.

Akhtar *et al.*, J. Pharm. exp. Therapeutics, 1978, 204, 655).
Antibodies prepared against PI diphosphate also react with
PI monophosphate (A.J. Greenberg *et al.*, Molecular Immunol.,
1979, 16, 193). The U.V. light-induced peroxidation of PI
diphosphate to a 'lysotriphosphoinositide' has been investigated (F. Hayashi *et al.*, Biochim. biophys. Acta, 1978, 510,
305).

Phosphatidyl Inositol Mannosides

Mono- and diacyl hexamannosides, a monoacyl tetramannoside,
a monoacyl trimannoside and mono- and diacyl dimannosides
of PI have been isolated by t.l.c. from *Mycobacterium tuberculosis* and the serological activities of these compounds
investigated (A. Sasaki, J. Biochem., 1975, 78, 547; B.
Banerjee and D. Subrahmanyan, Immunochem., 1978, 15, 359).
The mannophosphoinositides of *Corynebacterium aquaticum*
(J.A. Hackett and P.J. Brennan, Biochem. J., 1975, 148, 253),
Nocardia sp. (G.K. Khuller, Experientia, 1976, 32, 1371;
1977, 33, 1277; Ind. J. med. Res., 1977, 65, 657; 1978,
67, 734) and *Mycobacterium phlei* (K.R. Dhariwal *et al.*,
Experientia, 1975, 31, 776) have been studied and a t.l.c.
system for the separation of the deacylated derivatives has
been reported (G.K. Khuller and B. Banerjee, J. Chromatog.,
1978, 150, 518).

The monomannoside of PI has been synthesised {A.E. Stepanov
et al., Zhur. obshchei Khim., 1977, 47, 1653 (1515); Zhur.
org. Khim., 1977, 13, 1410 (1295); A.P. Kaplun *et al.*, *ibid.*,
1979, 15, 307 (266)}

Phosphatidyl Glycerol (PG)

The phosphonate analogue ('phosphotidyl glycerol' - shown
below) of PG has been synthesised from 3,4-dipalmitoxybutyl
1-phosphonic acid (D. Braksmayer *et al.*, Chem. Phys. Lipids,
1977, 19, 93).

$$RCOOCH_2CH(OCOR)\cdot CH_2CH_2\underset{\underset{O^-}{|}}{P}(O)\cdot O\cdot CH_2CH(OH)\cdot CH_2OH$$

3,4-Dihydroxybutyl 1-phosphonate is converted into the phosphonate analogue of PG phosphate (shown below) by *E. coli*
(R.J. Tyhach *et al.*, J. biol. Chem., 1976, 251, 6717).

$$\text{RCOOCH}_2\text{CH(OCOR)}\cdot\text{CH}_2\text{O}\cdot\overset{|}{\underset{\underset{\text{O}^-}{|}}{\text{P(O)}}}\cdot\text{O}\cdot\text{CH}_2\text{CH(OH)}\cdot\text{CH}_2\text{CH}_2\overset{\text{O}}{\underset{||}{\text{P}}}\text{(OH)}_2$$

The function of PG in lung surfactant (see G. Okano and T. Akino, Lipids, 1979, 14, 541) and its biosynthesis in lung from dihydroxyacetone phosphate has been studied (R.J. Mason, J. biol. Chem., 1978, 253, 3367).   The PG of *Acholeplasma laidlawii* has been extensively studied (see E.M. Bevers *et al.*, Biochim. biophys. Acta, 1978, 511, 509) and in the PG of the cholesterol-requiring *Mycoplasma sp.* the fatty acid distribution is unusual in that saturated fatty acids are predominantly on the 2-position of glycerol and unsaturated acids are on the 1-position (S. Rottem and O. Markowitz, FEBS Letters, 1979, 107, 379).   PG is the major phospholipid of the membrane of the bacteriophage PM2 (R.T. Ruettinger and G.J. Brewer, Biochim. biophys. Acta, 1978, 529, 181).

The enzymes in *Bacillus licheniformis* involved in the synthesis of PG from *sn*-glycerol 3-phosphate and CDPDG have been studied using affinity chromatography on a Sepharose column containing bound CDPDG (T.J. Larson *et al.*, Biochemistry, 1976, 15, 974).   PG synthesised from PC and glycerol by 'trans-phosphatidylation' in the presence of phospholipase D is a mixture of two isomers - enantiomeric in the free glycerol moiety (A. Joutti and O. Renkonen, Chem. Phys. Lipids, 1976, 17, 264).   The thermotropic behaviour of bilayers of PG and the influence of calcium and magnesium ions on this (see E.J. Findlay and P.G. Barton, Biochemistry, 1978, 17, 2400) and the [1]H-n.m.r. (D. Marsh and A. Watts, FEBS Letters, 1978, 85, 124) and ^{13}C-n.m.r. (J.M. Coddington *et al.*, Austral. J. Chem., 1981, 34, 357) spectra of PG have been studied.

Amino-acid Esters of Phosphatidyl Glycerol

The two isomeric lysyl esters of didodecanoyl PG (esterified on the 2[1]-or 3[1]-position of glycerol) have been synthesised. The lysyl group migrates from the 2[1]- to the 3[1]-position on treatment with hydrogen chloride.   The lysyl derivative isolated from *Staphylococcus aureus* has been shown to be the 3[1]-isomer (J.F. Tocanne *et al.*, Chem. Phys. Lipids, 1974, 13, 389).   The 3[1]-*O*-L-alanyl derivative of didodecanoyl PG has been synthesised and the effect of pH and ions on a monolayer of the lipid studied (M.M. Sacré, Chem. Phys. Lipids, 1977, 20, 305).

Acyl Derivatives of Phosphatidyl Glycerol

Monoacyl derivatives of PG are present in some bacteria (see M. Nishijima et al., Biochim. biophys. Acta, 1978, 528, 107) and diacyl derivatives of PG ('bis-phosphatidic acids') have been isolated from soya beans and a bacterium (see W.T. Morton et al., Lipids, 1977, 12, 451, 1083) and have been synthesised from glycerylphosphites and phosphoramidites {D.A. Predvoditelev et al., Bioorg. Khim., 1977, 3, 76 (58)}. The phospholipid, lyso-bis-phosphatidic acid {'(bis-monoacylglycero) phosphate'} (shown below) which accumulates in lysosomes treated with detergents and in BHK cells (see A. Joutti and O. Renkonen, J. lipid Res., 1979, 20, 230, 840) has an unusual configuration in the glycerol phosphate residues and appears to be synthesised, in lysosomes, from cardiolipin or PG by an unknown mechanism (Y. Matsuzawa et al., J. biol. Chem., 1978, 253, 6650). By contrast the newly formed '(bis-monoacylglycero) phosphate' in BHK cells have more sn-3-glycerophosphate residues than the bulk which contains mainly sn-1-glycerophosphate residues (A. Joutti, Biochim. biophys. Acta, 1979, 575, 10).

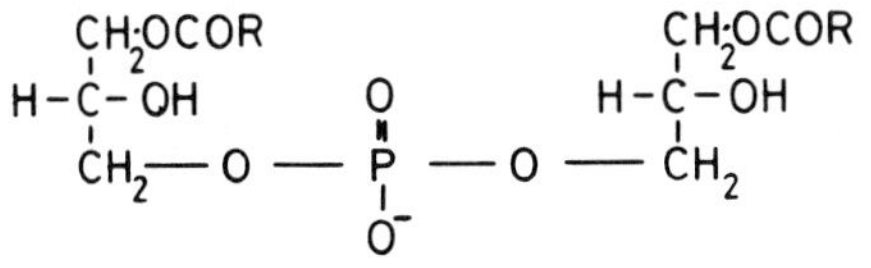

This phospholipid also accumulates in liver as a result of the ingestion of certain drugs (see Y. Matsuzawa et al., J. Biochem., 1977, 82, 1369) and its further degradation in lysosomes has also been studied (Y. Matsuzawa and K.Y. Hostetler, J. biol. Chem., 1979, 254, 5997; S. Huterer and J. Wherrett, J. lipid Res., 1979, 20, 966).

A glucosyl-derivative of PG has been isolated from moderately halophilic bacteria (see N. Stern and A. Tietz, Biochim. biophys. Acta, 1978, 530, 357; Y. Ohno et al., ibid., 1976, 424, 337) and a 2^1-O-α-glucosyl-derivative of PG has been

synthesised (C.A.A. van Boeckel and J.H. van Boom,
Tetrahedron Letters, 1979, 3561).

Cardiolipin

The chemistry and biochemistry of cardiolipins have been
reviewed (P.V. Ioannou and B.T. Golding, Progr. lipid Res.,
1979, 17, 279). Spin-labelled derivatives of cardiolipin
have been synthesised by incorporating a spin-labelled fatty
acid (M.B. Cable *et al.*, Proc. nat. Acad. Sci., 1978, 75,
1227) or by acylating the free hydroxy-group with a spin-
labelled acid (C. Landriscina *et al.*, Anal. Biochem., 1976,
76, 292). A derivative of cardiolipin containing an
α-glucosyl residue on the free hydroxy-group is present in
group B *Streptococci* (W. Fischer, Biochim. biophys. Acta,
1977, 487, 74, 89).

The biosynthesis of cardiolipin in rat liver mitochondria
from CDPDG and PG (K.Y. Hostetler *et al.*, *ibid.*, 1975, 380,
382) and in bacteria from two molecules of PG (see G.
Pluschke *et al.*, J. biol. Chem., 1978, 253, 5048; A.J.
De Siervo, Canad. J. Biochem., 1975, 53, 1031) has been
studied. The lysis of 'liposomes' containing cardiolipin
by antibodies to *Treponema pallidum* (the organism causing
syphilis) has been studied in relation to the Wasserman test
for syphilis (E. Rosenqvist and A.I. Vistnes, J. immunol.
Methods, 1977, 15, 147). ^{3}H-Labelled cardiolipin has been
prepared by oxidation of the free hydroxy-group and subsequent reduction with ^{3}H-sodium borohydride (E. Viola *et al.*,
Chem. Phys. Lipids, 1979, 25, 93).

Glycosyl Derivatives of Diacylglycerols

The bacterial glycolipids of this type have been reviewed
(N. Shaw in 'Lipids', Vol.1, Biochemistry, ed., R. Paoletti,
G. Porcellati and G. Jacini,Raven Press, New York, 1976,
p.253; Adv. microbial Physiol., 1975, 12, 141).

Galactosyl Diglycerides
The chemistry and biochemistry of these lipids have been
reviewed (H.C. van Hummell, Progr. Chem. org. Nat. Products,
1975, 32, 267). Di- and tri- galactosyl diglycerides (P.A.
Gent and R. Gigg, J. chem. Soc. Perkin I, 1975, 1521, 1779)
have been synthesised and a tetragalactosyl diglyceride has
been detected in plants (B.J.F. Hudson, Chem. Ind., 1976,

763; Y. Fujino and T. Miyazawa, Biochim. biophys. Acta, 1979,
572, 442).   The ^{13}C-n.m.r. spectra of mono- and digalactosyl
diglycerides (S.R. Johns *et al.*, Austral. J. Chem., 1977,
30, 823) and the fatty acid distribution in galactosyl diglycerides from plants (J. Rullkotter *et al.*, Z.
Pflanzenphysiol., 1975, 76, 163) and blue-green algae (H.D.
Zepke *et al.*, Arch. Microbiol., 1978, 119, 157) have been
investigated.   The biosynthesis of the galactolipids of
Bifidobacterium bifidum var. penn. has been studied (J.H.
Veerkamp, Biochim. biophys. Acta, 1976, 441, 403).

Glucosyl Diglycerides
A monoglucosyl diglyceride {3-*O*-(α-D-glucopyranosyl)-1,2-di-
O-hexadecanoyl-*sn*-glycerol} has been synthesised (R. Gigg *et
al.*, J. chem. Soc. Perkin I, 1977, 2014) and the phosphorylated derivatives of glucosyl diglycerides which occur in
Streptococci have been extensively investigated and their
relationship to the lipoteichoic acids determined (W.
Fischer in 'Lipids', Vol.1. Biochemistry, ed., R. Paoletti,
G. Porcellati and G. Jacini, Raven Press, New York, 1976,
p.255; R.A. Laine and W. Fischer, Biochim., biophys. Acta,
1978, 529, 250; W. Fischer *et al.*, Chem. Phys. Lipids, 1978,
21, 103).   A phosphatidyl α-glucosyl-diacyl-glycerol ('A'-
below) (C.A.A. van Boeckel and J.H. van Boom, Tetrahedron
Letters, 1980, 21, 3705) and a phosphatidyl α-diglucosyl
diacylglycerol ('B'- below) (C.A.A. van Boeckel and J.H.
van Boom, Chem. Letters, 1981, 581) have been synthesised.

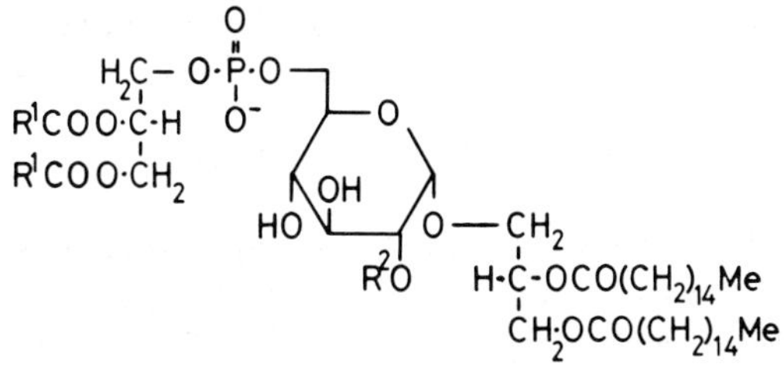

'A' R^1 = -(CH$_2$)$_7$CH=CH(CH$_2$)$_7$Me ; R^2=H   'B' R^1= -(CH$_2$)$_{16}$Me ; R^2= α-glucosyl

The glucose- and galactose-containing glycerophosphoglyco-
lipids of Gram-positive bacteria have also been studied in
detail (W. Fischer *et al.*, Biochim. biophys. Acta, 1978, 528,
298).   The glycolipid shown below is a major lipid of
Streptococcus lactis (W. Fischer *et al.*, *ibid.*, 1979, 575,
389).

gal(1α6) gal(1α3)⎤ sn-glycerol ⎤
gal(1α2)⎦ -1-phosphate⎦ (6)-gluc(1α2) 6-O-acyl (1α3) sn-
-gluc diglyceride

The glycolipid shown below, and more complex phosphoglyco-
lipids have been isolated from *Acholeplasma granularum* (P.F.
Smith *et al.*, *ibid.*, 1980, <u>617</u>, 419; 1981, <u>665</u>, 92).

gluc(1 β 3) gluc(1α2)gluc(1α1)diglyceride

Other Glycosyl Diglycerides

The biosynthesis of the glucuronosyl diglyceride of *Pseudomonas diminuta* has been studied (J.M. Shaw and R.A. Pieringer,
J. biol. Chem., 1977, <u>252</u>, 4391) and the structures of the
uronic-acid-containing glycolipids of *Streptomyces* have been
reviewed (S.G. Batrakov and L.D. Bergelson, Chem. Phys.
Lipids, 1978, <u>21</u>, 1). A heptosyl diacylglycerol (1,2-di-
O-acyl-3-O-α-D-*glycero*-D-*gluco*-heptopyranosyl glycerol) is
present in *Pseudomonas vesicularis* (S.G. Wilkinson and L.
Galbraith, Biochim. biophys. Acta, 1979, <u>575</u>, 244) and the
structure of the mannoheptose-containing pentaglycosyl diacylglycerol from *Acholeplasma modicum* has been established
as gal (1α2) gal (1α3)-D-*glycero*-D-*manno*-heptopyranose (1β3)
gluc (1α2) gluc (1α1) diglyceride (W.R. Mayberry *et al.*,
ibid., 1976, <u>441</u>, 115). The galactofuranose-containing
tetraglycosyl diglyceride from *Thermus thermophilus* ex
Flavobacterium thermophilum has been identified as gal(f)
(1β2) gal (1α6) *N*-15-methylhexadecanoyl glucosamine (1β2)
gluc (1α3)-*sn*-diglyceride (M. Oshima and T. Ariga FEBS
Letters, 1976, <u>64</u>, 440). Two other glucosamine-containing
glycolipids have been isolated from *Bacillus acidocaldarius*
(T.A. Langworthy *et al.*, Biochim. biophys. Acta, 1976, <u>431</u>,
550).

Plant Sulpholipid, 'Sulphoquinovosyl Diglyceride'

The biochemistry of this lipid has been reviewed (J.L.
Harwood and R.G. Nichols, Biochem. Soc. Trans., 1979, 7, 440).
The ^{13}C-n.m.r. spectrum (S.R. Johns *et al.*, Austral. J. Chem.
1978, 31, 65) and the fatty acid distribution in the lipid
from blue-green algae (H.D. Zepke *et al.*, Arch. Microbiol.,
1978, 119, 157) and plants (H.P. Siebertz *et al.*, Europ. J.
Biochem., 1979, 101, 429; M. Nishihara *et al.*, Biochim.
biophys. Acta, 1980, 617, 12) have been studied.   A
saturated sulphoquinovosyl diglyceride has been synthesised
(R. Gigg *et al.*, J. chem. Soc. Perkin I, 1980, 2490).

Plasmalogens

The chemistry and biochemistry of the plasmalogens have been
reviewed (H.K. Mangold, Angew. Chem. internat. Edn., 1979,
18, 493).   Plasmalogenic forms of PC and PE have been
prepared from the natural mixtures by hydrolysis of the
primary acyl group of the diacyl lipids with the lipase from
Rhizopus arrhizus and subsequent chromatography.   Some
alkyl-acyl-derivatives were still present as contaminants
(F. Paltauf, Lipids, 1978, 13, 165).   The molecular species
of PC in various tissues have been investigated by hydrolysis
with phospholipase C to give diglycerides and their analogues.
These are separated by chromatography into alkylacyl, alkenyl-
acyl and diacyl types which are separated into classes by
g.l.c. - mass spectrometry of their trimethylsilyl ethers
(T. Curstedt, Biochim. biophys. Acta, 1977, 489, 79).   The
quantification of the various species of PC in bovine heart
(diacyl-57%; alkyl-acyl-3%; alkenyl-acyl-39%; dialkyl-
0.2%) has also been achieved by hydrolysis and chromatography
(E.L. Pugh *et al.*, J. lipid Res., 1977, 18, 710) and similar
results were obtained for the PC and PE from rabbit heart
(A. Osanai and T. Sakagami, J. Biochem., 1979, 85, 1453).
Ether lipids have also been determined by phospholipase C
hydrolysis with subsequent saponification and periodate
oxidation of the 1-*O*-alkylglycerols (M.L. Blank *et al.*, Biochim. biophys. Acta, 1975, 380, 208).   The mass-spectra of
ether lipids {R.P. Evstigneeva *et al.*, Bioorg. Khim., 1977,
3, 83 (63), 1371 (1004); K. Satouchi and K. Saito, Biomed.
mass spectrometry, 1977, 4, 107 } and the chromatography of
plasmalogens and their degradation products (M.H. Hack and
F.M. Helmy, J. Chromatog., 1975, 107, 155; 1976, 128, 239;
1977, 135, 229; 1978, 151, 432;   V.E. Vaskovsky and V.M.

Dembitzky, *ibid.*, 1975, 115, 645) have been investigated.

The synthesis of plasmalogens has been reviewed (A.F.
Rosenthal, Methods in Enzymology, 1975, 35, 447; H.K.
Mangold, Angew. Chem. internat. Edn., 1979, 18, 493).
Phosphatidal choline and ethanolamine {A.V. Chebyshev *et al.*,
Bioorg. Khim., 1977, 3, 1362 (997)} and phosphatidal serine
{I.A. Vasilenko *et al.*, *ibid.*, 1976, 2, 1138 (823); see
also A.V. Chebyshev *et al.*, *ibid.*, 1979, 5, 628 (469), 918
for the synthesis of 1-O-alk-1'-enylglycerols} as well as
alk-1-enyl ethers of cholesterol (V.N. Klykov *et al.*, *ibid.*,
1979, 5, 918) have been synthesised. 1-Alkenyl ethers of
1-deoxy-1-thioglycerol have also been prepared (A.F. Hirsch,
Chem. Phys. Lipids, 1976, 17, 399).

Anaerobic *Mycoplasma* contain large quantities of the plasmalogenic form of PG (T.A. Langworthy *et al.*, J. Bacteriol,
1975, 122, 785) and the plasmalogens of anaerobic bacteria
have been reviewed (H. Goldfine in 'Lipids' Vol. 1., Biochemistry, ed. R. Paoletti, G. Porcellati and G. Jacini,
Raven Press, New York, 1976, p.11; see also N.G. Clarke *et*
al., Chem. Phys. Lipids, 1976, 17, 222). A complex lipid
containing a plasmalogenic glycolipid linked to a plasmalogenic phospholipid by esterification of the 2-hydroxy-
groups of the glycerol moieties with diabolic acid (29-
carboxy-15,16-dimethyl-nonacosanoic acid) has been
isolated from *Butyrivibrio sp.* (N.G. Clarke *et al.*, Biochem.
J., 1980, 191, 561).

The biosynthesis of the plasmalogens by desaturation of 1-
O-alkylglycerols has been reviewed (H.K. Mangold, Angew.
Chem. internat. Edn., 1979, 18, 493; R.L. Wykle and J.M.
Schremmer, Biochemistry, 1979, 18, 3512). The brain enzyme,
'plasmalogenase', responsible for the hydrolysis of the
vinyl ether group of plasmalogens, has been investigated in
connection with the defect in the neurological mutants,
'quaking' and 'jimpy' mice (R.V. Dorman *et al.*, J. Neurochem.,
1978, 30, 157; see also M.H. Hack and F.M. Helmy, J.
Chromatog., 1978, 145, 307).

Saturated Ether Lipids

The chemistry and biochemistry of these lipids have been
reviewed (H.K. Mangold, Angew. Chem. internat. Edn., 1979,
18, 493; see also ed. H.K. Mangold and F. Paltauf, 'Ether

Lipids: Biomedical applications', Academic Press, New York, 1981). 1-O-Alkyl-2-O-acyl-sn-glycerol 3-phosphoryl-ethanolamine and choline have been synthesised {A.E. Rozin et al., Bioorg. Khim., 1976, 2, 78(60); A.V. Chebyshev et al., ibid., 1979, 5, 628} and ³H-labelled 1-O-alkyl-2-O-acyl-sn-glycerol 3-phosphorylethanolamine has been synthesised for the investigation of plasmalogen biosynthesis (F. Paltauf, Chem. Phys. Lipids, 1976, 17, 148).

1-O-Alkyl-2-O-acetyl-sn-glycerol 3-phosphoryl choline has been shown to be a blood-platelet activating factor (PAF) (see K. Satouchi et al., J. biol. Chem., 1981, 256, 4425; Arch. biochem. Biophys., 1981, 211, 683; J.O. Shaw et al., Biochim. biophys. Acta, 1981, 663, 222; for reviews see R.N. Pinckard et al., J. Reticuloendothelial Soc., 1980, 28, 955; J.T. O'Flaherty et al., Amer. J. Pathol., 1981, 103, 70). Other derivatives of 1-O-alkyl-sn-glycerol 3-phosphorylcholine (e.g. the 2-O-benzyl-, 2-O-methyl-, 2-O-ethyl- and 2-deoxy-derivatives) have been investigated for their cytotoxic effects (D.J. Hanahan et al., Biochem. biophys. Res. Comm., 1981, 99, 183; R.L. Wykle et al., ibid., 1981, 100, 1651; H.U. Weltzein et al., Biochim. biophys. Acta, 1978, 530, 47; 1977, 466, 411; Chem. Phys. Lipids, 1976, 16, 267). The 1-O-alkyl-2,3-di-O-acylglycerols of the liver oil of the ratfish (Hydrolagus colliei) have been used as starting material for the synthesis of 'PAF' (T. Muramatsu et al., ibid., 1981, 29, 121).

Diether analogues of PC have been prepared from the glycerol diether and phosphorus oxychloride in the presence of choline toluene p-sulphonate (H. Brockerhoff and N.K.N. Ayengar, Lipids, 1979, 14, 88) and a phosphonate, diether analogue of PC (shown below) has been prepared for comparison with the pulmonary surfactant, dipalmitoyl lecithin (J.G. Turcotte et al., Biochim. biophys. Acta, 1977, 488, 235).

$$Me(CH_2)_{15}O \cdot CH_2CH[O(CH_2)_{15}Me] \cdot CH_2O \cdot P(O) \cdot CH_2CH_2CH_2\overset{+}{N}Me_3$$
$$\underset{O^-}{\quad}$$

The Raman spectra of the diether analogue of PC (S. Sunder et al., Chem. Phys. Lipids, 1978, 22, 279), the mass-spectra of alkoxy-lipids {A.E. Rozin et al., Bioorg. Khim., 1977, 3, 393 (297); 397 (300)} and t.l.c. of alkoxy-lipids (H.J. Liu and C.L.H. Lee, J. Chromatog., 1978, 147, 524) have been

studied.   The synthesis and mass-spectra of S-alkyl 1-deoxy-
1-thioglycerols (W.J. Ferrell *et al.*, Chem. Phys. Lipids,
1976, 16, 276; M.L. Blank *et al.*, *ibid.*, 1976, 17, 201) and
the autoxidation of alkoxy-lipids (N. Yanishlieva and H.K.
Mangold, *ibid.*, 1977, 20, 21) have been investigated.

The saturated ether lipids are synthesised biologically by
an exchange of a long-chain alcohol with the acyl-group of
1-O-acyl dihydroxyacetone phosphate and the mechanism and
stereochemistry of this reaction have been studied (see P.A.
Davis and A.K. Hajra, J. biol. Chem., 1979, 254, 4760; S.J.
Friedberg *et al.*, *ibid.*, 1980, 255, 1074).   The product is
reduced to 1-O-alkyl-sn-glycerol 3-phosphate which is converted into the alkyl phospholipids *via* the alkyl-acyl-glycerols
(see A. Radominska-Pyrek *et al.*, J. Lipid Res., 1977, 18,
53;   G. Goracci *et al.*, FEBS Letters, 1977, 80, 41).

The diether lipids of halophilic bacteria
These compounds have been reviewed (M. Kates in 'Lipids',
Vol. 1, Biochemistry, ed., R. Paoletti, G. Porcellati and G.
Jacini, Raven Press, New York, 1976, p.267; M. Kates and
S.C. Kushwaha in 'Energetics and Structure of the Halophilic
Microorganisms', ed. S.R. Caplan and M. Ginzburg, Elsevier-
North Holland, 1978, p.461).   Similar diphytanylglycerol
ether lipids are present in methanogenic and thermoacidophil-
ic bacteria (see T.G. Tornabene *et al.*, Science, 1979, 203,
51;   J. molecular Evolution, 1978, 11, 259).   Monophytanyl
ether analogues of lysophosphatidic acids and lysophospha-
tidyl glycerol (M. Kates and A.J. Hancock, Chem. Phys.
Lipids, 1976, 17, 155), 2,3-di-O-phytanyl-sn-glycerol 1-
phosphoryl choline (V.A. Vaver *et al.*, Bioorg. Khim., 1980,
6, 146) and gal (1β6) mann (1α2) gluc (1α1) 2,3-di-O-
phytanyl-sn-glycerol (C.A.A. van Boeckel *et al.*, Tetrahedron
Letters, 1981, 22, 2819) have been synthesised.   Related
glycolipids are present in *Halobacterium marismortui* which
grows in the Dead Sea (R.W. Evans *et al.*, Biochim. biophys.
Acta, 1980, 619, 533).   The ^{1}H- and ^{13}C-n.m.r. spectra of
the polar lipids of *Halobacterium halobium* have been recorded (H. Degani *et al.*, Biochemistry, 1980, 19, 626).   The
methanogenic bacterium *Methanospirillum hungatei* contains a
series of phospho- and glycolipids derived from the 'diphyt-
anyldiglycerol tetraether' shown below ($R^1=R^2=H$).   In these
lipids, sugars are glycosidically linked at R^1 and sn-
glycerol 3-phosphate is esterified at R^2 (S.C. Kushwaha *et
al.*, Science, 1981, 211, 1163; Biochim. biophys. Acta, 1981,

664, 156). The lipopolysaccharide from *Thermoplasma acido-phi̲lum* contains a polyglycosyl derivative of the 'diphytanyl-glycerol tetraether' (P.F. Smith, *ibid.*, 1980, 619, 367).

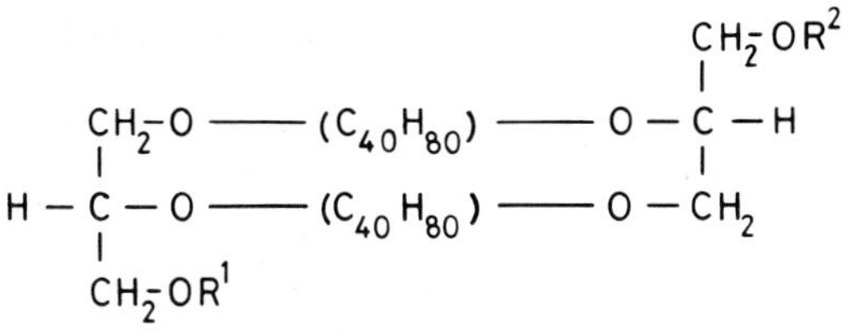

$$CH_2\text{-}O \underline{\quad\quad} (C_{40}H_{80}) \underline{\quad\quad} O\text{-}\overset{\displaystyle CH_2\text{-}OR^2}{\underset{\displaystyle O\text{-}CH_2}{C}}\text{-}H$$

$$H\text{-}\overset{\displaystyle CH_2\text{-}O}{\underset{\displaystyle CH_2\text{-}OR^1}{C}}\text{-}O \underline{\quad\quad} (C_{40}H_{80}) \underline{\quad\quad}$$

Seminolipid

The structure of seminolipid has been established as shown below (K. Ueno *et al.*, Biochim. biophys. Acta, 1977, 487, 61) and this structure has been confirmed by synthesis (R. Gigg, J. chem. Soc. Perkin I, 1979, 712).

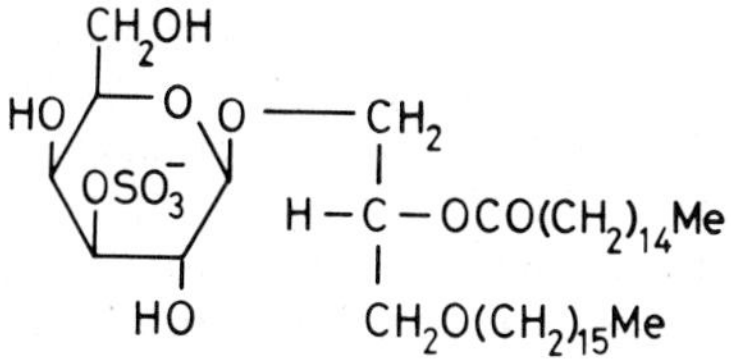

$$\text{HO}\overset{\displaystyle CH_2OH}{\underset{\displaystyle \underset{HO}{OSO_3^-}}{\bigcirc}}O\text{-}CH_2$$
$$H\text{-}C\text{-}OCO(CH_2)_{14}Me$$
$$CH_2O(CH_2)_{15}Me$$

The biochemistry of this lipid has been reviewed (R.K. Murray *et al.*, in 'Glycolipid Methodology', ed. L.A. Whitting, Amer. oil Chem. Soc., 1976, p.305; ACS Symposium Series, No. 128, 1980, p.105). HPLC has been used for the analysis of seminolipid (A. Suzuki *et al.*, J. Biochem., 1977, 82, 461) and it appears to be present only in testis and brain (C. Lingwood *et al.*, Canad. J. Biochem., 1981, 59, 556). The biosynthesis (M.J. Kornblatt, *ibid.*, 1979, 57, 255) and degradation (S. Reiter *et al.*, FEBS Letters, 1976, 68, 250; G. Fischer *et al.*, Z. physiol. Chem., 1978, 359, 863) of seminolipid have been studied. Seminolipid is located on the plasma membrane of rat testicular germinal cells (A. Klugerman and M.J. Kornblatt, Canad. J. Biochem., 1980, 58, 225). A similar lipid is present in brain where it occurs together with the corresponding diacyl-compound (see I. Ishizuka and M. Inomata, J. Neurochem., 1979, 33, 387; I. Ishizuka *et al.*, J. biol. Chem., 1978, 253, 898; J. Pieringer *et al.*, Biochem. J., 1977, 166, 421, 429).

A gluc-6-sulphate (1α6) gluc (1α6) gluc (1α3)-1-*O*-alkyl-2-*O*-acyl-*sn*-glycerol is a component of human gastric content (B.L. Slomiany *et al.*, Europ. J. Biochem., 1977, 78,33) and the corresponding tetraglucosyl glyceride (sulphated on the 6-position of the terminal glucose) is a major glycolipid of rabbit alveolar lavage (A. Slomiany *et al.*, *ibid.*, 1979, 98, 47).   Hexa- and octaglucosyl derivatives of alkyl-acyl-glycerols are present in human gastric content and saliva (B.L. Slomiany *et al.*, Biochemistry, 1977, 16, 3954;   Europ. J. Biochem., 1978, 84, 53).   Ether analogues of glucosyl-galactosyl diglycerides are also present in *Sulfolobus acidocaldarius* (T.A. Langworthy, J. Bacteriol., 1977, 130, 1326) and other oligoglycosyl derivatives of glycerol ethers are present in the mung bean *Phaseolus mungo* (Y. Kondo, Biochim. biophys. Acta, 1981, 665, 471).   An ether analogue of a monogalactosyl diglyceride has been synthesised (E. Heinz *et al.*, Chem. Phys. Lipids, 1979, 24, 265).

Phospholipids and Glycolipids derived from Sphingosine Bases

A terpenoid sphingosine base, 'apliadiasphingosine', which has antimicrobial and antitumour activity, with the structure shown below, has been isolated from a marine tunicate (G.T. Carter and K.L. Rinehart, J. Amer. chem. Soc., 1978, 100, 7441).

$$Me_2C=CH\cdot CH_2CH(OH)CHMe(CH_2)_3CMe=CH(CH_2)_2CHMeCH_2CH(OH)CH(NH_2)CH_2OH$$

Docosa-4,15-sphingadienine and 4-hydroxy-docosa-15-sphingenine have been isolated from the sphingophosphonolipids of *Turbo cornutus* (A. Hayashi *et al.*, Chem. Phys. Lipids, 1975, 14, 102) and other dienic bases are present in shell fish, (T. Matsubara, *ibid.*, 1975, 14, 247).   Branched-chain, long-chain bases (particularly *iso*-octadec-4-sphingenine and *anteiso*-nonadec-4-sphingenine) are present in the bivalve *Corbicula sandai* (M. Sugita *et al.*, *ibid.*, 1976, 16, 1). Dihydrosphingosine has been synthesised by the condensation of (S)-3-acetoxy-2-phthalimido-propanal with pentadecyl magnesium bromide and subsequent hydrazinolysis (Y. Saitoh *et al.*, Bull. chem. Soc., 1980, 53, 1783) and a synthesis of a mixture of isomers of methyl 2-benzamido-3-hydroxy-hex-4-enoate (a suitable intermediate for the preparation of the

isomers of the C6-homologue of sphingosine) by the hydroxy-
alkylation of the lithio dianion of methyl hippurate with
crotonaldehyde has been described (K.E. Harding *et al.*, J.
org. Chem., 1981, 46, 2809). Sphingosine (and its isomers)
have been synthesised by resolution of DL-*erythro*-or DL-
threo-ethyl-2-acetamido-3-hydroxy-octadec-*trans*-4-enoate
with the ester formed from L-(+)-acetyl-mandeloyl chloride
and separation of the diastereoisomers by chromatography.
The resolved esters,on hydrolysis to remove the *N*-acetyl
group, and reduction with ^{3}H-lithium aluminium hydride give
labelled bases (Y. Shoyama *et al.*, J. lipid Res., 1978, 19,
250).

The synthesis of phytosphingosine by the epoxidation of
sphingosine and reduction of the epoxide has been reinvestigated and shown to give a mixture of isomers (R.J. Kulmacz
et al., Chem. Phys. Lipids, 1979, 23, 291). Racemic phytosphingosine has been synthesised by a stereospecific hydroxylation of an isoxazoline 4-anion and subsequent reductive
cleavage of the isoxazole ring with lithium aluminium hydride
as shown below (W. Schwab and V. Jager, Angew. Chem. internat.
Edn., 1981, 20, 603).

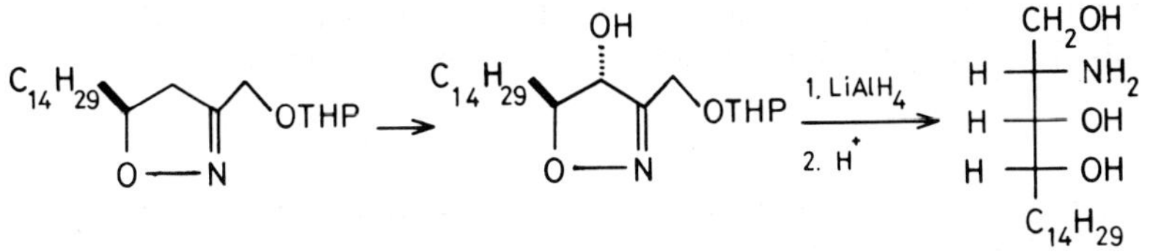

Hydroxylation of the double bond of sphingosine by osmium
tetroxide has been studied (B. Weiss and R.L. Stiller, Chem.
Phys. Lipids, 1975, 14, 59). Free phytosphingosines are
produced by *Candida intermedia* (1FO 0761) (A. Kimura, Ag.
Biol. Chem., 1976, 40, 239).

Oxidation of the allylic hydroxy-group of sphingosine, in
sphingolipids, by dichlorodicyanoquinone gives the corresponding ketone (M. Iwamori *et al.*, Biochim. biophys. Acta,
1975, 380, 308) and the uv-absorbtion of the unsaturated
keto-group was used for the estimation of sphingolipids.
Treatment of the 3-keto-derivatives of sphingosine-contain-
ing ceramides with base gives cyclic derivatives of the
following type (Y. Kishimoto and C. Costello, Chem. Phys.

460

Lipids, 1975, 15, 27.

$$Me(CH_2)_{12}$$

Me(CH₂)₁₂ — O ... O, NHCOR

The 3-keto-derivatives of sphinganine have been prepared by the condensation of a fully silylated derivative of serine with a Grignard reagent {B.I. Mitsner *et al.*, Bioorg. Khim., 1975, 1, 889 (672)} and the stereospecificity of the reduction of the product with sodium borohydride investigated {B.I. Mitsner *et al.*, Zhur. org. Khim., 1974, 10, 2518)}.

Sphingosine bases have been determined fluorometrically in the presence of pyridoxal (Y.S. Choi and K. Egawa, Jap. J. exp. Med., 1974, 45, 113) and dinitrophenyl- and dansyl-derivatives have been used for the separation of *threo*- and *erythro*-isomers {B.I. Mitsner *et al.*, Zhur. org. Khim., 1974, 10, 2514 (2523)}. Various boronate derivatives have been investigated for the g.l.c.-mass spectrometry of long-chain bases (S.J. Gaskell and C.J.W. Brooks, J. Chromatog., 1976, 122, 415). Sphingosine, dihydrosphingosine and phytosphingosine can be separated by chromatography on silica gel columns (Y. Barenholz and S. Gatt, Methods in Enzymology, 1975, 35, 529).

Biosynthesis of the Sphingosine Bases

In the biosynthesis of sphingosine from L-serine and palmit-oyl CoA, the proton at C-2 of serine and the carboxyl group are lost and replaced by a palmitoyl group and a proton from the medium (K. Krisnangkura and C.C. Sweeley, J. biol. Chem., 1976, 251, 1597) and in the catabolism of dihydro-sphingosine 1-phosphate to ethanolamine phosphate, a proton from the solvent is incorporated stereospecifically into the ethanolamine phosphate (T. Shimojo *et al.*, *ibid.*, 1976, 251, 4448; Biochim. biophys. Acta, 1976, 431, 433). In the biosynthesis of phytosphingosine the oxygen atom at C-4 originates from molecular oxygen, i.e. an oxygen- dependant hydroxylation of dihydrosphingosine gives phytosphingosine

R.J. Kulmacz and G.J. Schroepfer, J. Amer. chem. Soc., 1978, 100, 3963).

Ceramides

Ceramides have been synthesised, in good yield, from free fatty acids and sphingosine in the presence of triphenyl-phosphine and 2,2^1-dipyridyl-disulphide (Y. Kishimoto, Chem. Phys. Lipids, 1975, 15, 33). Radioactively-labelled ceramides have also been prepared (Y. Shoyama, J. lipid Res., 1978, 19, 250). The analysis of ceramides by HPLC has been investigated (M. Iwamori *et al.*, *ibid.*, 1979, 20, 86; M. Smith *et al.*, *ibid.*, 1981, 22, 714). The mass-spectrometry of the trimethylsilyl derivative of *N*-acetyl dihydrosphin-gosine has been studied (K. Krisnangkura and C.C. Sweeley, Chem. Phys. Lipids, 1974, 13, 415).

The biosynthesis of ceramides from fatty acids and sphingo-sine requires a pyridine nucleotide (I. Singh and Y. Kishimoto, Biochem. Biophys. Res. Comm., 1978, 82, 1287) and the effect of various synthetic compounds on the biosynthesis of ceramides has been investigated (I. Singh, J. biol. Chem., 1979, 254, 3840). Analogues of ceramide (particularly 2-decanoylamino-3-morpholino-1-phenylpropanol)inhibit gluco-cerebroside synthetase in mouse brain (R.R. Vunnam and N.S. Radin, Chem. Phys. Lipids, 1980, 26, 265). The enzyme 'ceramidase', responsible for the hydrolysis of ceramides, has been reviewed (N.S. Radin, Adv. Neurochem., 1975, 1, 51) and the *in vivo* metabolism of labelled ceramides (H. Okabe and Y. Kishimoto, J. biol. Chem., 1977, 252, 7068) and 3-keto-derivatives of ceramides (Y. Shoyama and Y. Kishimoto, J. Neurochem., 1978, 30, 377) in rat brain have been studied.

The conformations, mono-layer behaviour and thermotropic phase behaviour of ceramides have been studied by physical methods (see B. Dahlén and I. Pascher, Chem. Phys. Lipids, 1979, 24, 119).

Ceramide 1-phosphate has been synthesised (A.S. Bushnev *et al.*, Bioorg. Khim., 1979, 5, 1381; N.N. Karpyshev *et al.*, *ibid.*, 1980, 6, 1214) and 1-deoxy-1-sulpho ceramide has been isolated from the diatom *Nitzschia alba* (R. Anderson *et al.*, Biochim. biophys. Acta, 1978, 528, 89) and *Capnocytophaga sp.* (W. Godchaux and E.R. Leadbetter, J. Bacteriol., 1980, 144, 592) and has been synthesised (N.N.

Karpyshev *et al.*, Bioorg. Khim., 1977, **3**, 1374 (1007)}.

Glycosides of Ceramides

Several general reviews of these glycolipids have appeared
(C.C. Sweeley and B. Siddiqui in 'The Glycoconjugates', Vol.
1, ed. M.I. Horowitz and W. Pigman, Academic Press, New
York, 1977, p. 459; B.A. Macher and C.C. Sweeley, Methods
in Enzymology, 1978, **50**, 236; C.C. Sweeley *et al.*, ACS
Symposium Series, No. 80, 1979, p.47; ed. C.C. Sweeley,
ibid., No. 128, 1980; S. Hakomori in 'Internat. Rev.
Science, Org. Chem. Series 2, Vol. 7, ed. G.O. Aspinall,
Butterworths, 1976, p.223). The immunochemistry of the
glycosyl ceramides has been reviewed (D.M. Marcus in 'Glycolipid Methodology', ed. L.A. Witting, Amer. oil Chem. Soc.,
1976, p.233; D.M. Marcus and G.A. Schwarting, Adv. Immunol.,
1976, **23**, 203; B. Niedieck, Progr. Allergy, 1975, **18**, 353).

The HPLC of peracylated neutral glycosyl ceramides (M.D.
Ullman and R.H. McCluer, J. lipid Res., 1977, **18**, 371; 1978,
19, 910; S.K. Gross and R.H. McCluer, Anal. Biochem., 1980,
102, 429; T. Yamazaki *et al.*, J. Biochem., 1979, **86**, 803)
and non-derivatised neutral glycosyl ceramides (K. Watanabe
and Y. Arao, J. lipid. Res., 1981, **22**, 1020), the t.l.c. of
neutral glycosyl ceramides (V.P. Skipski, Methods in Enzymology, 1975, **35**, 396), the structure determination of
glycosyl ceramides by chromic acid oxidation, (R.A. Laine
and O. Renkonen, J. lipid Res., 1975, **16**, 102), ^{1}H-n.m.r.
(K.-E. Falk *et al.*, Arch. Biochem. Biophys., 1979, **192**, 164,
177, 191; J. Dabrowski *et al.*, Biochemistry, 1980, **19**,
5652) and mass-spectrometry (H. Egge, Chem. Phys. Lipids,
1978, **21**, 349; K.A. Karlsson in 'Glycolipid Methodology',
ed. L.A. Witting, Amer. oil Chem. Soc., 1976, p.97; Progr.
Chem. Fats Lipids, 1978, **16**, 207; S. Ando *et al.*, J.
Biochem., 1977, **82**, 1623; P. Hanfland *et al.*, Arch.
Biochem., Biophys., 1981, **210**, 383, 396, 405; Biochemistry,
1981, **20**, 5310; M.E. Breimer *et al.*, FEBS Letters, 1981,
124, 299) of glycosyl ceramides have been studied.

The labelling of the galactose or *N*-acetylgalactosamine
moieties of glycolipids by oxidation with galactose oxidase
and subsequent reduction of the aldehyde formed with ^{3}H-
labelled sodium borohydride (C.G. Gahmberg, Methods in
Enzymology, 1978, **50**, 204) and labelling by reduction of
unsaturated linkages with ^{3}H-sodium borohydride-palladium

chloride-sodium hydroxide (G. Schwarzmann, Biochim. biophys. Acta, 1978, 529, 106) have been studied. The attachment of glycosyl ceramides to solid supports for immunological studies (W.W. Young *et al.*, Methods in Enzymology, 1978, 50, 137; J. lipid Res., 1979, 20, 275) and the biosynthesis of glycosyl ceramides (M. Basu in 'Glycolipid Methodology, ed. L.A. Witting, Amer. oil Chem. Soc., 1976, p.123) have been reviewed.

Monoglycosyl Ceramides

The preparative isolation of glucosyl and galactosyl ceramides have been improved (N.S. Radin, J. lipid Res., 1976, 17, 290) and monoglucosyl ceramides have been analysed at less than the nanomole level by HPLC (F.B. Jungalwala, J. lipid Res., 1977, 18, 275; G. Nonaka and Y. Kishimoto, Biochim. biophys. Acta, 1979, 572, 423). Monoglucosyl ceramides have been specifically labelled at the 3-position (of the sphingosine base) by oxidation of the allylic hydroxy-group with dichlorodicyanoquinone and reduction of the keto-group with ^{3}H-sodium borohydride (M. Iwamori *et al.*, J. lipid Res., 1975, 16, 332). Treatment of the keto-intermediate with base causes elimination of the sugar residue (M. Iwamori and Y. Nagai, Chem. Phys. Lipids, 1977, 20, 193). The 6-position (of glucose) in a glucosyl ceramide has been labelled with tritium after chemical oxidation of a derivative and reduction with ^{3}H-borohydride (M.C. McMaster and N.S. Radin, J. labelled Compd. and Radiopharm., 1977, 13, 353). The chemical ionisation mass-spectrometry of monoglycosyl ceramides (T. Murata *et al.*, J. lipid Res., 1978, 19, 370) and the X-ray crystallographic analysis of a galactosyl ceramide (I. Pascher and S. Sundell, Chem. Phys. Lipids, 1977, 20, 175) have been studied. The ^{13}C-n.m.r. spectrum of a glucosyl ceramide has confirmed the β-linkage of the glucosyl residue (T.A.W. Koerner, *et al.*, J. biol. Chem., 1979, 254, 2326) and high resolution ^{1}H- and ^{13}C-n.m.r. spectra of galactosyl ceramides have been recorded (J. Dabrowski *et al.*, Chem. Phys. Lipids, 1980, 26, 187). Spin-labelled galactosyl ceramides have been prepared for use in membrane studies (F.J. Sharom and C.W.M. Grant, J. supramol. Struct., 1977, 6, 249) and the unnatural isomer, L-glucosyl ceramide,has been synthesised and used as a model compound for the accumulation of glucosyl ceramides in Gaucher's disease, since it is not hydrolysed by gluco-cerebrosidase (A.E. Gal *et al.*, Proc. nat. Acad. Sci., 1979,

<u>76</u>, 3083).   For a review of Gaucher's disease see R.O. Brady
in 'Glycolipid Methodology', ed. L.A. Witting, Amer. oil
Chem. Soc., 1976, p.295.   A β-D-mannosyl ceramide is present
in the bivalve *Hyriopsis schlegelii* (T. Hori *et al.*, Biochim.
biophys. Acta, 1981, <u>665</u>, 170).

Cerebroside Sulphate, 'Sulphatide'

A large scale isolation procedure for cerebroside sulphate
has been briefly described (N.S. Radin, Federation Proc.,
1979, <u>38</u>, 405).   HPLC has been performed on this lipid
after benzoylation and desulphation with perchloric acid in
acetonitrile (G. Nonaka and Y. Kishimoto, Biochim. biophys.
Acta, 1979, <u>572</u>, 423).   The desulphated, benzoylated
material has been resulphated to give a ^{35}S-labelled sulphatide (G. Nonaka and Y. Kishimoto, J. Neurochem., 1979, <u>33</u>,
23).   Saponification of sulphatide gives a mixture of the
deacylated material, $3^1,6^1$-anhydro-galactosyl sphingosine
and sphingosine (G. Nonaka *et al.*, J. Biochem., 1979, <u>85</u>,
511).   Reacylation of the deacylated sulphatide with
labelled fatty acids has also been used to prepare labelled
sulphatides (G. Nonaka and Y. Kishimoto, Biochim. biophys.
Acta, 1979, <u>572</u>, 423;   J. Neurochem., 1979, <u>33</u>, 23).   The
critical micelle concentration (0.01mM in water) has been
studied (H.J. Jeffrey and A.B. Roy, Austral. J. exp. Biol.
Med. Sci., 1977, <u>55</u>, 339).

Diglycosyl Ceramides

Sulpholactosyl ceramide is a major glycolipid of the testis
of mature salmon and trout (M. Levine *et al.*, Biochim.
biophys. Acta, 1976, <u>441</u>, 134).   The diglycosyl ceramide
                    Mann (1β4) gluc (1β1) ceramide
is present in *Corbicula sandai* (O. Itasaka *et al.*, J. Biochem., 1976, <u>80</u>, 935).

Triglycosyl Ceramides

The triglycosyl ceramide which accumulates in Fabry's
disease
            gal (1α4) gal (1β4) gluc (1β1) ceramide
has been synthesised (D. Shapiro and A.J. Acher, Chem. Phys.
Lipids, 1978, <u>22</u>, 197) and the ^{13}C-n.m.r. spectrum of the
oligosaccharide portion investigated (D.D. Cox *et al.*,

Carbohydrate Res., 1978, 67, 23). Antibodies to the Fabry
glycolipid have been prepared (M.C. Hamers *et al.*, Immunochem., 1978, 15, 353; J.T.R. Clarke and J.A. Embil, Biochim.
biophys. Acta, 1979, 582, 283). The triglycosyl ceramide
       *N*Acgal (1β4) gal (1β4) gluc (1β1) ceramide
(asialo GM_2) accumulates in transformed mouse fibroblasts
and antibodies against this glycolipid have been prepared
(G. Rosenfelder *et al.*, Cancer Res., 1977, 37, 1333; K.
Uemura *et al.*, J. Biochem., 1978, 83, 1199). The triglycosyl ceramide
       *N*Acgluc (1β3) gal (1β4) gluc (1β1) ceramide
is present in human erythrocyte membranes and is a probable
precursor of the blood group-active glycolipids (S. Ando *et
al.*, *ibid.*, 1976, 79, 625) and the triglycosyl ceramide
       mann (1α4) mann (1β4) gluc (1β1) ceramide
is present in *Corbicula sandai* (O. Itasaka *et al.*, *ibid.*,
1976, 80, 935).

Tetraglycosyl Ceramides

The tetraglycosyl ceramide
    *N*Acgal (1β3) gal (1α3) gal (1β4) gluc (1β1) ceramide
is the main glycolipid of granuloma (E. Hanada *et al.*, J.
Biochem., 1978, 83, 85) and is also a major glycolipid of
rat spleen (H. Arita and J. Kawanami, *ibid.*, 1977, 81, 1661).
The glycolipid
    Gal (1β4) *N*Acgluc (1β3) gal (1β4) gluc (1β1) ceramide
('paragloboside') accumulates in transformed cells (J.S.
Sundsmo and S. Hakomori, Biochem. Biophys. Res. Comm., 1976,
68, 799) and antibodies prepared against this glycolipid
react with human erythrocytes and lymphocytes (G.A. Schwarting
and D.M. Marcus, J. Immunol., 1977, 118, 1415). 'Paragloboside' has been synthesised (M.M. Ponpipom *et al.*,
Tetrahedron Letters, 1978, 1717). A sulphate ester of
'paragloboside' (sulphated on the 6-position of *N*-acetyl
glucosamine) has been isolated from hog gastric mucosa (B.L.
Slomiany and A. Slomiany, J. biol. Chem., 1978, 253, 3517).
The tetraglycosyl ceramide
    *N*Acgal (1α3) *N*Acgal (1β3) gal (1β4) gluc (1β1) ceramide,
which contains the antigenic determinant of the Forssman
antigen (see below), is present in hamster fibroblasts (C.G.
Gahmberg and S. Hakomori, *ibid.*, 1975, 250, 2438) and the
glycolipid
    *N*Acgal (1β2) mann (1β3) mann (1β4) gluc (1β1) ceramide
is present in the sperm of the fresh-water bivalve *Hyriopsis*

schlegelii (T. Hori *et al.*, J. Biochem., 1977, $\underline{81}$, 107).
Antibodies to the glycolipid 'asialo GM_1'
  gal (1β3) *N*Acgal (1β4) gal (1β4) gluc (1β1) ceramide
have been prepared by the technique of affinity chromatography, using a derivative of 'asialo GM_1' coupled to
Sepharose-4B (T. Taki *et al.*, *ibid.*, 1981, $\underline{89}$, 503).

Pentaglycosyl Ceramides

The structure of the Forssman antigen has been confirmed by
permethylation studies (S.R. Kundu *et al.*, Carbohydrate Res.,
1975, $\underline{39}$, 179) and the ^{13}C-n.m.r. spectrum assigned (T.A.W.
Koerner *et al.*, Biochem. J., 1981, $\underline{195}$, 529). The penta-
saccharide portion of the Forssman antigen has been synthesised (H. Paulsen and A. Bünsch, Angew. Chem. internat. Ed.,
1980, $\underline{19}$, 902) and the biosynthesis of the glycolipid from a
trihexosyl ceramide studied (S. Kijimoto-Ochiai *et al.*, J.
biol. Chem., 1980, $\underline{255}$, 9037). A more complex fucose-
containing glycolipid with Forssman antigenic activity has
been isolated from dog gastric mucosa (B.L. Slomiany and A.
Slomiany, Europ. J. Biochem., 1978, $\underline{83}$, 105).

The P_1-antigen of red blood cells has been characterised as
the pentaglycosyl ceramide
  gal(1α4)gal(1β4)*N*Acgluc(1β3)gal(1β4)gluc(1β1)ceramide
(M. Naiki *et al.*, Biochemistry, 1975, $\underline{14}$, 4831; W.M. Watkins
and W.T.J. Morgan, J. Immunogenetics, 1976, $\underline{3}$, 15).

Fucose-containing Glycosyl Ceramides

These compounds have been extensively investigated since
they include the blood group substances and several reviews
have been published (S. Hakomori and K. Watanabe in 'Glycolipid Methodology', ed. L.A. Witting, Amer. oil Chem. Soc.,
1976, p.13; K.A. Karlsson, *ibid.*, p.97; A. Slomiany *et al.*,
ibid., p.49; J.M. McKibbin, *ibid.*, p.77; S. Hakomori,
Progr. Biochem. Pharmacol., 1975, $\underline{10}$, 167; K. Watanabe and
S. Hakomori, J. exp.Med., 1976, $\underline{144}$, 644; J.M. McKibbin, J.
lipid Res., 1978, $\underline{19}$, 131; S. Hakomori, Methods in Enzymology, 1978, $\underline{50}$, 207; J. Koscielak *et al.*, *ibid.*, p.211).
The following references are to work by each group active
in the isolation and structural determination of blood group
active glycolipids (P. Hanfland, *et al.*, Chem. Phys. Lipids,
1978, $\underline{22}$, 141; J. Koscielak *et al.*, Europ. J. Biochem.,
1978, $\underline{91}$, 517; 1979, $\underline{96}$, 331; N. Sharon *et al.*, *ibid.*,

1978, 83, 363; S. Hakomori *et al.*, J. exp. Med., 1979, 149,
975; B.L. Slomiany *et al.*, Biochem. Biophys. Res. Comm.,
1979, 88, 1092).

Extensive investigations on the chemical synthesis of the
antigenic determinants of these complex glycolipids have led
to the synthesis of several tri- and tetrasaccharides (for
leading references see R.U. Lemieux, Chem. Soc. Rev., 1978,
7, 423; S. David *et al.*, Nouveau J. Chim., 1980, 4, 547;
A. Veyrières *et al.*, J. chem. Soc. Perkin I, 1979, 1825;
1981, 1626; Carbohydrate Res., 1981, 92, 310; P. Sinaÿ
et al., Tetrahedron, 1979, 35, 365; Angew. Chem. internat.
Edn., 1979, 18, 464; Carbohydrate Res., 1981, 92, 183; J.
chem. Soc. Perkin I, 1981, 326; H. Paulsen *et al.*, Tetrahedron Letters, 1981, 22, 1387; Ber., 1981, 114, 306, 333).

The pentaglycosyl ceramide shown below

gal(1β4)⎤
          ⎬ NAcgluc(1β3)gal(1β4)gluc(1β1)ceramide
L-fuc(1α3)⎦

is a major neutral oligoglycosyl ceramide of human brain
(M.T. Vanier *et al.*, FEBS Letters, 1980, 112, 70) and the
glycolipid shown below is the main glycolipid of the macro-
phage and is probably the receptor for the migration inhi-
bition factor of rat macrophages (T. Miura *et al.*, J.
Biochem., 1979, 86, 773).

gal(1α3)⎤
         ⎬ gal(1β3)NAc gal (1β3)gal(1β4)gluc(1β1)ceramide
L-fuc(1α2)⎦

Glycosyl Ceramides containing Sialic Acids. Gangliosides.

Several reviews of these glycolipids have appeared (R.W.
Ledeen and R.K. Yu in 'Research Methods in Neurochemistry',
Vol. 4, ed. N. Marks and R. Rodnight, Plenum Press, New
York, 1978, p.371; R.W. Ledeen, J. supramol. Struct., 1978,
8, 1; several chapters in 'Glycolipid Methodology', ed. L.A.
Witting, Amer. oil Chem. Soc., 1976; ed., G. Porcellati,

B. Ceccarelli and G. Tettamanti, 'Ganglioside Function: Biochemical and Pharmacological Implications', Plenum Press, New York, 1976; P.H. Fishman and R.O. Brady, Science, 1976, 194, 906; C.C. Sweeley and B. Siddiqui in 'The Glyco-conjugates', Vol. 1, ed., M.I. Horowitz and W. Pigman, Academic Press, New York, 1977, p.459; Ed. L. Svennerholm, P. Mandel, H. Dreyfus and P.-F. Urban, 'Structure and Func-tion of Gangliosides', Plenum Press, New York, 1980).

Methods for the isolation, chromatography and analysis of gangliosides have been discussed by the following: N. Kawamura and T. Taketomi, J. Biochem., 1977, 81, 1217; S. Harth et al., Anal. Biochem., 1978, 86, 543; S. Ando et al., ibid., 1978, 89, 437; C.C. Irwin and L.N. Irwin, ibid., 1979, 94, 335; M. Iwamori and Y. Nagai, Biochim. biophys. Acta, 1978, 528, 257; L. Svennerholm and P. Fredman, ibid., 1980, 617, 97; E.G. Bremer et al., J. lipid Res., 1979, 20, 1028. The ^{13}C- and ^{1}H-n.m.r. spectra of gangliosides have been studied (L.O. Sillerud et al., Biochemistry, 1978, 17, 2619; P.L. Harris and E.R. Thornton, J. Amer. chem. Soc., 1978, 100, 6738) and the critical micelle concentration and dialysis behaviour have been reinvestigated (S. Formisano et al., Biochemistry, 1979, 18, 1119; R. Ghidoni et al., Lipids, 1978, 13, 820; H. Rauvala, Europ. J. Biochem., 1979, 97, 555; M. Corti et al., Chem. Phys. Lipids, 1980, 26, 225).

The interaction of cholera toxin with 'liposomes' containing GM$_1$ has been studied (P.H. Fishman et al., Biochemistry, 1979, 18, 2562) and a large scale purification of cholera toxin on silica beads derivatised with lyso-GM$_1$ has been achieved (J.-L. Taylor et al., Europ. J. Biochem., 1981, 113, 249). The interaction of the oligosaccharide portion of GM$_1$ with cholera toxin has also been investigated by ^{13}C-n.m.r. (L.O. Sillerud et al., J. biol. Chem., 1981, 256, 1094). The phase behav-iour of hydrated gangliosides has been studied by X-ray diffraction, calorimetry and polarised light microscopy (W. Curatolo et al., Biochim. biophys. Acta, 1977, 468, 11).

Several reviews of the ganglioside storage diseases have been published: ed., B.W. Volk and L. Schneck, 'Gangliosidoses', Plenum Press, New York, 1975; ed., M.M. Kaback, 'Tay-Sachs Disease: Screening and Prevention', A.R. Liss Inc., New York, 1977; H.J. Barker et al., Federation Proc., 1976, 35, 1193; see also the general references on lipids at the beginning of the chapter.

Several fucose-containing gangliosides have been characterised: the heptaglycosyl ceramide ('A'-shown below) has been
isolated from pig cerebellum (S. Sonnino *et al.*, J. Neurochem., 1978, 31, 947; see also P. Fredman *et al.*, Europ. J.
Biochem., 1981, 116, 553) and a similar ganglioside containing only one NANA residue is present in bovine thyroid (G.A.F.
van Dessel *et al.*, J. biol. Chem., 1979, 254, 9305; B.A.
Macher *et al.*, Biochim. biophys. Acta, 1979, 588, 35) and
the decaglycosyl ceramide ('B'-shown below) has been isolated
from human erythrocytes (K. Watanabe *et al.*, J. biol. Chem.,
1978, 253, 8962).

L-fuc (1∝2) gal (1β3) NAc gal (1β4)⎤
⎥ gal(1β4)gluc (1β1)ceramide   'A'
NANA (2∝8) NANA (2∝3)⎦

L-fuc (1∝2)gal(1β4)NAc gluc (1β6)⎤
⎥ gal(1β4)NAcgluc(1β3)gal(1β4)gluc
NANA (2∝3)gal(1β4)NAc gluc (1β3)⎦   1   'B'
                                                    β
                                                    1
                                                 ceramide

NANA = N-Acetylneuraminic acid

A new trisialoganglioside (S. Ando and R.K. Yu, *ibid.*, 1977,
252, 6247) and a new monosialoganglioside (M. Iwamori and Y.
Nagai, J. Biochem., 1978, 84, 1601) from brain, a blood
group I-active ganglioside (K. Watanabe *et al.*, J. biol.
Chem., 1979, 254, 3221) and a novel ganglioside with a NANA
(2α3)*N*Acgal linkage (K. Watanabe and S. Hakomori, Biochemistry, 1979, 18, 5502), from erythrocytes, have been
isolated and characterised.

The tetrasialoganglioside ('A'-below) is present in mammalian
brain whereas the tetrasialoganglioside ('B'-below) is present
in cod fish brain (see S. Ando and R.K. Yu, J. biol. Chem.,
1979, 254, 12224; P. Fredman *et al.*, FEBS Letters, 1980, 110,
80).

The ganglioside patterns of the erythrocytes from horses and
cattle have been extensively examined (S. Hamanaka *et al.*,
J. Biochem., 1980, 87, 639) and the *N*-glycolylneuraminic
acid-containing gangliosides of animal serum have been
characterised as the 'serum sickness antigens' (M. Nakai and

H. Higashi, in 'Structure and Function of Gangliosides', ed. L. Svennerholm, P. Mandel, H. Dreyfus and P.-F. Urban, Plenum Press, New York, 1980, p.359).

NANA(2α8)NANA(2α3)gal(1β3)NAcgal(1β4) ⎤
                                        ⎥ gal(1β4)gluc(1β1)ceramide
        NANA(2α8)NANA (2α3) ⎦

`A´

NANA(2α3)gal (1β3)NAcgal(1β4) ⎤
                              ⎥ gal(1β4)gluc(1β1)ceramide
NANA(2α8)NANA(2α8)NANA (2α3) ⎦

`B´

Sphingomyelin

The properties of sphingomyelin in bilayers and biological membranes have been reviewed (Y. Barenholz and T.E. Thompson, Biochim. biophys. Acta, 1980, 604, 129). The reversed phase and 'argentation' HPLC separation of molecular species of sphingomyelins has been described (F.B. Jungalwala *et al.*, J. lipid Res., 1979, 20, 579; M. Smith *et al.*, *ibid.*, 1981, 22, 714) and the molecular species of sphingomyelins have also been determined after enzymic hydrolysis to ceramides which are analysed by t.l.c., g.l.c. and mass spectrometry (M.E. Breimer, *ibid.*, 1975, 16, 189). A hemolytic toxin from the sea anemone *Stoichactis helianthus* binds sphingomyelin (R. Linder *et al.*, Biochim. biophys. Acta, 1977, 467, 290).

Oxidation of the allylic hydroxy-group of sphingomyelin to a ketone with 2,3-dichloro-5,6-dicyanobenzoquinone and treatment of the product with base (pH 11.5) results in β-elimination to give a diene (M. Iwamori and Y. Nagai, Chem. Phys. Lipids, 1977, 20, 193). Sphingomyelin is hydrolysed to ceramide by 40% hydrogen fluoride at 40°C for 72h. (P.V. Reddy, *ibid.*, 1976, 17, 373).

Sphingomyelin has been synthesised by methylation of the corresponding dimethylaminoethyl derivative {E.N. Zvonkova *et al.*, Bioorg. Khim., 1975, 1, 1746 (1261)}. DL-3-O-Benzoylsphingenine has been resolved with the aid of (+)-

tartaric acid. The phosphonic acid analogue of sphingo-myelin, shown below, has been synthesised (V.M. Kapoulas and M.C. Moschidis, Chem. Phys. Lipids, 1981, 28, 357).

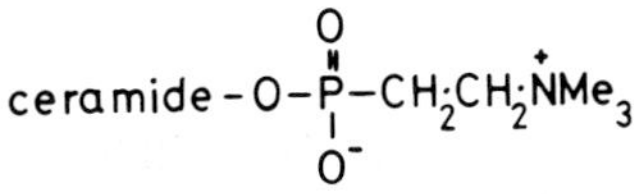

$$\text{ceramide}-O-\overset{\overset{O}{\|}}{\underset{\underset{O^-}{|}}{P}}-CH_2CH_2\overset{+}{N}Me_3$$

The Raman spectra (see R. Faiman, *ibid.*, 1979, 23, 77) circular dichroism (B.J. Litman and Y. Barenholz, Biochim. biophys. Acta, 1975, 394, 166), X-ray diffraction (R.S. Khare and C.R. Worthington, *ibid.*, 1978, 514, 239), n.m.r. (see P.L. Yeagle *et al.*, Biochemistry, 1978, 17, 5745) and thermotropic behaviour (R. Cohen and Y. Barenholz, Biochim. biophys. Acta, 1978, 509, 181; W.I. Calhoun and C.G. Shipley, *ibid.*, 1979, 555, 436) of sphingomyelins have been studied.

Ceramide 1-(2-Aminoethyl phosphonate)

This compound has been isolated from the fungus *Pythium prolatum* (M.K. Wassef and J.W. Hendrix, Biochim. biophys. Acta, 1977, 486, 172) and from the protozoa *Entamoeba invadens* (H.H.D.M. van Vliet *et al.*, Arch. Biochem. Biophys., 1975, 171, 55) and its content (together with that of the *N*-methyl derivative) in various shell fish (T. Matsubara, Chem. Phys. Lipids, 1975, 14, 247; Biochim. biophys. Acta, 1975, 388, 353) and in *Tetrahymena pyriformis* (M. Sugita *et al.*, J. Biochem., 1979, 86, 281) has been investigated. A 'sphingophosphonoglycolipid', containing galactose, glucose, galactosamine, 3-*O*-methylgalactose and 2-aminoethyl phosphonic acid has been isolated from the skin of the marine gastropod *Aplysia kurodai* (S. Araki *et al.*, J. Biochem., 1980, 87, 503).

Sphingoethanolamine

This lipid has been isolated from cultured mosquito cells
(T.K. Yang *et al.*, Lipids, 1974,9, 1009) and its biosynthesis
in the anaerobic bacterium *Bacteroides melaninogenicus* (M.
Lev and A.F. Milford, Arch. Biochem. Biophys., 1978, 185, 82)
and the anaerobic protozoa *Entodinium caudatum* (T.E. Broad
and R.M.C. Dawson, Biochem. J., 1975, 146, 317) studied.
It has been synthesised from 3-*O*-benzoyl ceramide 1-phosphate
and *N*-tritylaziridine (A.S. Bushnev *et al.*, Chem. Phys.
Lipids, 1975, 14, 263). Ceramide 1-(serine phosphate) has
also been synthesised (N.N. Karpyshev *et al.*, Bioorg. Khim.,
1979, 5, 1422).

Ceramide 1-(Glycerol 1-phosphate)

The racemic compound and its 3-amino-1,2-propane-diol
analogue have been synthesised {R.P. Evstigneeva *et al.*,
Bioorg. Khim., 1979, 5, 234 (167); 238 (170)} and its bio-
synthesis in *Bacteroides melaninogenicus* has been studied
(M. Lev and A.F. Milford, J. Bacteriol., 1977, 130, 445;
Arch. Biochem. Biophys., 1978, 185, 82).

Inositol Phosphate-containing Sphingolipids

These compounds have been reviewed (P.J. Brennan, Adv.
Microbial Physiol., 1978, 17, 48; R.L. Lester in 'Cyclitols
and Phosphoinositides', ed. W.W. Wells and F. Eisenberg,
Academic Press, New York, 1978, p.83) and methods of
extraction of these lipids from *Saccharomyces cerevisiae*
and *Neurospora crassa* have been investigated in detail (B.A.
Hanson and R.L. Lester, J. lipid Res., 1980, 21, 309). Six
more inositol phosphate-containing sphingolipids have been
isolated from tobacco leaves (K. Kaul and R.L. Lester,
Biochemistry, 1978, 17, 3569) and the major compound was
shown to have the structure below (T.C.-Y. Hsieh *et al.*,
Biochemistry, 1978, 17, 3575) which is similar (but not
identical) to the structure reported for a portion of the
phytoglycolipid molecule (see also R.A. Laine *et al.*, ACS
Symposium Series, No. 128, 1980, p.65 and T.C.-Y. Hsieh *et
al.*, J. biol. Chem., 1981, 256, 7747).

$$\underset{\substack{\text{D-glucuronic}\\ \text{acid}}}{\text{NAc gluc}(1\alpha 4)} \quad (1\alpha 2) \quad \underset{\substack{\text{myo-}\\ \text{inositol}}}{} -(1)-O\cdot\overset{\overset{\displaystyle O}{\parallel}}{\underset{\underset{\displaystyle O^-}{|}}{P}}\cdot O\cdot\text{ceramide}$$

Ceramide phosphoryl inositol derivatives have also been isolated from *Aspergillus niger* (J.A. Hackett and P.J. Brennan, FEBS Letters, 1977, 74, 259; P.F.S. Byrne and P.J. Brennan, Biochem. Soc. Trans., 1976, 4, 893) and are also present in the plasma membranes of *Acanthamoeba castellanii* (D.G. Dearborn *et al.*, J. biol. Chem., 1976, 251, 2976) and *Trypanosoma cruzi* (R.M. Lederkremer *et al.*, Biochem. Biophys. Res. Comm., 1978, 85, 1268).

Miscellaneous Phospholipids and Glycolipids

Cord Factor has been reviewed (J. Asselineau and C. Asselineau, Progr. Lipid Res., 1978, 16, 59; E. Lederer, Chem. Phys. Lipids, 1976, 16, 91) and new methods for the synthesis of cord factor and its analogues have been developed (J. Polonsky *et al.*, Carbohydrate Res., 1978, 65, 295; R. Toubiana *et al.*, J. Carbohydrates Nucleosides, Nucleotides, 1978, 5, 127; E. Durand *et al.*, Europ. J. Biochem., 1978, 93, 103; M. Kato *et al.*, *ibid.*, 1978, 87, 497; A. Liav and M.B. Goren, Chem. Phys. Lipids, 1980, 27, 345). 2^1-Sulphate esters of acylated cord factor have been isolated from *Mycobacterium tuberculosis* (M.B. Goren *et al.*, Biochemistry, 1976, 15, 2728; Nouveau J. Chim., 1978, 2, 379). A phenolic glycolipid, containing 2,3-di-*O*-methylrhamnose, 3-*O*-methylrhamnose and 3,6-di-*O*-methylglucose is present in high concentration in *Mycobacterium leprae* (isolated from infected armadillo liver) and a similar lipid containing 2,4-di-*O*-methylrhamnose, 2-*O*-methylrhamnose and 2-*O*-methylfucose is present in *Mycobacterium kansasii* (S.W. Hunter and P.J. Brennan, J. Bacteriol., 1981, 147, 728).

474

Torulopsis Glycolipids. The glycolipids produced extra-
cellularly by yeasts have been reviewed (A.P. Tulloch in
'Glycolipid Methodology', ed. L.A. Witting, Amer. oil Chem.
Soc., 1976, p.329).

Polyisoprenoid Lipid Intermediates have been reviewed (W.J.
Lennarz, Science, 1975, 188, 986; C.J. Waechter and W.J.
Lennarz, Ann. Rev. Biochem., 1976, 45, 95; J.L. Lucas and
C.J. Waechter, Molecular Cell Biochem., 1976, 11, 67; W.T.
Forsee and A.D. Elbein in 'Glycolipid Methodology', ed. L.A.
Witting, Amer. oil Chem. Soc., 1976, p.369; F.W. Hemming,
Biochem. Soc. Trans., 1977, 5, 1682; D.K. Struck and W.J.
Lennarz in 'The Biochemistry of Glycoproteins and Proteo-
glycans, ed. W.J. Lennarz, Plenum Press, New York, 1979,
p.35). The synthesis of these compounds has also been
reviewed (C.D. Warren and R.W. Jeanloz, Methods in Enzymology, 1978, 50, 122; see also C.D. Warren *et al.*, Carbohydrate Res., 1981, 92, 85).

Lipid A. The chemistry and biochemistry of the lipid A's
from various bacteria have been reviewed (O. Luderitz *et al.*,
Naturwissenschaften, 1978, 65, 578; in Int. Rev. Biochemistry. Biochem. of Lipids II, Vol. 14, ed. T.W. Goodwin, 1977,
p.239) and the structures of the lipopolysaccharides from a
heptose-less mutant of *E. coli* have been studied in detail
(M.R. Rosner *et al.*, J. biol. Chem., 1979, 254, 5906, 5918,
5926; for a review see H.G. Khorana, Bioorg. Chem., 1980,
9, 363). Various syntheses of the acylated glucosamine
(1β6) glucosamine disaccharide, which constitutes the main
structural feature of lipid A, have been described (M.
Inage *et al.*, Chemistry Letters, 1980, 1373; Tetrahedron
Letters, 1980, 21, 3889; 1981, 22, 2281; M. Kiso *et al.*,
Carbohydrate Res., 1981, 88, C5, C10; M.A. Nashed and L.
Anderson, *ibid.*, 1981, 92, C5).

Ornithine Lipids of Streptomyces have been reviewed (S.G.
Batrakov and L.D. Bergelson, Chem. Phys. Lipids, 1978, 21,
1) and the chlorosulpholipids in the flagellar membrane of
Ochromonas danica have been investigated (L.L. Chen *et al.*,
J. biol. Chem., 1976, 251, 1835). $2^1,3^1$-Di-O-acyl-D-
glucopyranosyl (1α2) D-glyceric acid has been isolated from
Nocardia caviae (M.-T. Pommier and G. Michel, Europ. J.
Biochem., 1981, 118, 329).

Guide to the Index

This index is constructed in a similar manner to the volume indexes of the first edition of the Chemistry of Carbon Compounds. However, to make the index easier to use, more descriptive entries have been made for the commonly occurring individual, and groups of chemicals.

The indexes cover primarily the chemical compounds mentioned in the text, and also include reactions and techniques, where named, and some sources of chemical compounds such as plant and animal species, oils, etc.

Chemical compounds have been indexed alphabetically under the names used by authors, editing being restricted to ensuring uniformity of entries under the same heading. In view of the alternative nomenclature that can often be used, a limited amount of cross-referencing has been done where it is considered to be helpful, but attention is particularly drawn to Convention 2 below.

For this and the succeeding volumes, the indexing conventions listed below have been adopted.

1. *Alphabetisation*
   (a) The following prefixes have not been counted for alphabetising:

n-	*o-*	*as-*	*meso-*	D	*C*
sec-	*m-*	*sym-*	*cis-*	DL	*O-*
tert-	*p-*	*gem-*	*trans-*	L	*N-*
	vic-				*S-*
		lin-			*Bz-*
					Py-

Some prefixes and numbering have been omitted in the index, where they do not usefully contribute to the reference.

   (b) The following prefixes have been alphabetised:

Allo	Epi	Neo
Anti	Hetero	Nor
Cyclo	Homo	Pseudo
	Iso	

(c) A letter by letter alphabetical sequence is followed for entries, firstly for the main entry, followed by the descriptive entry. The only exception to this sequence is the placing of plural entries in front of the corresponding individual entries to prevent these being overlooked by a strict alphabetical sequence which could lead to a considerable separation of plural from individual entries. Thus "butanes" will come before *n*-butane, "butenes" before 1-butene, and 2-butene, etc.

2. *Cross references*

In view of the many alternative trivial and systematic names for chemical compounds, the indexes should be searched under any alternative names which may be indicated in the main body of the text. Only a limited amount of cross-referencing has been carried out, where it is considered that it would be helpful to the user.

3. *Esters*

In the case of lower alcohols esters are indexed only under the acid, e.g. propionic methyl ester, not methyl propionate. Ethyl is normally omitted e.g. acetic ester.

4. *Derivatives*

Simple derivatives are not normally indexed if they follow in the same short section of the text.

5. *Collective and plural entries*

In place of "– derivatives" or "– compounds" the plural entry has normally been used. Plural entries have occasionally been used where compiunds of the same name but differing numbering appear in the same section of the text.

6. *Main entries*

The main entry of the more common individual compounds is indicated by heavy type. Multiple entries, such as headings and sub-headings over several pages are shown by "–", e.g., 67–74, 137–139, etc.

488